PRACTICAL HORTICULTURE

A Guide to Growing Indoor and Outdoor Plants

LAURA WILLIAMS RICE
ROBERT P. RICE, JR.

1980 SAUNDERS COLLEGE
Philadelphia

This book is dedicated with respect and affection to the late

Dr. John Carew,
Dr. D.C. Kiplinger, and
Dr. Phil C. Kozel,

who were our friends, our teachers, and outstanding horticulturists.

Saunders College
West Washington Square
Philadelphia, PA 19105

Library of Congress Cataloging in Publication Data

Rice, Laura Williams.
Practical horticulture.

Includes index.
1. Horticulture. 2. Gardening. 3. Indoor gardening. I. Rice, Robert P., joint author.
II. Title.
SB318.R48 635 79-17985
ISBN 0-03-041455-5

Printed in the United States of America
0123 032 987654321

Preface

This text was conceived and developed as a guide for students enrolled in continuing education, attending evening college, and majoring in fields other than horticulture. The material focuses on those aspects of horticulture which would be useful to an average gardener or homeowner. We have attempted to organize a range of horticultural information in a readable fashion and present it to the reader at a level intermediate between popular gardening books and a textbook for horticulture majors.

We use botanical and horticultural terms where simpler phrases are inadequate or misleading, but we do not use jargon for its own sake. For example, gynoecious flowers are referred to as female flowers, since the latter term conveys the same information in a simpler manner. All botanical names are used in accordance with "Hortus III" by L. H. Bailey, the recognized definitive work on cultivated plants in the United States.

Because the book covers so many aspects of indoor and outdoor horticulture, certain explanations and procedures have necessarily been generalized. However, exceptions can be found in many instances. Likewise, cultivation practices are also generalized to be applicable throughout the United States and Canada, with the expectation that the instructor will discuss variations in cultural practices due to local climate.

The general text is organized into three parts. Part I contains the core chapters on anatomy, physiology, taxonomy, and the growth and development of plants, as well as a chapter on climate and another on propagation.

Part II deals with the cultivation of outdoor plants. Specific instructions are given for the cultivation of fruits and vegetables but not for ornamentals because of the diversity of species which are grown in different regions.

Part III covers indoor plant growing and can be used independently of Part II. In this section numerous plant examples and cultivation specifics are provided since the species of indoor plants vary only slightly in different regions.

Selected references are listed at the end of each chapter for students who wish to do further reading on a particular topic.

A glossary for quick reference of botanical and horticultural terms and an appendix on plant societies are included at the back of the book.

San Obispo, California
September, 1979

Laura Williams Rice
Robert P. Rice, Jr.

Acknowledgements

We wish to thank the following individuals for reviewing the manuscript and offering their suggestions for its improvement: William E. Barrick, University of Florida; Richard Biamonte, Agricultural Research and Education Center; Charles A. Conover, University of Florida; Marge Coon, California State Polytechnic University; David Cox, Diablo Valley College; John W. Day, University of Tennessee Institute of Agriculture; F.H. Emerson, Purdue University; William C. Fonteno, North Carolina State University; Ted Hirayama, Los Angeles Unified School District; the late Phil C. Kozel, Ohio State University; Herbert V. Marsh, Jr., University of Massachusetts at Amherst; Dexter McDonald, Ventura College; Edward L. McWilliams, Texas A&M University; Quentin Nakagawara, Butte Community College; Clifford R. Parks, University of North Carolina at Chapel Hill; Robert L. Rucker, Professor Emeritus, Texas A&M University; Carol Sherman, North Carolina State University; Jeanne Waller, Mount Hood Community College; David Williams, University of Illinois at Urbana; and Frank Yu, Cerritos College.

Special thanks also to Marge Coon, Rick Smith, Kirk Zirion, and the United States Department of Agriculture for supplying numerous photographs which appear in this book.

We are particularly grateful to the staff of the California Polytechnic State University Library, the staff of Holt, Rinehart and Winston, and to Lisa Asato, Heidi Howard, and Sandy Weidlich, who typed the manuscript.

L.W.R.
R.P.R., Jr.

CONTENTS

PRACTICAL HORTICULTURE

Introduction

The word "horticulture" has its origins from Latin. The Latin word "hortus," meaning garden, is the root of the first part. "Culturea," which composes the second part, means cultivation. Together they give a fairly accurate translation: garden cultivation.

Today horticulture is only one branch of the broad field of agricultural plant sciences. But it is among the most important, dealing with the growing of fruits, vegetables, and both indoor and outdoor ornamental plants. Two other branches of agricultural plant sciences, forestry and crop science, encompass the remaining plants in cultivation. Trees grown for other than ornamental use are usually covered under forestry. Grains, pasture grasses, and fiber crops such as cotton are discussed under agronomy or crop science.

Plant pathology (the study of plant diseases), entomology (the study of insects), and plant breeding are allied to all branches of agriculture. Their contribution to the successful cultivation of plants is immeasurably important.

Contrasted to these applied plant sciences is botany, the pure sciences of plants. Botanists frequently deal with noncultivated plants, researching plant classification, and internal plant processes among other topics. However, they also sometimes work with food or ornamental plants as do horticulturists.

THE DIVISIONS OF HORTICULTURE

Because plants are grown for a variety of uses, the art and science of horticulture has many divisions. The most familiar are listed here, along with types of employment opportunities found in each area.

Fruit Production

Fruit production, called *pomology,* covers the growing of tree fruits such as apples and citrus. It also deals with the "small fruits," including blueberries, grapes, and strawberries.

In the pomology industry, orchard supervisors establish and maintain fruit orchards throughout the country. Vineyard supervisors specialize in the cultivation of grapes, and their crops can be destined for either fresh eating or winemaking. In universities, research pomologists dedicate their efforts to improving fruit quality or yield. Other pomologists produce young fruit trees for sale to commercial orchards and have extensive knowledge of propagation.

Vegetable Production

Vegetable production is called *olericulture.* Olericulture covers the cultivation of all of the vegetables and also such crops as melons and rhubarb.

Because most vegetables are cultivated outdoors, most vegetable growers work in field vegetable production. They are responsible for the crop from planting through harvest, and their tasks may include weed control, insect and disease control, and irrigation.

In some areas of the country vegetables are grown in greenhouses. Vegetable growers in this specialization must understand both greenhouse operation and vegetable cultivation to perform their job.

Landscape Horticulture

The landscaping facet of horticulture is called *environmental horticulture* or *landscape horticulture.* The landscape horticulture industry employs thousands of people in hundreds of job specialties. Workers in nursery production propagate and grow millions of ornamental trees, shrubs, and groundcovers annually and ship them throughout the country. Horticulturally trained designers then develop landscape plans for residence and commercial buildings which are implemented by landscape installers, or *crews.* Later, landscape maintenance personnel maintain the plants in a healthy condition by watering, pruning, fertilizing, and controlling pests.

In garden centers, nursery salespeople furnish plants and gardening supplies to the public. They in turn are supplied with plants by salespeople representing the wholesale plant growers.

The field of landscape horticulture also includes nonindustry occupations. Botanic gardens and arboretums employ landscape horticulturists for teaching and maintaining the gardens. City beautification programs may require people having landscape horticulture training.

Turf

The branch of horticulture dealing with turfgrass culture is also an environmentally oriented science. People who work in this field can be involved in breeding new turfgrass varieties, controlling insects and diseases, or maintaining quality turf for golf courses, parks, and home lawns.

Floriculture

The production and sale of greenhouse-grown flowers and plants is a branch of horticulture known as *floriculture.*

Floriculturists can be involved in either the wholesale or the retail parts of the industry. Many wholesale growers are in charge of greenhouses where they raise flowers and indoor plants throughout the winter and bedding plants (seedlings of flowers and vegetables) in the spring. But in warm areas like Florida, California, and Texas, outdoor production of cut flowers and foliage plants is common.

Much of the wholesale growers' output reaches the public through retail florists, where floral designers create arrangements, bouquets, and corsages.

Indoor plant shops are dynamic small businesses in the retail portion of floriculture. They are one way for a floriculturist to start his own business. Some indoor plant shops expand into plant rental businesses and employ people to care for leased plants in offices and hotels.

RECREATIONAL HORTICULTURE

Two aspects of horticulture emphasize the emotional and recreational value of working with plants: horticulture therapy and home horticulture.

Horticulture Therapy

The value of horticulture as therapy for those with physical, intellectual, or emotional disabilities has only recently been recognized. The blind, deaf, wheelchair confined, or otherwise physically handicapped can all participate in some horticultural activity because of the many senses that are involved when growing a plant. A blooming flower can be appreciated by sight by many people, but through touch by others, and through smell by nearly all.

For the emotionally disturbed, intellectually handicapped, or imprisioned, horticulture has been proven to be a constructive and rewarding activity which teaches patience and responsibility.

Home Horticulture

As a recreational activity, horticulture is the nation's most popular pastime. The satisfaction of working with soil and plants to produce home-grown vegetables, lush houseplants, or a beautiful landscape is enjoyed by millions of people. Balcony gardens made up of dozens of small pots flourish in the cities, dispelling the notion that gardening is a suburban or farm activity.

There are only two prerequisites one must have to assure success in gardening endeavors: a sustained interest in growing plants and information on how to grow them. Contrary to popular belief, green thumbs are not born. They are made by experience and observation.

A local library or bookstore is one source of horticultural information. Because of the popularity of home horticulture, many books are currently available. They include information from basic gardening to specialized orchid and bonsai growing. Local nurseries are a second source of gardening information. Their employees are usually very willing and qualified to answer gardening questions.

Perhaps the most overlooked source of gardening help is the Cooperative Extension Service. This government agency is a federal, state, and county cooperative whose purpose is to offer free agricultural and home economics information. Almost every county in the United States has a Cooperative Extension Office, listed under county government in the phone book. The office title will vary among states but will usually be County Extension Service, Cooperative Extension Service, Agriculture Extension Service, or Farm Advisor.

Some of the many services provided by the Extension Service are free gardening publications, phone advice regarding gardening questions, and diagnosis of disease and insect problems. If the county office is large enough, it may provide additional services such as soil testing, public lectures, and recorded dial-a-garden-tip messages.

The County Extension Service satisfies the information requests of most gardeners. However, each state also has at least one Agricultural Experiment Station. The Experiment Station conducts research on agricultural topics including those of interest to home gardeners. The results are then published and made available to the public free or for a small fee. Lists of these publications are issued regularly and can be ordered by writing directly to the Agriculture Experiment Station of the state (see Table I-1).

The United States Department of Agriculture issues many publications for gardeners which are either free or inexpensive. A list may be obtained by writing to the Superintendent of Documents, U.S. Government Printing Office, Washington, D.C. 20402.

Those whose interest has narrowed to a specific group of plants should note that there are over 60 plant societies ranging from The African Violet Society of America to The Terrarium Association. These societies usually issue newsletters dealing with their area of specialization, and members frequently cooperate in exchanging plants. See Appendix A for the addresses of these plant societies.

Table I-1 State Agricultural Experiment Stations

ALABAMA
Agricultural Experiment Station
Auburn University
Auburn, AL 36830

ALASKA
Institute of Agricultural Sciences
University of Alaska
Fairbanks, AK 99701

ARIZONA
Agricultural Experiment Station
University of Arizona
Tucson, AZ 85721

ARKANSAS
Agricultural Experiment Station
University of Arkansas
Fayetteville, AR 72701

CALIFORNIA
Universitywide Administration
Agricultural Experiment Station
University of California
Berkeley, CA 94720

COLORADO
Agricultural Experiment Station
Colorado State University
Fort Collins, CO 80521

CONNECTICUT
Agricultural Experiment Station
P.O. Box 1106
New Haven, CT 06504

Agricultural Experiment Station
University of Connecticut
Storrs, CT 06268

DELAWARE
Agricultural Experiment Station
University of Delaware
Newark, DE 19711

FLORIDA
University of Florida
Institute of Food and Agricultural Sciences
Gainesville, FL 32601

GEORGIA
Agricultural Experiment Station
University of Georgia
Athens, GA 30602

GUAM
Resource Development Center
University of Guam
P.O. Box EK
Agana, GU 96910

HAWAII
Agricultural Experiment Station
University of Hawaii
Honolulu, HI 96822

IDAHO
Agricultural Experiment Station
University of Idaho
Moscow, ID 83843

ILLINOIS
Agricultural Experiment Station
University of Illinois
109 Mumford Hall
Urbana, IL 61801

INDIANA
Agricultural Experiment Station
Purdue University
West Lafayette, IN 47907

IOWA
Agricultural & Home Economics Experiment Station
Iowa State University
Ames, IA 50010

KANSAS
Agricultural Experiment Station
Kansas State University
113 Waters Hall
Manhattan, KS 66506

KENTUCKY
Agricultural Experiment Station
University of Kentucky
Lexington, KY 40506

LOUISIANA
Agricultural Experiment
Station
Louisiana State University
and A&M College
Drawer E University Station
Baton Rouge, LA 70803

MAINE
Agricultural Experiment
Station
University of Maine
105 Winslow Hall
Orono, ME 04473

MARYLAND
Agricultural Experiment
Station
University of Maryland
College Park, MD 20742

MASSACHUSETTS
Agricultural Experiment
Station
University of Massachusetts
Amherst, MA 01002

MICHIGAN
Agricultural Experiment
Station
Michigan State University
East Lansing, MI 48823

MINNESOTA
Agricultural Experiment
Station
University of Minnesota
St. Paul Campus
St. Paul, MN 55101

MISSISSIPPI
Agricultural and Forestry
Experiment Station
Mississippi State University
P.O. Drawer ES
Mississippi State, MS 39762

MISSOURI
Agricultural Experiment
Station
University of Missouri
Columbia, MO 65201

MONTANA
Agricultural Experiment
Station
Montana State University
Bozeman, MT 59715

NEBRASKA
Agricultural Experiment
Station
University of Nebraska
Lincoln, NB 68503

NEVADA
Agricultural Experiment
Station
University of Nevada
Reno, NV 89507

NEW HAMPSHIRE
Agricultural Experiment
Station
University of New Hampshire
Durham, NH 03824

NEW JERSEY
Agricultural Experiment
Station
Rutgers University
P.O. Box 231
New Brunswick, NJ 08903

NEW MEXICO
Agricultural Experiment
Station
New Mexico State University
P.O. Box 3BF
Las Cruces, NM 88003

NEW YORK
Agricultural Experiment
Station
Cornell University
Cornell Station
Ithaca, NY 14850

NEW YORK
Agricultural Experiment
Station
State Station
Geneva, NY 14456

NORTH CAROLINA
Agricultural Experiment
Station
North Carolina State
University
Box 5847
Raleigh, NC 27607

NORTH DAKOTA
Agricultural Experiment
Station
North Dakota State University
State University Station
Fargo, ND 58102

Table I-1 *(continued)*

OHIO
Ohio Agricultural Research
and Development Center
Ohio State University
Columbus, OH 43210

OKLAHOMA
Agricultural Experiment
Station
Oklahoma State University
Stillwater, OK 74074

OREGON
Agricultural Experiment
Station
Oregon State University
Corvallis, OR 97331

PENNSYLVANIA
Agricultural Experiment
Station
Pennsylvania State University
229 Agricultural Administration Building
University Park, PA 16802

PUERTO RICO
Agricultural Experiment
Station
University of Puerto Rico
P.O. Box H
Rio Piedras, PR 00928

RHODE ISLAND
Agricultural Experiment
Station
University of Rhode Island
Kingston, RI 02881

SOUTH CAROLINA
Agricultural Experiment
Station
Clemson University
Clemson, SC 29631

SOUTH DAKOTA
Agricultural Experiment
Station
South Dakota State University
Brookings, SD 57006

TENNESSEE
Agricultural Experiment
Station
University of Tennessee
P.O. Box 1071
Knoxville, TN 37901

TEXAS
Agricultural Experiment
Station
Texas A&M University
College Station, TX 77843

UTAH
Agricultural Experiment
Station
Utah State University
Logan, UT 84322

VERMONT
Agricultural Experiment
Station
University of Vermont
Burlington, VT 05401

VIRGINIA
Agricultural Experiment
Station
Virginia Polytechnic Institute
and State University
Blacksburg, VA 24061

VIRGIN ISLANDS
Agricultural Experiment
Station
P.O. Box 166
College of the Virgin Islands
Kingshill,
St. Croix,
VI 00850

WASHINGTON
Agricultural Experiment
Station
Washington State University
Pullman, WA 99163

WEST VIRGINIA
Agricultural Experiment
Station
West Virginia University
Morgantown, WV 26506

WISCONSIN
Agricultural Experiment
Station
University of Wisconsin
Madison, WI 53706

WYOMING
Agricultural Experiment
Station
University of Wyoming
University Station
Box 3354
Laramie, WY 82070

Selected References for Additional Reading

American Horticulture Society. *Directory of American Horticulture.* 1978 ed. Mt. Vernon, VA: The American Horticultural Society, 1978.

Baker, H. G. *Plants and Civilization.* Belmont, CA: Wadsworth, 1965.

Edlin, H. L. *Man and Plants.* London: Aldus Books, 1967.

Janick, J. *Horticultural Science.* 2d ed. San Francisco: W.H. Freeman, 1972.

United States Department of Agriculture. *Yearbook of Agriculture 1972: Landscape for Living.* Washington, D.C.: U.S. Government Printing Office, 1972.[1]

United States Department of Agriculture. *Yearbook of Agriculture 1977: Gardening for Food and Fun.* Washington, D.C.: U.S. Government Printing Office, 1977.

[1]Yearbooks of agriculture are issued by the USDA in the early part of the year. A copy may be obtained free by writing your congressman or woman.

PART I

FUNDAMENTALS OF HORTICULTURE

Chapter 1

Climate and Plant Growth

All plants outdoors and even those indoors are affected by the climate of the area in which they grow. Humans can alter growing conditions to a limited extent (such as by irrigating to compensate for scanty rainfall), but for the most part plants grow within natural climate restrictions. Therefore a knowledge of how basic climate factors affect plant growth is of prime concern for the gardener.

CLIMATE

The elements combining to make up a climate include primarily (1) temperature, (2) precipitation, (3) humidity, (4) light, and (5) wind. Each factor has a wide range of variation and can have a dramatic effect on plant growth.

Temperature

Temperature, and particularly minimum winter temperature, largely determines the geographic range over which a plant will grow. The lowest temperature which a plant can withstand is called its *cold hardiness* or *cold tolerance* and is usually expressed in degrees Fahrenheit or centigrade.

For many plants a temperature of about 28°F (−2.2°C) is critical for survival. At this temperature the liquid contents of the plant cells freeze and the plant dies. These plants are designated as *frost-tender,* and included in this group are many vegetables and flowers.

Plants which are able to survive temperatures less than freezing are called *frost-hardy.* They vary widely in their tolerance to subfreezing temperatures. Some plants will live to 10°F (−12.2°C), others to −5°F (−20.6°C) or even −30°F (−34.6°C). In some cases the woody portions or root system may be frost-hardy but not the flowers and leaves. For example, a late spring frost may kill the blossoms and young leaves of a fruit tree, but the branches will leaf out again and continue growing. Some garden flowers die to the ground in fall but grow back from the roots the following spring.

Although "frost-hardy" and "frost-tender" are rough classifications of cold tolerance, some houseplants and many tropical plants such as bananas incur "chilling injury" at temperatues less than 50°F (10°C).

Minimum winter temperature limits the areas where many plants can live, but for some plants a lack of cold prevents survival. Fruit trees such as apple and cherry and flowers such as peony and tulip require a period of cold temperatures to grow normally. They cannot be grown in tropical areas because of the comparatively warm winters. (The requirement of plants for cold is discussed in Chapter 3 under "Vernalization.")

FROST

Frost and the weather conditions which favor it are important because they signal an end to the growing season of frost-tender plants.

A frost can be one of two types. The first type is called *radiation frost,* which occurs when the air is cool and skies are clear. With this frost, solar energy accumulated by the soil and plants during the day is lost at night as heat radiating upward. If the day is warm and the night only slightly below freezing, the reservoir of heat in the soil and plants may last through the night. If not, frost damage will result.

When plants are covered by a barrier that blocks the flow of heat to the sky, radiation frost can often be prevented. In nature, clouds form this barrier, and radiation frosts seldom occur on cloudy nights. On clear nights when a radiation frost is likely, protection of the plants by layers of newspaper, cloth, or plastic film can serve the same function.

The second type of frost injury is caused by a cold air mass moving into an area. The air sweeps over the earth, carrying off heat from the plants, and the plants become frosted. This type of frost occurs irrespective of cloud cover.

Frost often leaves ice crystals on plants and the ground after its occurrence. These crystal-depositing frosts are called *hoar frosts.* They occur when moisture from humid air condenses onto the plants because of the drop in temperature.

If the moisture content of the air is low, the air temperature can reach freezing without moisture condensation, but plants will still be injured. These frosts are called *black frosts,* because the first sign of their occurrence is the blackening of injured plants.

The appearance of frost in some areas but not in others close by is usually due to elevation differences. Cold air is heavier than warm air and will flow downward and settle in the lowest area. Accordingly, a low-lying "frost pocket" may experience a frost 2 to 3 weeks earlier.

Precipitation

Precipitation can take many forms including rain, snow, hail, and sleet. It is the precipitation that falls as rain which is of greatest value to outdoor plants.

RAIN

Lack of rainfall all year or when other weather conditions favor growth is the limiting factor to plant growth in much of the western United States. In these areas irrigation water is essential.

Excessive rain or rain occurring out of season can be as detrimental to plants as lack of rainfall. Excessive rain can kill species adapted to dry areas, particularly if the water cannot drain away from the roots. Unseasonable rain near fruit harvest can make strawberries watery-tasting and can split unpicked apples and grapes. Frequent rains are also known to spread plant diseases by splashing microorganisms from one leaf to the next and to trigger the growth of dormant disease organisms present on plants.

SNOW

Precipitation in the form of snow also affects plant growth. During northern winters, snow insulates plants against continual freezing and thawing, conditions more harmful than continuously subfreezing temperatures. By covering the surface of the soil, snow also reduces evaporation.

SLEET, HAIL, AND FREEZING RAIN

Sleet, hail, and freezing rain can be very injurious to plant health. Sleet and hail injure plants by tearing young leaves and bruising or knocking off developing fruits. These forms of precipitation can fall with such force that tender plants will be completely beaten to the ground. Freezing rain frequently causes branch breakage on trees. As the rain hits cold plants, it quickly freezes. Layers of ice multiply the weight of branches, causing them to bend and eventually to break. Heavy snowfall can also cause breakage from its weight.

DEW

Dew occurs between sunset and sunrise. Although its effect on the growth of most plants is not great, it can be a factor in the spread of diseases in turfgrass. It is also an important water source for plants which absorb water readily through their leaves.

Dew is most likely to occur when the air is warm and humid. After sunset, the air temperature drops, and the atmosphere is unable to contain the amount of water it held during the daytime. The water is then precipitated out over the area in the small droplets called dew.

Humidity

Humidity also determines to some extent how well a plant will grow in an area. Often higher humidity improves plant growth by reducing the rate at which plants lose water from their leaves. But low humidity will not seriously hinder the growth of most outdoor plants, provided adequate water reaches the roots.

Plants growing indoors in cold-winter climates may commonly be damaged by insufficient humidity. Heated indoor air has an average relative humidity of only 10 to 25 percent. Since many plants used indoors are native to the humid tropics, the drop from the accustomed 50 to 90 percent to 25 percent can cause problems (see Chapter 17).

FOG AND MIST

Two other forms of humidity, fog and mist, can greatly influence plant growth. In dry Mediterranean climates the moisture from fog and mist is absorbed by plants and supplements the meager rainfall (Figure 1-1). At the same time this water vapor retards loss of water from leaves.

Light

Light duration and intensity affect the growth of outdoor and indoor plants, often controlling their flowering and growth rate. Light duration is dependent on distance from the

Figure 1-1 Fog in a redwood forest of coastal California. This moisture-laden air is the only source of water during the dry summer months.

equator and season. Directly at the equator, day and night are equal in length all year. Further north (the distance being marked by latitude lines), days become longer in summer and shorter in winter. In the northernmost areas of the United States 16-hour days in the summer are normal.

The duration of daily light a plant receives is extremely important. It determines how long the plant can manufacture the carbohydrate necessary for growth. This process, called *photosynthesis,* is dependent on the presence of light. The longer the duration of light, the more carbohydrate will be produced during photosynthesis.

Light intensity or brightness also controls photosynthesis, with moderate to high intensities most beneficial to the majority of plants. Latitude will alter intensity. At the equator the sun passes directly overhead all year. But in the northern United States it is nearly overhead in summer but low in the southern sky in winter. Thus the winter intensity is lowered because of the distance through the atmosphere which the rays must travel.

The lesser amount of light in winter does not seriously affect outdoor plants. Some enter a resting phase at the onset of cooler temperatures. For others the light intensity is sufficient to continue photosynthesis and growth continues. However, indoor plants can be seriously retarded by the decrease in light because many are growing at subsistence light levels already. The dual handicaps of short light duration and low light intensity combine to make the climatic factor of light critical for indoor plants.

Wind

Wind is the fifth climatic element governing plant growth. Ocean wind limits the number of plants that can be grown near the shoreline, both by the salt deposits, which brown plant leaves, and by the intensity, which whips and breaks the aboveground portions. In dry areas hot desert winds increase the rate of moisture loss from the soil and plant leaves, intensifying existing drought conditions.

Even in cold northern areas wind is an inhibiting factor to plant health. Subzero winds damage evergreens by removing moisture from the foilage. In frozen ground the plant cannot replace the water and injury results.

FACTORS MODIFYING CLIMATE

Natural factors modifying climate include elevation, terrain, and the nearby presence of large bodies of water.

Elevation

Changes in elevation can give areas only a small distance apart completely different climates for plant growth. The higher the elevation of an area, the colder the average year-round temperature, with every rise of 300 feet (91.4 meters) causing an average temperature decrease of 1°F (0.56°C).

Terrain (Topography)

Changes in terrain also alter climate, particularly rainfall. Most rain storms move from west to east. In a narrow range of hills, the west-facing slopes often receive considerably more rainfall than the east-facing slopes. The heaviness of the water in rain clouds prevents them from passing over the mountains until most of the precipitation has been released.

Sloping terrains can also vary several degrees in temperature from the top portion of the slope to the lowest part or valley. Cool air, which is heavier than warm air, will flow down a slope and collect at the bottom, a phenomenon called *air drainage* (Figure 1-2).

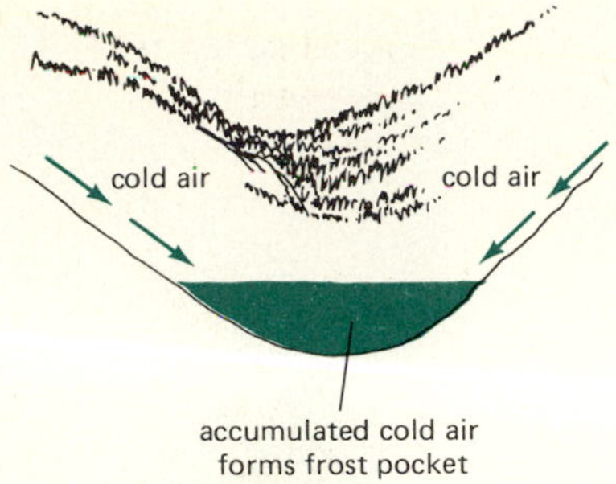

Figure 1-2 Typical cold-air drainage on sloping land.

Consequently, plants growing higher on the slope are less likely to be damaged by cold than those at the bottom.

Bodies of Water

Large bodies of water such as the Great Lakes and oceans exert strong effects on the climate of nearby land. They moderate temperatures all year, making the summer cooler and the winter warmer than in inland areas. The enormous volume of water absorbs heat in the summer, making adjacent land cooler, and gives it off in the winter, raising the temperature.

The influence of large bodies of water on agriculture is considerable. Warmer winter temperatures found along the Great Lakes make these regions large commercial growing areas for fruits such as grapes, cherries, and peaches. Similarly, most of the Pacific coast is warm enough to grow cool-season vegetables all winter, but inland these crops would freeze.

Climate Modification by Humans

People can modify climate either intentionally or unintentionally. Among the most widespread unintentional causes of climate change are air pollution and smog.

Air pollution can be composed of both gases (carbon monoxide, for example) and small airborne pieces of matter called *particulates*. Combined together they become smog, a persistent cloud that hangs over metropolitan areas.

One way smog affects plant growth is by temperature modification. A smog cloud acts like an insulation layer over an area and prevents heat from escaping to the atmosphere. Thus areas with smog cover have warmer summers than nearby areas without smog, and plants may be damaged by summer heat.

Smog also decreases the intensity of sunlight which reaches the ground, giving plants less light from which to manufacture carbohydrate. Particulates settle onto leaves and further decrease the light intensity reaching the plants.

Another way smog affects plant growth is by the toxic gases it contains. The gases enter the leaves, causing injury and sometimes death. Plants vary in their susceptibility to smog injury, ranging from smog-tolerant to highly smog-intolerant. Plants classified as intolerant (many pines, for example) cannot be grown in smoggy areas.

MICROCLIMATES

Microclimates (Figure 1-3) are small pieces of land which have slightly different climate characteristics than the surrounding area. They may be less windy, shadier, moister, warmer, or in any other way different from the typical climate. These differences affect plants growing in the microclimate, sometimes helping and sometimes hindering growth.

Microclimates may be either artificially made or naturally occurring. They can be formed by the natural terrain and vegetation or unintentionally created by the erection of structures such as buildings, fences, and roadways. Although they often go unnoticed, they can be used by the gardener to provide the specialized growing conditions favored by different plants. In some cases they make it pos-

Figure 1-3 A microclimate for growing ferns. The north side of the house provides shelter from excess sun and wind.

sible to grow plants not normally considered cold-hardy in that area.

Natural Outdoor Microclimates

One example of a natural microclimate is a frost pocket. In addition to being colder, this microclimate might also be moister due to water runoff and would be suitable for moisture-loving plants not easily damaged by cold.

Another natural microclimate can be found under trees. The area is shadier and cooler in summer and would be good for shade-loving plants. It would also provide ideal conditions for summering indoor foliage plants.

Artificial Outdoor Microclimates

Many microclimates are created by people. The area under the eaves of a house is a microclimate found around most homes; plants there live under different environmental conditions than plants living away from the house. For one thing, they are shielded from rainfall by an eave, so the soil will be drier. Moreover, the wall affords protection from cold or drying winds, and it radiates warmth from the heated dwelling. During cool periods this additional warmth speeds up growth, but in summer the heat can become excessive. A parking-lot tree can also be subjected to excess heat because of radiation from the pavement, and scorch can result (Figure 1-4).

At night when temperatures drop, objects which absorbed heat during the day begin losing it to the cooler night air. Heat radiating from the ground and buildings can protect plants from frost and extend the growing season.

Wind can be intensified or lessened in a microclimate. A plant by a wall will generally receive wind protection unless it is directly in the path of prevailing winds. However, when two walls are parallel, a "wind tunnel" can be formed. Wind conditions will be worse there than in a completely unprotected area. Such wind tunnels are frequently formed between closely spaced buildings.

Indoor Microclimates

Microclimates are usually thought of as being outdoors. However, indoor microclimates with varying temperature, light, and humidity levels can be found throughout a house or apartment.

Microclimates varying by temperature exist in most homes. Because heat rises, upstairs rooms may be several degrees warmer than downstairs ones. A basement may be up to 20°F (11.2°C) cooler than the remainder of the home on a hot day, and the north side of the house may stay considerably cooler than the south side.

Within a room, cool microclimates will be found adjacent to windows in winter because of heat loss through the glass.

Microclimates with varying humidity levels are also common. In the kitchen and bath, where water is used, evaporation raises the humidity. A basement is usually also a humid place.

Light microclimates are easy to detect, being governed by the size and location of room windows. Obviously the further from a window a plant is, the less light it will receive. However, each foot of distance can be thought of as a different microclimate. In addition, south-facing windows are maximum-light microclimates, east and west moderate, and north-facing low-light microclimates.

Figure 1-4 Scorch on a maple leaf due to excess heat.

PLANT-GROWING ZONES OF THE UNITED STATES

The question of where in the country a specific plant can be grown is important to every gardener. There is little to worry about when plants are bought at local nurseries because nurserymen seldom carry plants which do not grow in their area. But when a plant is ordered from a mail-order dealer, the consumer must check to make sure the plant he has selected will grow in his area.

Since minimum winter temperature is often what determines where a plant can be grown, knowing its minimum survival temperature is essential. This and other valuable information is usually given in the written descriptions of the plant. It can also be looked up in a gardening book from a library.

United States Department of Agriculture Plant Hardiness Zones

After determining the minimal survival temperature of a plant, it is then necessary to determine the minimum winter temperature of one's own area. To simplify this procedure, the United States Department of Agriculture has prepared maps showing zones of minimum winter temperatures throughout the United States and Canada (see Figure 1-5). Ten zones are outlined, ranging from completely frost-free areas in Florida, Texas, and California to a −50°F (−45.8°C) or colder zone in Canada. The map takes into account cold mountain climates in warmer zones, as shown in Colorado. But for the most part it is made up of bands which correlate minimum winter temperature with distance from the equator.

The USDA zone map is not the only plant hardiness map that exists, but it is the most widely used. Some states whose climate is influenced by many factors have designed more detailed maps for use by its inhabitants. California is one such state, and the University of California at Davis lists 21 separate climate zones in the state due to the influences of the Pacific Ocean and mountain ranges.

Zone Maps Based on Overall Climate Characteristics

Several zone maps divide the country into general climate types. Such maps are helpful because they include such climatic factors as precipitation and light intensity as well as temperature. They give a more general view of the basic climate in large sections of the country. Such information is particularly valuable today when people move so frequently.

One such general climate classification developed by Dr. J. B. Edmond, Professor Emeritus, Mississippi State University, divides the continental United States into five regions (Figure 1-6).

1. *Northeast and Great Lakes Region.* This area is characterized by comparatively cool summers, long cold winters, and moderate rainfall. Light intensity is moderate in the summer and low in the winter owing to a persistent cloud cover. Summer daytime temperatures may reach 90–95°F (32.2–35°C), but they are for the most part cool enough to grow deciduous fruits (from trees which drop their leaves) like apples, pears, and cherries. At the same time, temperatures in summer are high enough to grow most warm-season vegetables (like tomatoes and beans). But summers are not long enough to mature varieties of watermelons and sweet potatoes which require a long maturation period. The rainfall throughout this region is usually sufficient and no irrigation is needed.

2. *Southeast and Gulf Coast Region.* This region has long warm summers, short mild winters, high light intensity, and moderate to heavy rainfall. The temperature of this region is quite variable: in the northern half yearly temperatures are cool enough for deciduous fruits; in the southern half they are warm enough for semitropical fruits like citrus and avocado. Winter temperatures along the coast are mild enough to grow most vegetables, and much of the winter produce for the East is shipped from this region.

3. *Great Plains Region.* Wide fluctuation in temperature, low rainfall, and high light intensity are typical of this region. Summer temperatures in the northern part of this large region may equal those in the southern section, but winter temperatures differ greatly. In the north they can drop as low as −30°F (−1.1°C), whereas near the Gulf they are quite mild. The short growing season in summer limits the commercial production of crops in the north, but parts of Oklahoma and Texas produce vegetable crops throughout the year with the aid of irrigation.

4. *Intermountain Region.* This region has variable temperatures, low rainfall, and high light intensity. The area is characterized by a wide range of elevations which, along with

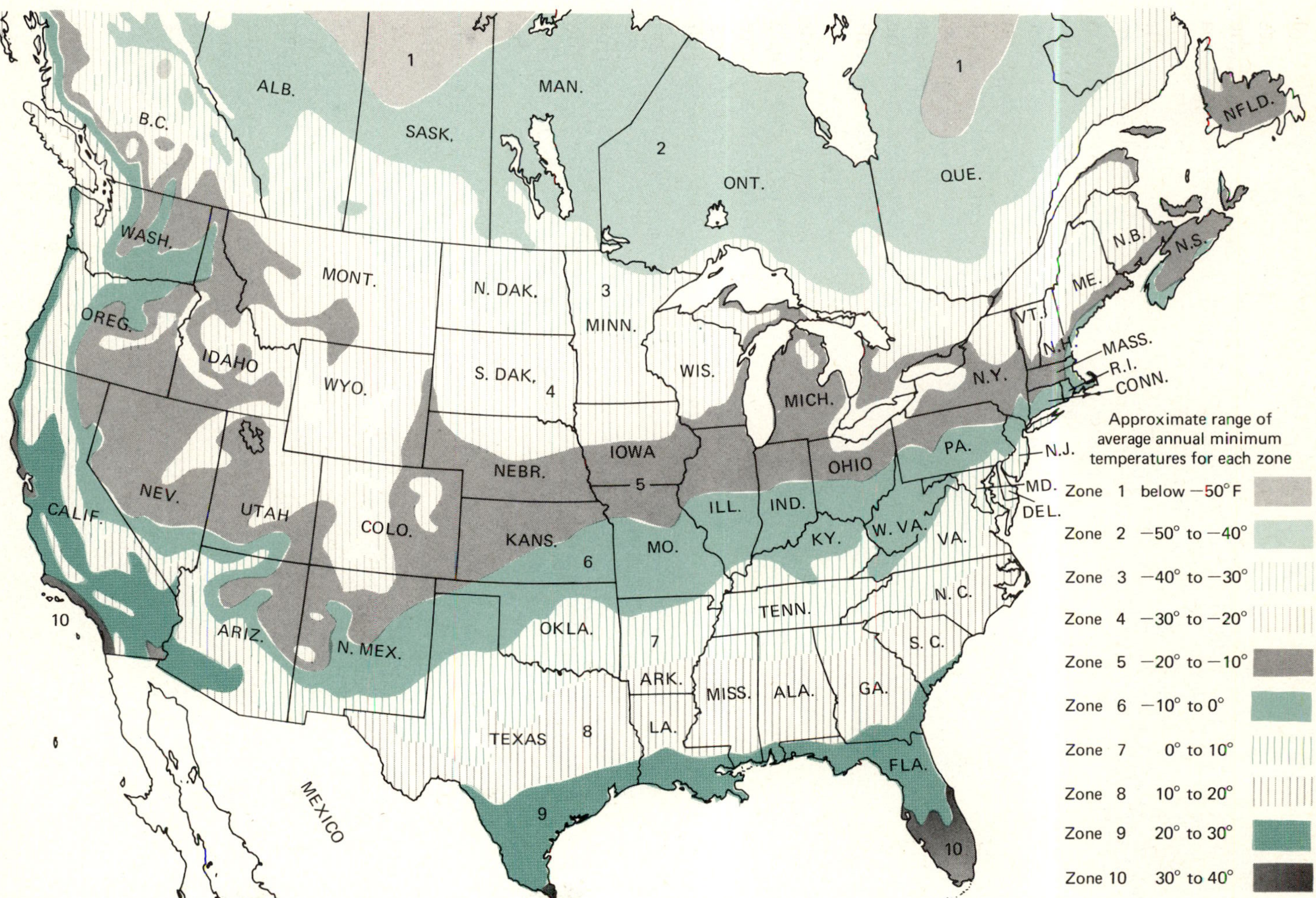

Figure 1-5 Plant hardiness zone map showing minimum winter temperatures throughout the continental United States and Canada.

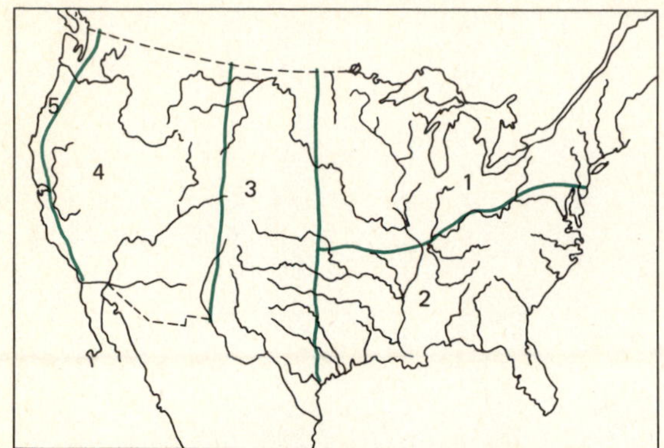

Figure 1-6 Climates of the continental United States. (1) Cool-humid of Northeast and Great Lakes. (2) Warm-humid of Southeast and Gulf Coast. (3) Subhumid of Great Plains. (4) Arid of Intermountain region. (5) Summer-dry of Pacific Coast.

latitude, gives this region variable temperatures. In the southern part, temperatures are high, and the growing season is long. In the remainder, temperatures are generally low, and the growing season is of moderate length. Snowfall in the winter provides water for summer irrigation.

5. *Pacific Coast Region.* The Pacific Coast divides into two sections lengthwise: the shoreline and the intermountain valleys directly over the coastal mountains. Summer temperatures are cool along the coast due to the Pacific Ocean but high in the valleys east of the coastal mountains. Winter temperatures are mild throughout the region but particularly so along the ocean where frost is infrequent. Winter is the rainy season in this area, with seasonal rainfall ranging from very high in the northern half to very low toward the Mexican border. The amount of rainfall also influences light intensities, making them moderate in the north but high through the south. This region is very productive agriculturally, producing fruits and vegetables all year.

Selected References for Additional Reading

Bannister, P. *Introduction to Physiological Plant Ecology.* New York: Halsted Press, 1976.

Daubenmire, R. F. *Plants and Environment.* 3d ed. New York: Wiley, 1974.

Gates, D. M. *Man and His Environment: Climate.* New York: Harper and Row, 1972.

Ortho Books. *Weather-Wise Gardening.* San Francisco: Chevron Chemical Company, 1974.[2]

Treshow, M. *Environment and Plant Response.* New York: McGraw-Hill, 1970.

Trewartha, G. I. *An Introduction to Climate.* New York: McGraw-Hill, 1968.

United States Department of Agriculture. *Yearbook of Agriculture 1941: Climate and Man.* Washington, D.C.: U.S. Government Printing Office, 1941.

[2]Different editions for separate regions of the country.

Chapter 2

Botanical Nomenclature, Anatomy, and Physiology

PLANT NOMENCLATURE AND CLASSIFICATION

Botanical nomenclature is the orderly classification and naming of plants. It is a subject which unnecessarily intimidates people unfamiliar with it because they believe one must know Latin in order to use the system. But the botanical naming system is not overly complex, and it does not require any background in Latin. In fact, a surprising number of common names are the same as botanical names, such as iris, fuchsia, and citrus.

There are several reasons for using botanical names in place of common names. The first reason is the universality of botanical names. Latin names are the standard worldwide system for communicating the identity of plants. The words *Pyrus communis* would be recognized as the common pear by people in the Orient, Africa, the Americas, or Europe, provided they were familiar with botanical nomenclature.

Consistency is the second reason. Common names such as "daisy" or "ivy" can refer to any number of similar plants. If only common names were used to identify these plants, confusion would result.

Knowing the botanical name of a plant can also give clues to its growing requirements. For example, the common names bunny ears, beaver tail, and prickly pear do not indicate that these plants have anything in common, but their botanical names are similar and so is their culture.

The Origin and Construction of Botanical Names

The branch of botany which deals with the naming of plants is called *taxonomy,* and people doing the work are taxonomists. The naming system used dates back 250 years to the Swedish botanist Carolus Linnaeus, the "father of botany." He named and published the first references to many plants, using a naming formula called the "binomial system." This system specifies that a plant name must have at least two parts.

For example, in the botanical name *Tagetes patula* (French marigold), *Tagetes* is called the *genus* (*genera,* plural). The second word, *patula,* is called the *specific epithet*. When combined, these two words form the plant *species*.

For ease of understanding, the genus can be thought of as the surname of a plant. It indicates that the plant is related to other plants in that same genus. The specific epithet can be thought of as the first name of the plant, the name that distinguishes it from its relatives in the same genus.

In the two plants *Tagetes patula* and *Tagetes erecta,* for example, *Tagetes patula* is a small marigold with thumbnail-sized blossoms. *Tagetes erecta,* on the other hand, is a tall marigold with fist-sized blossoms. It is apparent that both plants are marigolds. But since they grow so unlike each other, they are given different specific epithets and hence become separate species.

For many wild plants, a two-part name is sufficient to distinguish them from similar related plants. However, new plants which are mutations or offspring of existing species are constantly being developed. To distinguish them from their parents, a third part called the *variety,* or *cultivar,* is added to the name. Common marigold varieties, for example, would be *Tagetes patula* 'Lemondrop' and *Tagetes patula* 'Petite Harmony.'

The botanical names given to plants are not assigned haphazardly. A set of rules called the International Code of Botanic Nomenclature is adhered to strictly when new species are named. Although the genera and specific epithets must be in Latin-like form, the words are not always derived from that language. The names of prominent botanists, statesmen, and even explorers were often Latinized and incorporated in plant names. Frequently though, a Latin name conveys information about the plant it represents. The specific epithets *rubra, alba, atropurpurea,* and *variegata* are used with many plant names to connote colors. Table 2-1 defines some common specific epithets.

Writing Botanical Names

The writing of botanical names follows a prescribed pattern. The genus is always capitalized, followed by the specific epithet beginning with a lowercase letter. Any variety name follows and may be set off from the species by "v.," "var.," "cv.," or single quotes. For example, the variety of ash known as Modesto ash may be written as:

Fraxinus	*velutina*	'Modesto'
F.	*v.*	cv. Modesto
F.	*v.*	v. Modesto
F.	*v.*	var. Modesto.

Note that in a listing, the genus or specific epithet is written out once and abbreviated thereafter.

Another botanical abbreviation is the use of the genus name followed by the word "species," "sp.," or "spp." The sp. indicates an unknown species in that genus. Spp. is plural and is used to refer collectively to all species of that genus. The latter abbreviation technique is useful when discussing a broad topic such as the diseases of oaks (*Quercus* spp.) or the culture of different philodendrons (*Philodendron* spp.).

Pronouncing Botanical Names

Unlike the standardized method of assigning and writing botanical names, pronouncing the names is somewhat arbitrary. Botanists themselves disagree on certain pronunciations. Common usage has become the basis for pronunciation.

Botanical Classification of Plants

SEED- AND SPORE-BEARING PLANTS

Ninety percent of the plants grown are flowering plants which reproduce by seed, but a few common ones are not. These *Pteridophytes* have a completely different reproductive system based on spores (see Chapter 3). Ferns are the most widely known exception, and they belong to a branch of the plant kingdom which emerged early in plant evolution (Figure 2-1).

The seed plants are further divided into

Table 2-1 English Translations of Common Specific Epithets and Varieties

Latin epithet	English translation	Plant example
alba	White	White mulberry (*Morus alba*)
atropurpurea	Dark purple	Red-leaf Japanese barberry (*Berberis thunbergii* 'Atropurpurea')
aureum	Golden	Golden pothos (*Epipremnum aureum*)
carnosa	Fleshy	Wax plant (*Hoya carnosa*)
compacta	Compact	Dwarf Oregon grape (*Mahonia aquifolium* 'Compacta')
esculentus	Edible	Okra (*Abelmoschus esculentus*)
fastigiata	Erect	Upright English oak (*Quercus robur* 'Fastigiata')
floribunda	Free-flowering	Showy crab apple (*Malus floribunda*)
glauca	With white or gray coating	Blue fescue (*Festuca ovina* 'Glauca')
grandiflora	With large or showy flowers	Glossy abelia (*Abelia* × *grandiflora*)
horizontalis	Horizontal	Creeping juniper (*Juniperus horizontalis*)
nanus	Dwarf	Dwarf European cranberry bush (*Viburnum opulus* 'Nanum')
nidus	Nest	Bird's nest fern (*Asplenium nidus*)
occidentalis	From the Western Hemisphere	American plane tree (*Platanus occidentalis*)
officinalis	Medicinal	Rosemary (*Rosmarinus officinalis*)
pendula	Hanging	Wandering Jew (*Zebrina pendula*)
rotundifolia	Round-leaved	Button fern (*Pellaea rotundifolia*)
rubrum	Red	Red maple (*Acer rubrum*)
sativus	Cultivated	Cucumber (*Cucumis sativus*)
semperflorens	Ever (semper)-flowering (florens)	Wax begonia (*Begonia* × *semperflorens-cultorum*)
stellata	Starlike	Star magnolia (*Magnolia stellata*)
sylvestris	Of the forest	Scotch pine (*Pinus sylvestris*)
tuberosum	Bearing tubers	Potato (*Solanum tuberosum*)
variegata	Variegated	Variegated rubber plant (*Ficus elastica* 'Variegata')
vulgaris	Common	Common bean (*Phaseolus vulgaris*)

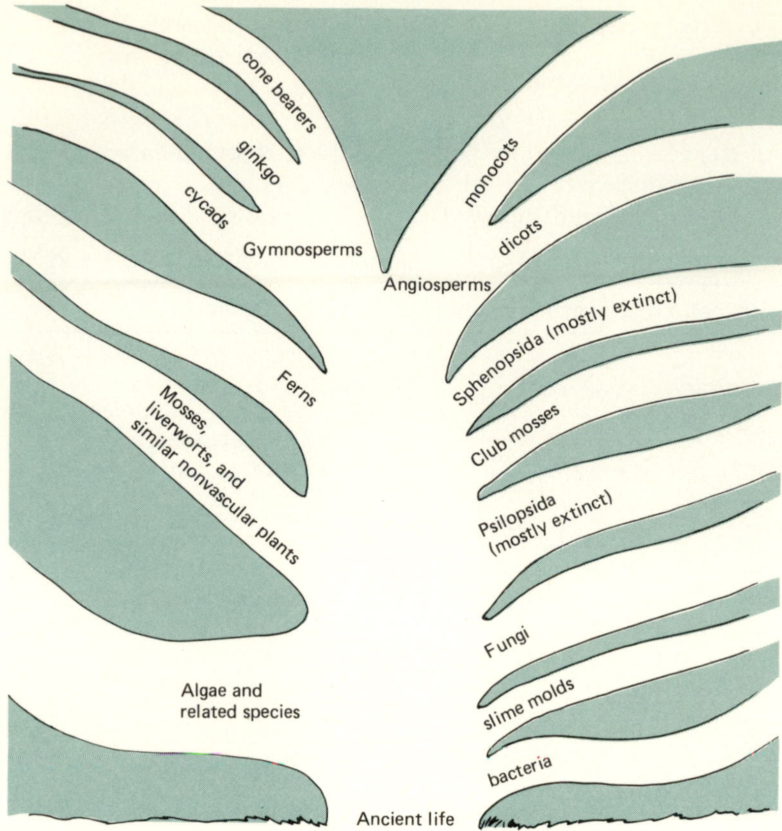

Figure 2-1 Evolutionary history of plants.

two groups. Gymnosperms are the smaller of these two groups, consisting primarily of evergreen cone-bearing plants such as pines, spruces, junipers, and yews. Their foliage is generally needlelike, and they do not produce showy flowers. Of the indoor plants, cycad, podocarpus, and Norfolk island pine are the common gymnosperms.

Angiosperms compose the majority of cultivated plants, including such diverse plants as rose, cabbage, and palm. It is within the angiosperm group that all ornamental flowering plants and nearly all food plants are found. The primary identifying characteristic of angiosperms is the growth of seeds completely within the plant ovary, which swells to become the fruit (see Chapter 3).

MONOCOTS AND DICOTS

The angiosperms are further separated into the divisions Monocotyledoneae and Dicotyledoneae, or monocot and dicot for short. The monocots are the smaller of the two groups and are usually nonwoody plants which have a short stem and overlapping, whorled leaves, a form called a *rosette*. Their leaves are frequently long and narrow, with parallel veins running the length. Other identifying characteristics include flower petals fused at the base to form a bell and, often, a bulb or fleshy root system (see "Modified Roots").

The dicots, on the other hand, are frequently woody plants which may grow to a large size. Their leaves have a branching vein pattern, and their flowers have a variety of shapes. Most trees and shrubs are dicots, as well as most fruits and garden vegetables. Included among the monocots are all grasses, lilies, irises, onions, cattails, and most flowering bulbs.

Being able to recognize the difference between monocots and dicots is not of academic interest only. Whether a plant is a monocot or dicot can also determine its method of propagation and its susceptibility to weed killers.

PLANT FAMILIES

The division and subdivision of plants lead ultimately to the smallest division, variety (Figure 2-2). For the most part the intermediate subdivisions are of interest to botanists, with the exception of plant families, the division prior to genus. Each family links a number of genera having like characteristics together. These families have both Latin and common names. Some families are listed in Table 2-2 with a sampling of the genera each includes as well as a key identifying characteristic.

Basic knowledge of plant families can help identify an unknown plant. For example, by remembering that all pea family members have pod-type fruits, you could deduce that such diverse plants as mimosas and snapbeans are both in that family.

In addition, plants in the same family are often susceptible to the same diseases. If a gardener has had a recurring problem with a disease, he could avoid buying plants with similar susceptibilities through familiarity with the family characteristics.

Horticultural Classification of Plants

Botanists are interested primarily in classifying plants by their evolutionary relationship to each other. Horticulturists, on the other hand, frequently classify plants according to their use.

Most cultivated plants are valued either for their ornamental or for their edible value, with a few multipurpose plants such as herbs and ornamental vegetables. Within the edible group, plants are commonly classified as the following.

- Tree fruits: apple, peach, plum, orange, avocado, and so on
- Nuts: pecans, walnuts, filberts, almonds, and so on
- Small fruits: blackberries, raspberries, strawberries, figs, grapes, and so on
- Vegetables: cauliflower, tomatoes, sweet corn, peppers, squash, and so on
- Herbs: rosemary, basil, dill, thyme, and so on
- Grains: wheat, rice, corn, oats, and so on

The last category, grains, is generally not covered under horticulture. Grains are classified as field crops and fall under the study of crop science or agronomy.

Within ornamentals, the classification

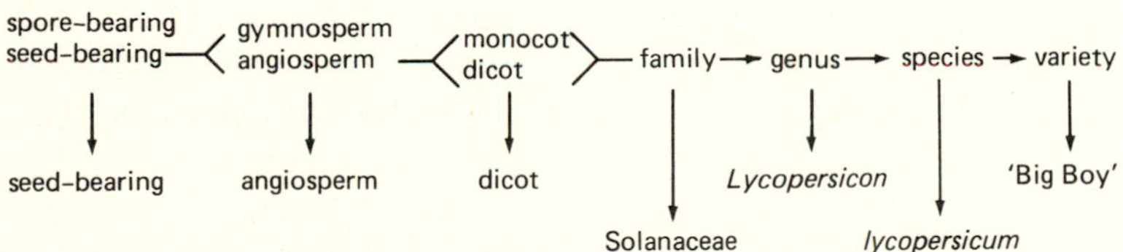

Figure 2-2 A botanical classification of the garden tomato variety 'Big Boy.'

Table 2-2 Plant Families and Their Key Identifying Characteristics

Family name		Key identifying characteristics	Genera	Plants
Latin	Common			
cactaceae	Cactus family	Fleshy growth habit and spines	*Mammillaria, Schlumbergera, Opuntia*	Pincushion cactus, Christmas cactus, prickly pear
compositae	Aster family	Daisy or buttonlike flower heads composed of many tiny flowers which resemble petals	*Senecio, Chrysanthemum, Rudbeckia*	Florist cineraria, chrysanthemum, blackeyed Susan
cruciferae	Mustard family	Flowers with four petals arranged crosswise	*Brassica, Alyssum*	Cabbage, alyssum
gramineae	Grass family	Narrow, parallel-veined leaves encircling the stem at the base	*Poa, Avena*	Bluegrass, oats
labiatae	Mint family	Square stems; flowers have two lips	*Coleus, Mentha, Plectranthus*	Coleus, mint, creeping Charlie
leguminosae	Pea family	Long, podlike fruits, characteristic flowers	*Lathryus, Phaseolus, Gleditsia*	Sweet pea, green bean, honey locust
liliaceae	Lily family	Thin, grasslike leaves; bell-shaped flowers with six petals	*Chlorophytum, Lilium, Yucca*	Spider plant, Easter lily, yucca
rosaceae	Rose family	Flower petals in multiples of five; flowers have many stamens	*Malus, Prunus, Pyracantha*	Apple, flowering cherry, firethorn
umbelliferae	Carrot family	Flowers in an umbrellalike cluster	*Daucus, Anethum Petroselinum*	Carrot, dill parsley

systems link the plant to its form and use.

- Trees: oak, magnolia, pine, eucalyptus, and so on
- Shrubs: juniper, yew, lilac, azalea, and so on
- Vines: clematis, ivy, and so on
- Groundcovers: bugleweed, periwinkle, and so on
- Garden flowers: tulip, marigold, daylily, and so on
- Turfgrasses and lawn substitutes: bluegrass, St. Augustine grass, dichondra, and so on
- Houseplants: umbrella plant, African violet, spider plant, and so on

Plant Identification

"What plant is that?" is one of the most frequently asked questions among gardeners. Determining the identity of a plant may be relatively easy if the plant is commonly grown, but it becomes difficult and time-consuming if the plant is a wild flower or other uncultivated species. As a first step, cut off a stem with foliage and preferably flowers and/or fruits. Without the latter it will be more difficult, unless the plant is a foliage houseplant. Enclose the sample in a plastic bag to lessen wilting, having sprinkled it previously with water if possible. Make note of the relative size of the plant, whether it is woody or nonwoody, and where it was growing (sun, shade, marsh, landscape, and any other relevant conditions). All this information can be helpful.

To have the unknown plant identified, begin by showing the sample to a knowledgeable local nurseryman. If he is unfamiliar with it, consult mail-order plant catalogs with color pictures or library books with photographs. If the plant was growing wild, botanical references may be necessary. However, these are usually complicated to use (unless you have a background in botany) because they include a considerable amount of technical terminology that can become confusing.

As a last resort, dry the sample by pressing it flat between several layers of newspaper and weighting it with a book. When the sample is completely dry (several weeks), it can be mounted on cardboard with tape and mailed to a university department of botany or horticulture along with the information on size, growing location, and so on, previously noted.

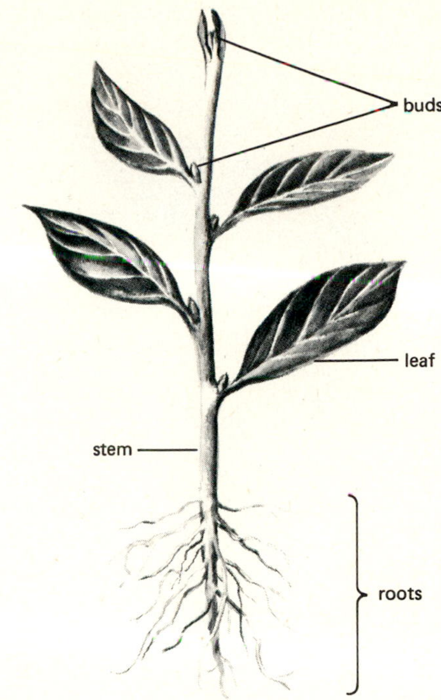

Figure 2-3 Vegetative parts of a typical plant.

PLANT ANATOMY

Understanding how the parts of a plant form the whole, as well as the structure of those parts, is necessary if you are to grow plants successfully.

Vegetative Plant Organs and Their Function

Nearly all cultivated plants are composed of a limited number of basic parts, or organs: leaves, stems, buds, and roots. Collectively these are termed *vegetative organs* because they are not part of the sexual reproductive system of the plant (Figure 2–3). In its simplest form, a young plant would have the organs discussed in the next section.

LEAVES

Leaves are the most obvious part of most plants. They are usually made up of two main parts: the wide section termed the *blade* and the *petiole,* or *leaf stem* (Figure 2-4). The angle formed between the petiole and its supporting stem is the *leaf axil,* and in that axil will be a *bud.*

Figure 2-4 is a diagram of what is called a simple leaf. In contrast to this simple leaf, Figure 2-5 shows several compound leaves. A compound leaf is composed of individual leaflets but will only have a bud at the base of the entire leaf.

Leaves of the types shown in Figures 2-4 and 2-5 are common leaf forms on the angiosperm plants. But the leaves of cone-bearing gymnosperms are often scalelike or needlelike. Figure 2-6 shows sample leaves of gymnosperms.

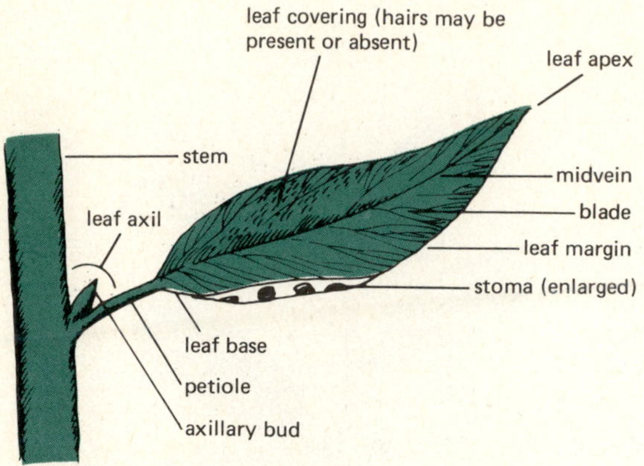

Figure 2-4 Parts of a simple leaf.

Parts of the Leaf Blade. Looking closely at a simple leaf, the visible parts are as follows.

Margin. The leaf margin is its outside edge. The margin may be toothed, barbed, lobed, or in any other way different from smooth (Figure 2-7). Leaf margin characteristics are a means of identifying plants. Barbed or spined margins can protect a plant from being eaten by grazing animals.

Figure 2-6 Leaves of gymnosperm plants.

Veins. The patterns of veins in leaves can be seen by looking closely or holding the leaf up to the light. Dicot plants have a central vein (midvein) with many branch veins. Monocots have parallel veination running the length of the leaf (Figure 2-8).

Figure 2-5 Compound leaves.

Figure 2-7 Barbed (left), toothed, (middle), and smooth (right) margins of leaves.

Leaf apex. The apex is the tip of the leaf. The apex may be pointed, blunt, notched, or a number of other shapes.

Leaf base. This is the part of the blade which attaches to the petiole or directly to the stem (if the leaf has no petiole).

Leaf covering. Any hair, scales, or film on the leaf blade can be considered a leaf covering. Almost all leaves have an invisible wax layer, called the *cuticle*, which prevents water loss from the leaf surface. A leaf may have hair or scales in addition to this cuticle (Figure 2-9).

Stomata. Stomata are minute openings found primarily on the undersides of leaves (Figure 2-9). They are flanked on either side by *guard cells* which open and close the opening according to environmental conditions. Stomata regulate the flow of gases and water vapor in and out of the leaves.

Modified Leaves. Modified leaves have evolved to perform functions not customarily associated with leaves. The "petals" of poinsettia are modified leaves called *bracts.* Twining tendrils and spines are also thought by some botanists to be modified leaves.

STEMS

Growth of stems increases plant height or breadth. The stem transports carbohydrates and water. It is also the site of leaf and flower attachment.

Modified Stems. Not all stems grow aboveground. Underground bulbs contain short

Figure 2-8 Venation types. (a) Elm leaf with midvein and branch veins. (b) Maple leaf with several main veins and numerous branch veins. (c) Amaryllis leaf with parallel veins.

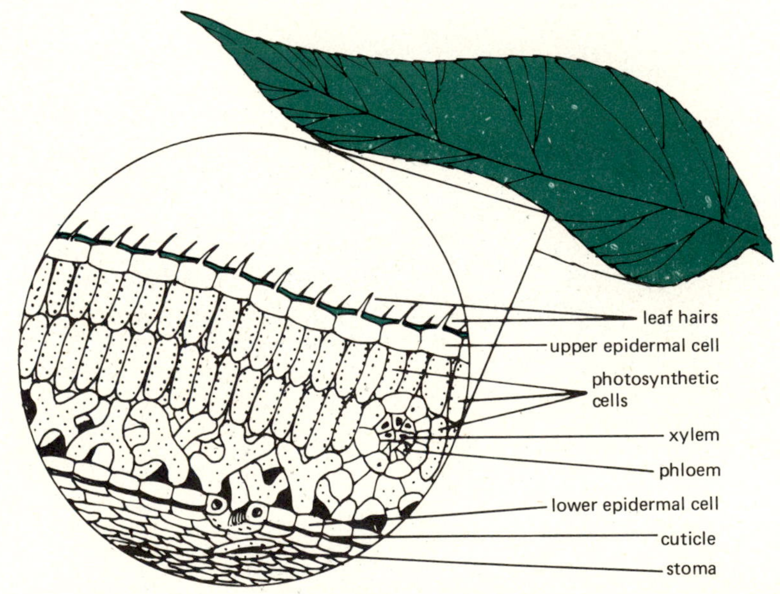

Figure 2-9 Cross section of the internal parts of a leaf.

stems and attached scales (onion layers, for example) which are modified leaves. The potato is an underground stem thickened to serve as a storage site for carbohydrate. Frequent use is made of modified stems in plant propagation, discussed in more detail in Chapter 4.

Shortened stems found in the aboveground portions of plants create a rosette plant form (Figure 2-10). Rosette plants are easily recognized because of the shortened stem; the leaves appear to grow from one central point and radiate outward like the overlapping petals of a rose. Many plants grow in rosette form, including African violets, cabbages, strawberries, and daylilies.

Arrangement of Leaves on the Stem. Leaves can be arranged in an opposite or an alternate pattern on a stem. The pattern of leaf arrangement is important and is used in identifying plant species.

As the names imply, leaves arranged oppositely grow in pairs, and those arranged alternately are attached stepwise down a stem (Figure 2-11). The site at which a leaf is or was once attached is called a *node,* and the section of stem between nodes is the *internode.*

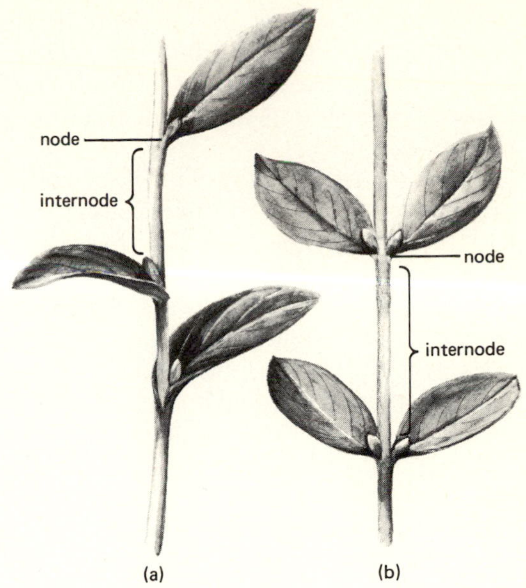

Figure 2-11 Alternately (a) and oppositely (b) arranged leaves attached at nodes on the stem.

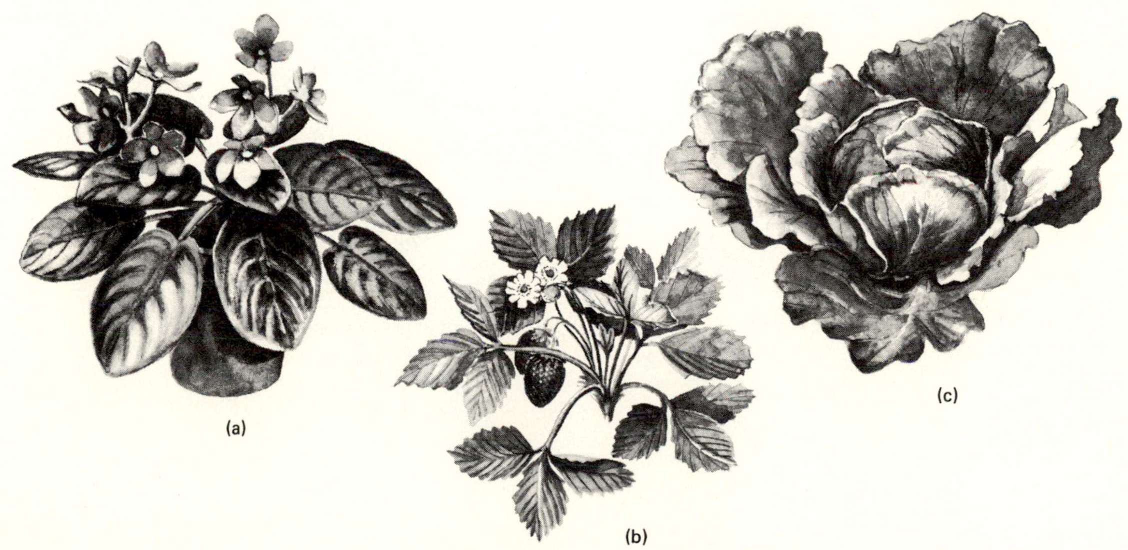

Figure 2-10 Typical plants with a rosette form. (a) African violet. (b) Strawberry. (c) Cabbage.

A bud will be present at each node, although it can be undeveloped and difficult to distinguish. Knowing the locations of nodes and internodes on a stem is important in pruning and propagation.

Internal Stem Structure. The life of a plant is dependent on the correct functioning of its vascular system, the network of pathways that moves carbohydrates, minerals, and water within the plant. These pathways, which can be thought of as a plant's circulatory system, are divided into xylem and phloem. Water and nutrients are moved up in the xylem from the roots, through the stem, and out to the leaves. Carbohydrate manufactured in the leaves moves down the phloem to the roots.

Dicots have their xylem and phloem arranged in adjacent rings inside the stem (Figure 2-12). The phloem is always found outside the xylem. Monocots, however, have xylem and phloem together in "vascular bundles" scattered throughout the stem. Both xylem and phloem extend not only through the stem of a plant but also up the petiole and into the leaf blade.

The vascular cambium is the third component of a stem. The cambium manufactures new xylem and phloem cells to increase the transporting abilities of the vascular system. The cambium is a thin layer of rapidly dividing cells found between the xylem and phloem areas in the stem.

A second type of cambium is found in woody stems. This cambium, called the *cork cambium,* produces bark cells and is found directly beneath the bark surface (Figure 2-12).

BUDS

A bud contains immature plant parts and can be of three types: a vegetative, flower, or mixed bud. A vegetative bud contains a leaf or leaves and sometimes an embryonic shoot. A flower bud includes the rudiments of one or more flowers and is frequently larger than a vegetative bud. A mixed bud contains the potential for both shoot and flower tissues.

The arrangement of buds on a stem places them in either *axillary* or *terminal* positions (Figure 2-13). Axillary buds are found in the leaf axils, usually directed above the point at which the petiole attaches to the stem. They may be either vegetative, flowering, or mixed. Terminal buds occur on the tips of stems.

In some cases buds form at other sites, such as along the veins of a leaf or at the petiole and leaf blade junction. Buds produced in these unusual places are called *adventitious.*

ROOTS

Roots, the fourth vegetative plant organ, anchor the plant in its growing position. They also absorb and transport water and nutrients for use in photosynthesis. This absorption

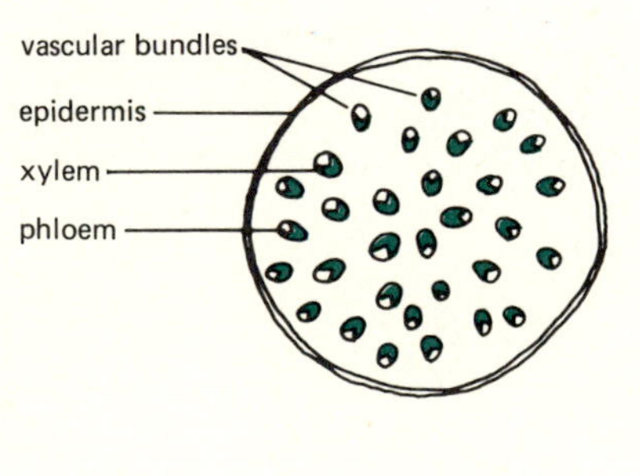

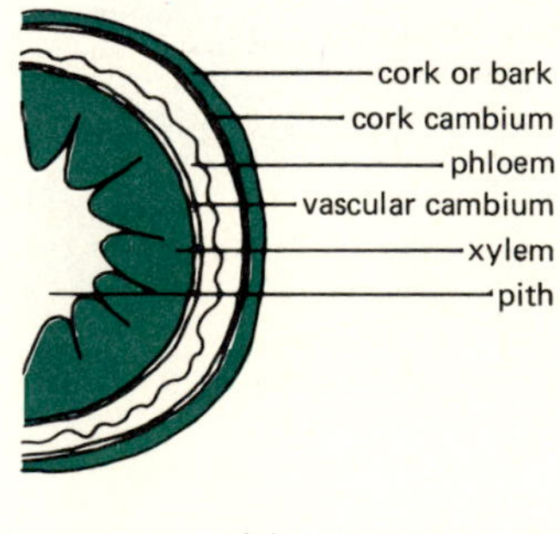

Figure 2-12 Cross sections showing arrangement of vascular tissue inside monocot (a) and dicot (b) stems.

takes place mainly through the young tips of roots, rather than through the older, larger parts of the root system. In addition, carbohydrates are stored in the older parts of the root system. This storage capacity of roots is very important because carbohydrates are the source of energy for generating new shoots if the top of the plant is damaged.

Root Types. In the development of a plant from seed, a single "primary root" is first produced. In many plants this primary root does not branch and remains the only site of anchorage and absorption. Such roots, called *taproots*, are found typically on citrus, dandelions, and carrot plants (Figure 2-14).

In most plants the primary root branches

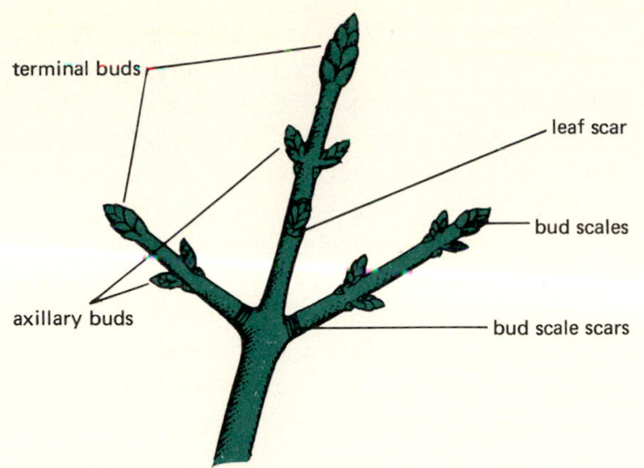

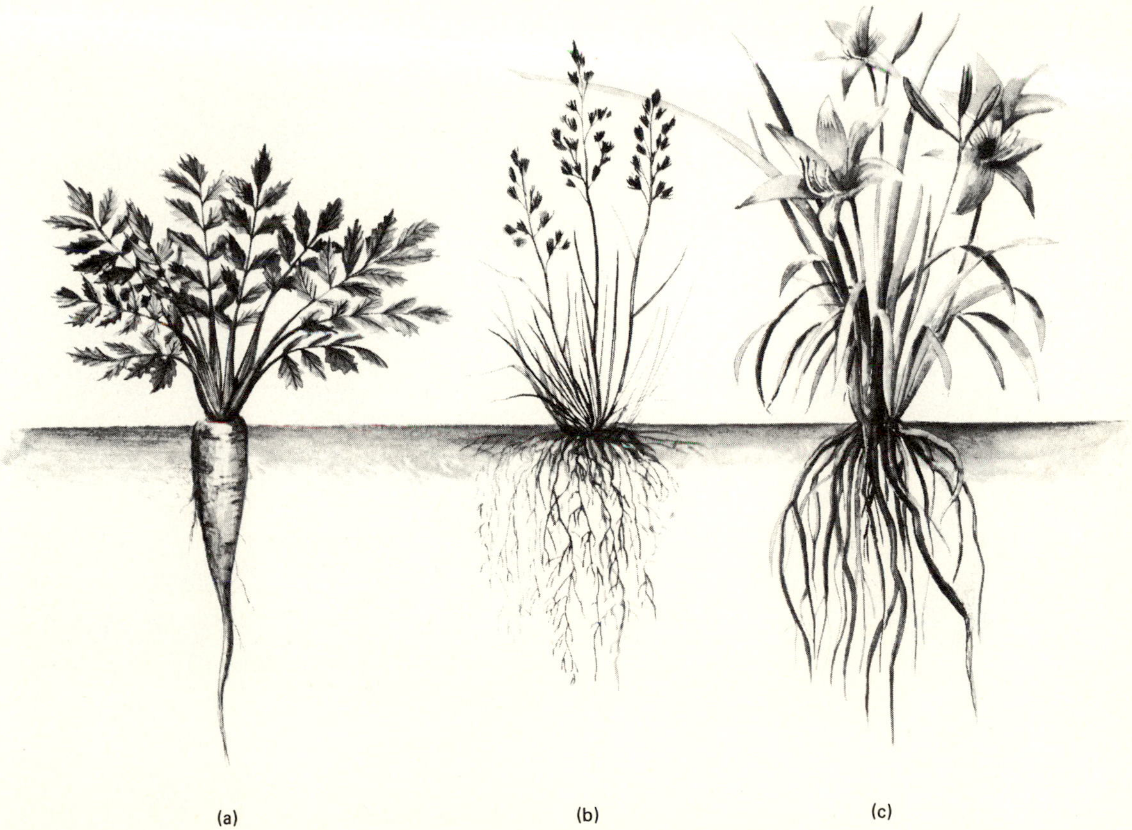

Figure 2-13 Stem with axillary and terminal buds.

(a) (b) (c)

Figure 2-14 (a) Primary taproot of parsnip. (b) Fibrous roots of grass. (c) Fleshy roots of daylily.

to form a netlike mass called a *fibrous root system* (Figure 2-14). Grasses, African violets, and squash are examples of plants with fibrous roots.

A third type of root system is made up of roots called *fleshy roots* (Figure 2-14) which are thick like tap roots but branch like fibrous roots.

Parts of the Root Tip. Regardless of whether a plant has tap, fibrous, or fleshy roots, the structure of the root tips will be the same. At the tip of every root is the root cap. The cap is made up of a layer of cells that prevents damage to the rest of the root as it pushes through the soil (Figure 2-15). Directly behind is the root meristem, an area which produces new cells to lengthen the root and replace cells scraped off as the cap pushes against the soil.

Beyond the meristem is the zone of elongation. Cells produced in the meristem are deposited here, grow larger, and lengthen the root. Cell lengthening in this zone pushes the root through the soil.

Water and nutrients are absorbed in the fourth zone. Cells in this area form fragile root hairs (Figure 2-16) about 1/16 inch (1–2 millimeters) long through which water and nutrients enter the plant.

Root hairs often live only a few weeks, to be replaced by hairs developing on younger cells. In this way root hairs constantly contact fresh soil, assuring a steady supply of nutrients.

Modified Roots. Although most roots resemble those in Figure 2-14, modified roots exist, much the same as do modified leaves and stems.

Roots fashioned for storing large quantities of carbohydrates are the most common modified roots. Sweet potatoes and beets are examples of this modification.

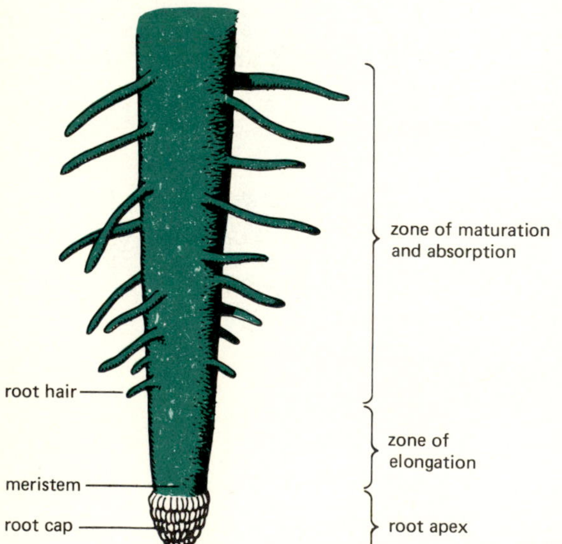

Figure 2-15 Parts of a root tip.

Figure 2-16 Root hairs of a radish seedling.

(a)

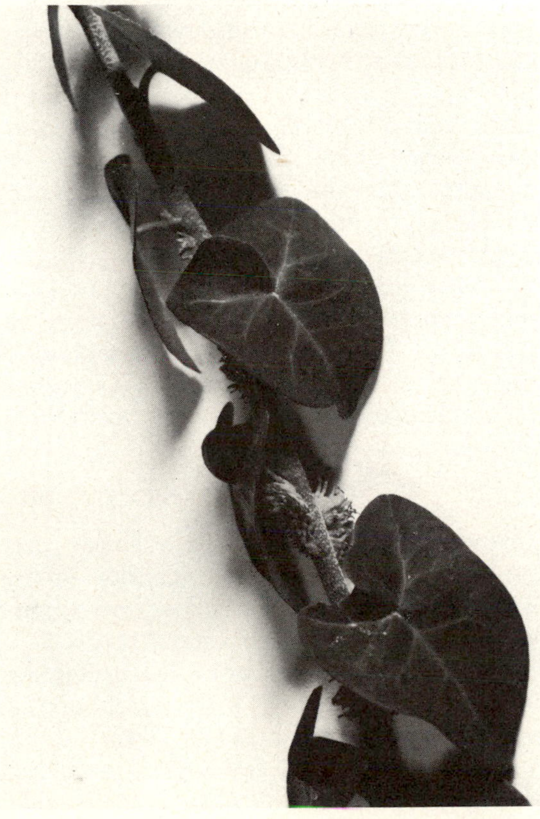

(b)

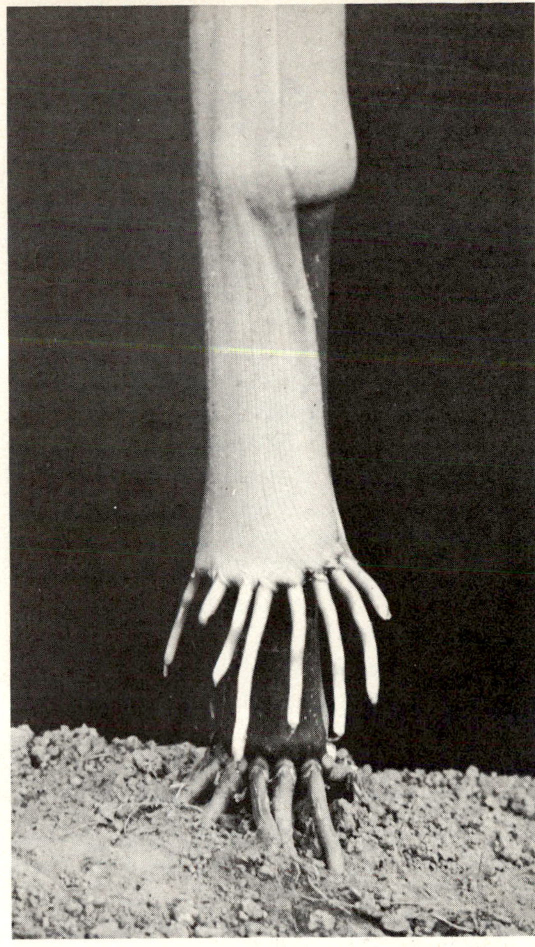

(c)

Figure 2-17 Modified roots. Aerial roots of orchid (a) and English ivy (b). (c) Prop roots of corn.

Adventitious roots grow from a stem instead of from other root tissues. The adventitious roots of ivy and orchids (Figure 2-17) attach them to trees. Adventitious prop roots of corn (Figure 2-17) anchors it against uprooting by wind.

Sexual Plant Organs

Flowers and fruits are found on most cultivated plants at some time during their life cycle. They may appear continuously, seasonally, or only once, depending on the species and its growing conditions. Although the biological function of flowers is seed production, flowers are not essential for the continuation of the species and can be of only ornamental interest to a gardener.

FLOWER TYPES

Flowers are classified by their sexual parts. A flower with both male and female parts is called *perfect;* most species produce perfect flowers. However, a flower can be exclusively male or female, with blooms of opposite sexes needed to complete seed formation. Some species have male and female blooms on one plant, as is the case with cucumber, sweet corn, and pecan. Others have male flowers on one plant and female on another, with plants of opposite sexes needed to produce seed on the female. These *dioecious* species include holly, date palm, bittersweet, and ginkgo.

Occasionally flowers are sterile, that is, they have neither male nor female parts. Sterile flowered plants are frequently prized by gardeners because they often have double sets of petals. The doubleness is caused by sexual parts mutating into petals.

FLOWER STRUCTURE

A typical flower is made up of a central female structure, the pistil, and numerous surrounding male parts called stamens (Figure 2-18).

The pistil contains three main parts: the knoblike top called the *stigma*, the thin vertical shaft called the *style,* and the bulblike base called the *ovary.* It is the ovary that contains the eggs which will develop into seeds.

Stamens are composed of the *anther* at the top, which produces pollen (functionally similar to sperm), and the supporting stalk, called the *filament.*

Surrounding or beneath the sexual parts are the petals and the sepals (the green, leaflike base beneath the petals). Under the sepals is the receptacle, the enlarged base on which the flower rests. The colored petals attract

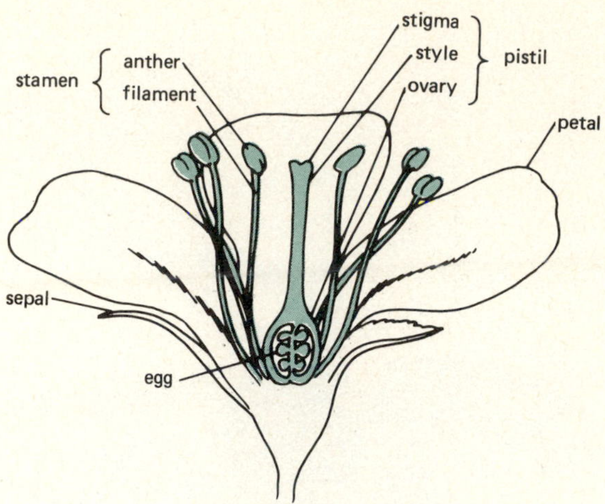

Figure 2-18 Parts of a complete flower.

insects, which transfer the pollen to the female flower parts. They also shield the pistil and stamens from the weather. The sepals form the covering for the developing flower bud, protecting it from damage.

Figure 2-18 shows a typical form for a flower, but not all flowers have such easily distinguished parts. Instead the parts will be in various forms due to the divergent evolution of plant families and genera. Figure 2-19 illustrates other typical flower forms.

Moreover, not all flowers are produced singly. Often they occur in clusters, and even though each flower may be tiny, when grouped together they form a showy head. Flower clusters are named on the basis of the positioning of the blooms (Figure 2-20).

FRUIT PRODUCTION AND TYPES

After flowering is complete, the flower parts shrivel and drop, with the exception of the ovary. The ovary continues developing, producing a swollen fruit with seeds.

Fruits are either fleshy or dry, depending on whether the ovary is juicy and plump at maturity or dry and shrunken. Fruits of cucumber, orange, and crabapple are fleshy, whereas pea pods, ears of corn, and pine cones are dry.

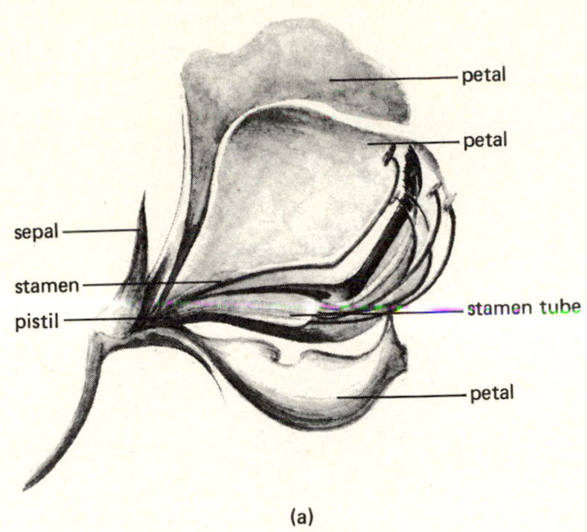

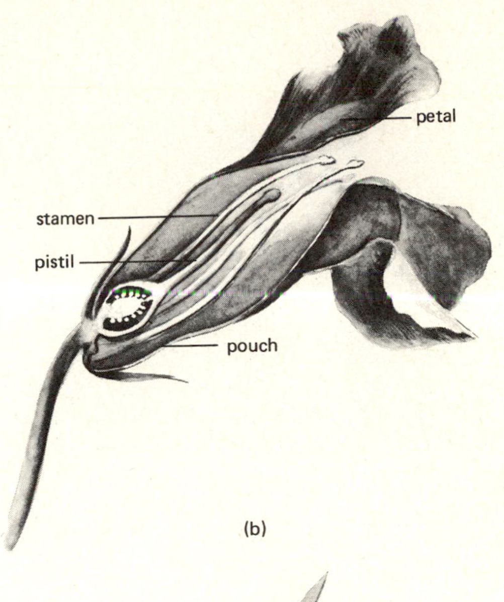

Figure 2-19 Three flower forms. (a) Pea. (b) Snapdragon. (c) Stonecrop.

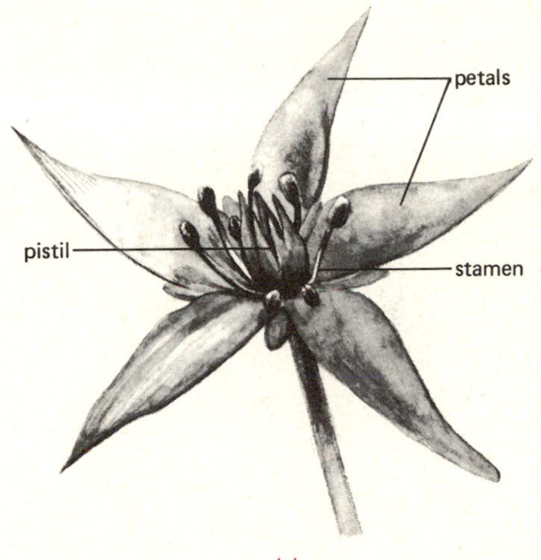

PHYSIOLOGICAL PROCESSES

Photosynthesis, respiration, translocation, absorption, and transpiration are complex chemical reactions which occur constantly in plants. Understanding these processes will equip the gardener with the information needed to make decisions on proper plant care.

Photosynthesis

Photosynthesis is the process by which plants manufacture the carbohydrates they require to live and grow. The process is made possible by the green chemical chlorophyll. In the presence of water, light, and carbon dioxide, chlorophyll produces carbohydrates (starches and sugars) and, as a by-product, oxygen gas (Figure 2-21).

Light, the first essential for the photosynthetic process, generally comes from the sun. With the exception of plants which naturally grow in shady habitats, bright light intensity and long light duration maximize photosynthesis and increase growth.

Water, the second essential, is drawn from the soil by a process called *absorption.* Any lack of water will slow photosynthesis, and a severe deficiency will stop it completely.

Carbon dioxide, the third ingredient, is taken in from the air through leaf stomata. It is through this same pathway that the excess oxygen by-product is released into the air. Lack of carbon dioxide is seldom a cause of low photosynthesis. However, photosynthesis

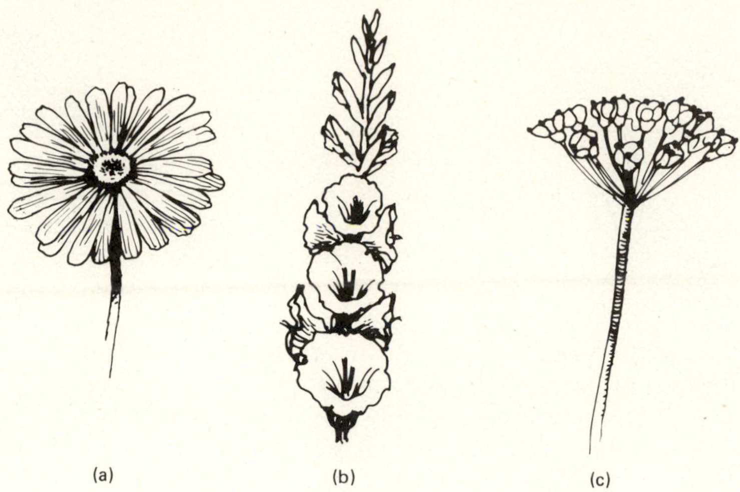

Figure 2-20 Three types of inflorescences. (a) Head. (b) Spike. (c) Umbel.

can be increased through the addition of supplemental carbon dioxide, which is occasionally used in commercial greenhouse flower production.

Respiration

Respiration can be thought of as an opposite reaction to photosynthesis. Whereas photosynthesis makes carbohydrates out of energy, respiration breaks carbohydrates down into energy.

Respiration requires two ingredients—carbohydrates and oxygen—and yields three products—energy, water, and carbon dioxide gas (Figure 2-22). The carbohydrates are already present in the plant. Oxygen is taken from the air primarily through the stomata on the leaves and also through the roots. Of the three products resulting from respiration, energy is the primary one used by the plant. It is needed for activities such as absorption of water and nutrients, chlorophyll formation, and development of flowers or fruits. However, the carbon dioxide and water released in the reaction can also be reused in photosynthesis.

The photosynthetic process requires light but respiration does not. Respiration continues constantly in every cell of the plant, even at the same time photosynthesis is taking place. It does, however, generally occur at a slower rate than photosynthesis so that more carbohydrates will not be respired than are being photosynthesized.

The relative rates of respiration and photosynthesis govern the health and growth rate of plants. To grow a successful crop of vege-

$$\text{water + light + carbon dioxide gas} \xrightarrow[\text{presence of chlorophyll}]{\text{in the}} \text{carbohydrates + oxygen gas}$$

Figure 2-21 A simple equation for photosynthesis.

$$\text{carbohydrate + oxygen gas} \longrightarrow \text{energy + water + carbon dioxide gas}$$

Figure 2-22 A simple equation for respiration.

tables, it is estimated that the photosynthesis rate should be 8 to 10 times higher than the respiration rate. This assures that the majority of the carbohydrates will accumulate in the plant or be used for new growth. Correct environmental conditions are crucial for obtaining a high photosynthesis/low respiration ratio. Maximum photosynthesis can be encouraged by moderate to high light intensity, sufficient water in the soil, and warm temperatures. Respiration, on the other hand, can be slowed by cooler temperatures. Since lowering the temperature also decreases photosynthesis, the best growing conditions are warm days with cool nights. Under these conditions the respiration rate is decreased at night, but daytime photosynthesis is not affected.

When photosynthesis and respiration rates are equal, there is no new plant growth. The carbohydrates being photosynthesized are entirely used up for the respiration of existing cells.

Occasionally a plant will have a higher rate of respiration than of photosynthesis. An example would be a small tree in a forest of larger trees. Shading from the larger trees reduces photosynthesis, while respiration remains moderate and steady. All the sugar being produced, plus part of the stored reserve, is used for respiration. As a result, the tree stops growing and slowly dies.

Translocation

The movement of carbohydrates, minerals, and water through the plant is called *translocation.* As discussed earlier, the xylem is responsible for water and mineral movement and the phloem for sugar movement throughout the plant.

The forces which cause upward movement of water and minerals through the xylem are a matter of debate among plant scientists. It is sufficient to think of the movement as caused by upward pressure from the roots combined with pressure deficit when water vaporizes from the leaves.

The rate of water movement in the xylem is surprisingly rapid. This can be demonstrated by noting the time required to revive a wilted flower after it is placed in water. Complete revival of a flower on a 6-inch (15-centimeter) stem in 15 to 30 minutes is not unusual.

Movement of sugars in the phloem is called a "source to sink" movement because the sugars are taken from their site of manufacture (source) to a storage or utilization site (sink). One of the primary sinks for carbohydrate is the roots. It is stored there as starch compounds made up of many sugar molecules combined together. The starches make compact, high-energy packages which can be efficiently stored.

A second sink is developing flowers, fruits, and seeds. In order to survive, seeds must be well supplied with stored carbohydrate to provide energy for respiration after separation from the parent plant. The constant sugar flow to these sites also accounts for the continual increasing in size of fruits from bloom until maturity and explains why fruits are often sweet.

A third sink for photosynthesized sugars is the growing regions (called *meristems*) at the stem and root tips. They require the constant supply of sugars to grow and to manufacture compounds such as pigments for flowers and fruits, waxes for cuticle formation, and cellulose fibers to strengthen the stems.

Absorption

Water and mineral absorption is the fourth internal plant process. It is primarily a function of the roots; however, it can also occur through the leaves by a process called *foliar absorption.*

Absorption of materials through root cells is a complex topic, which is still being heavily researched today. To date, research has explained absorption as consisting of two processes: active absorption and passive absorption. Active absorption is an energy- and oxygen-requiring process, the primary process responsible for the absorption of mineral compounds. Passive transport, on the other hand, is a result of natural physical forces and is the main way in which water is absorbed.

Transpiration

Transpiration, the fifth process, is the opposite of absorption in the sense that it involves the loss of water from the plant. Water in liquid form in the plant is released into the air as vapor from the leaves, stems, and flowers. Of all the water taken from the soil by absorption, a relatively small amount is used in photosynthesis; the rest is lost through transpiration.

As mentioned earlier, the main parts of the leaf through which water is lost are the stomata. These are opened and closed by the bordering cells called *guard cells.* An increase in the amount of water in the leaf and guard cells forces the stomata open and transpiration occurs. As the leaf loses water, guard cells go limp and the stomata close, decreasing transpiration.

Transpiration can be increased or decreased by environmental factors such as temperature, humidity, and wind. High temperature increases transpiration. High humidity decreases transpiration because humid air contains almost its capacity of water vapor, thus suppressing the transpiration of more water from the leaf. Wind usually increases transpiration by drawing newly transpired water vapor from the leaf. The dry air surrounding the leaf then causes rapid transpiration again.

Summary

These five processes—photosynthesis, respiration, translocation, absorption, and transpiration—are basic concepts in plant science. Knowledge of each and how each is related to the others is crucial to a thorough understanding of plant growth.

Selected References for Additional Reading

Bell, C. R. *Plant Variation and Classification.* Belmont, CA: Wadsworth, 1967.

Bonnor, J., and A. W. Galston. *Principles of Plant Physiology.* San Francisco: W. H. Freeman, 1971.

Devlin, R. M. *Plant Physiology.* 3d ed. New York: Van Nostrand Reinhold, 1975.

Esau, K. *Anatomy of Seed Plants.* 2d ed. New York: Wiley, 1977.

Lawrence, G. H. M. *Taxonomy of Vascular Plants.* New York: Macmillan, 1951.

———. *An Introduction to Plant Taxonomy.* New York: Macmillan, 1955.

Noggle, G. R., and G. J. Fritz. *Introductory Plant Physiology.* Englewood Cliffs, N J: Prentice-Hall, 1976.

Salisbury, F. B., and C. Ross. *Plant Physiology.* Belmont, CA: Wadsworth, 1969.

Smith, A. W. *A Gardener's Dictionary of Plant Names.* London: Cassell, 1972.

Thomas, M., S. L. Ranson, and J. A. Richardson. *Plant Physiology.* 5th ed. London: Longman, 1973.

Chapter 3

Plant Growth and Development

PLANT LIFE CYCLES

In a typical life cycle a plant will pass through two states, growth and rest. The growth phase occurs with suitable environmental conditions (temperature, rainfall, and the like) and is characterized by activities such as flowering, shoot lengthening, and leaf production. The rest phase, often called *dormancy*, also reflects conditions imposed by the environment such as cold, drought, or inadequate light. It is characterized by slowed or stopped growth, leaf drop, or death of the entire aboveground parts of the plant.

Dormancy should not be thought of solely as an outdoor plant phase. Many indoor plants also undergo rest periods, induced by the shorter days of winter or by the plant adhering to its normal outdoor growth cycle.

To understand the importance of growth and dormancy in the life of a plant, let us first classify plants according to the length of time they live (Figure 3-1). Plants fall into one of four botanical groups: annual, biennial, perennial, and monocarp.

Annual

An annual plant grows to maturity, flowers, produces seeds, and dies during one growing season. It experiences dormancy only as a seed.

To most people an annual is a frost-tender flowering or vegetable plant grown in summer. In mild climates there are also winter annuals which are planted in fall to bloom through the winter. These common-usage definitions usually do not clash with the botanical definition of an annual. However, exception must be made for plants which would live more than 1 year in their mild native climate but which are not able to survive winter when planted in cold regions. These plants are not true annuals but function as annuals because of climate. Included in this group are vegetables such as tomato and flowers such as alyssum and impatiens.

Biennial

A biennial plant grows for one season, becomes dormant during winter, resumes growth

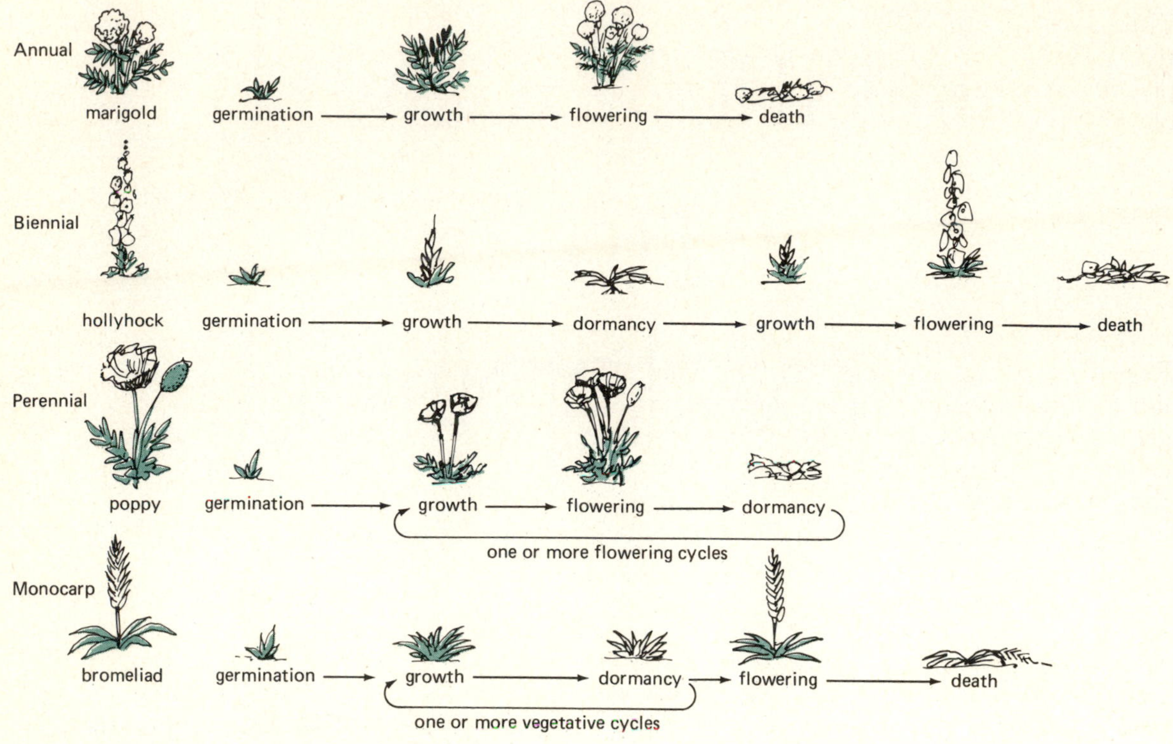

Figure 3-1 The four life-cycle types of plants.

in spring, flowers, and dies. The life cycle is completed in 2 years, hence the term "biennial." Biennials are relatively uncommon among cultivated plants. Some of the better-known examples are hollyhocks, parsley, carrots, and sweet William.

Perennial

Perennials, or plants which live for more than two growing seasons, make up the third and largest botanical division. Perennials are further subdivided into woody and herbaceous types. Woody perennials have sturdy, bark-covered stems which remain alive over the winter. The plants increase in size each year, producing additional new wood, and may grow to a large size. Tree and shrubs are examples of woody perennials.

The herbaceous perennial is the type of plant most often referred to as "perennial." Herbaceous perennials only produce tender leaves and stems. In most climates their aboveground portions die in fall, and only the belowground portions survive the winter. But in frost-free areas they may continue active growth all year.

Because they do not produce woody aboveground parts capable of overwintering, herbaceous perennials seldom are over about 5 feet (1.5 meters) high. They produce flowers and seeds yearly after attaining a minimum size or age. Common examples of herbaceous perennial plants are poppies, rhubarb, and strawberries.

Monocarp

A monocarp can live for many years but will flower only once in its lifetime and die afterward. Bromeliads and century plants (*Agave* spp.) are two examples of monocarpic plants.

These plants bloom after several years of active growth. The tops then die, and new plants are produced by the root system of the old plant.

STAGES OF PLANT MATURATION

Regardless of how long a plant lives, it will usually pass through four stages of maturation: germination, juvenility, maturity, and senescence. These stages and the role each has in the life of the plant are discussed in the following sections.

Germination

Many plants begin life as germinating seeds. Germination starts when the seed absorbs water and ends when the primary root emerges (Figure 3-2). After germination, the seedling goes through a period called *establishment*, which lasts until the seedling is independent and photosynthesizing. Both these stages are exceedingly important in the life of a plant, since it is during these phases that plants are most susceptible to adverse environmental conditions and the greatest number of plants die.

PARTS OF THE SEED

A seed must contain an embryo to develop into the new plant and a source of carbohydrates to supply energy to the embryo during germination and establishment. In addition, a covering called the *seed coat* helps prevent injury and drying (Figure 3-2).

In many seeds the carbohydrates required for the seed to germinate are stored in two *cotyledons*. These cotyledons resemble leaves and are attached to the stem when the seedling emerges from the ground. They are frequently called "seed leaves," as opposed to the true leaves of the plant which are produced later (Figure 3-2).

In other species carbohydrates are stored in an *endosperm*. Whereas the endosperm is a single organ (Figure 3-3) used only for carbohydrate storage, cotyledons may serve not only as storage sites but also as photosynthesizing organs.

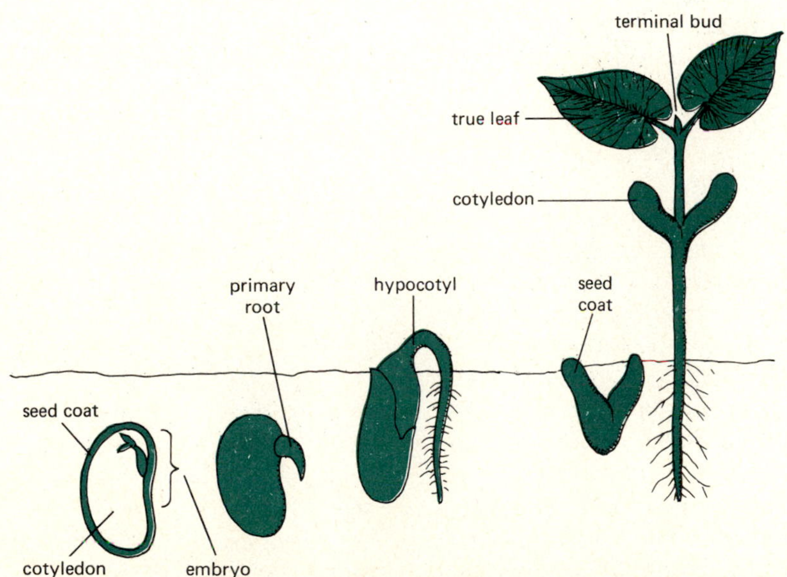

Figure 3-2 Germination of a bean seedling.

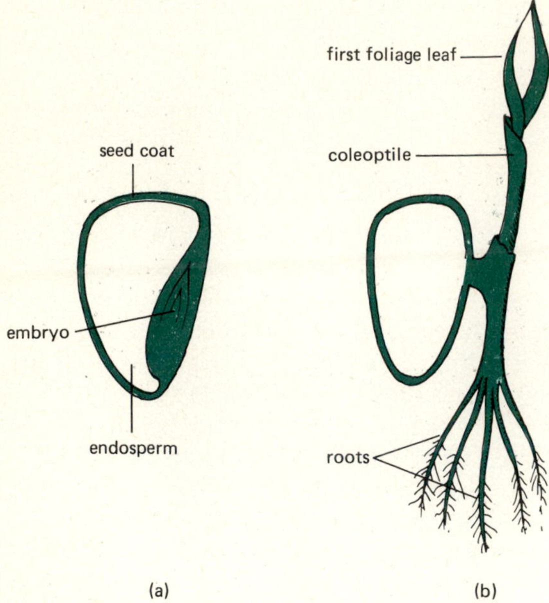

Figure 3-3 (a) A corn grain showing the enlarged endosperm. (b) The germinated corn grain. Note the absence of the cotyledon leaves found on the bean.

ENVIRONMENTAL REQUIREMENTS FOR GERMINATION

For a seed to germinate and establish a new plant, correct environmental conditions are necessary. First, water must be present. Second, oxygen must be available. Third, the temperature must be within the range acceptable to that species.

Uptake of water is the first process in germination. The uptake causes the seed to swell and triggers many chemical reactions, including an increase in the respiration rate. As carbohydrates are respired, energy is freed for the growth of the embryo.

Oxygen is necessary for the respiratory process; without it, respiration will not begin and the seed will not germinate. This fact can be proven by the following example. Seeds placed in a glass of water will absorb the water and swell, a sign that the germination process is proceeding. If taken out of the water a day later and planted, germination will continue. But if the seeds are left in the water, the process will stop after the swelling stage.

Suitable temperature is the third essential for germination. Temperatures either too high or too low will kill the embryo or prevent germination from proceeding. For most seeds, 70 to 80°F (21.1–26.7°C) temperatures promote vigorous germination, although plants native to cold climates will germinate at considerably cooler temperatures.

After sufficient swelling and respiration have taken place, the *radicle,* or primary root, will emerge. It grows down into the soil, beginning the absorption process, and soon afterward the shoot emerges.

Many times the shoot will arise from the soil as an arch, with the stem of the young plant forming a loop (Figure 3-2). The tip of the shoot is eventually pulled out of the soil by the growth of the stem, and the plant then straightens to an upright position. The pulling of the shoot tip from the soil prevents injury that could occur if it were pushed straight through the soil.

After the shoot emerges, the plant begins photosynthesis. The shoot tip develops leaves, and the cotyledons usually wither and drop from the stem.

Juvenility

After germination, most plants enter a period of juvenility which may last from several weeks to years. The plant will grow rapidly vegetatively but will not begin its reproductive process.

TRAITS OF JUVENILITY

In most cases the juvenile appearance of a plant resembles the mature form. However, some plants have recognizable traits which signal that the plant is in a juvenile state.

Leaf Form. Leaf form different from that found on the mature plant is a common juvenile trait. The leaves of some citrus species, for example, are frequently simple in the juvenile stage but compound on mature trees. Other plants including sassafras have deeply lobed and divided leaves when they are young, but when they mature, unlobed leaves will predominate.

Growth Form. Variation in form also distinguishes juvenile from mature plants. The most well-known example is ivy (*Hedera* spp.). Juvenile ivy forms are trailing or climbing and deeply lobed. Mature ivy will produce upright shoots which grow without support and have leaves without lobes.

Juvenile branch structure is sometimes found on young fruit trees in the form of whiplike vertical shoots. Mature trees occasionally produce such juvenile branches. They usually arise from the base of the trunk and are called *suckers,* or water sprouts.

Thorns. Thorns are a third common characteristic of juvenility. Citrus and locust trees are frequently thorny when young, but branches produced after the tree reaches maturity are thornless.

Leaf Retention. A tendency for the juvenile parts of a tree to hold leaves throughout the winter is the fourth characteristic of juvenility. Many trees demonstrate this phenomenon (Figure 3-4).

Plants may indicate juvenility other than by visible characteristics. In propagation, juvenile parts almost always root faster than mature portions.

The exact causes of juvenility are not completely known. Pruning, temperature modification, and applications of growth-regulating chemicals (p. 56) can change a plant from the juvenile to the mature state and back again.

Maturity

Maturity is the third stage in plant growth. It is during this stage that sexual reproduction takes place. The phases of this process are discussed here.

FLOWER INDUCTION

Flower induction is the first indication of maturity and the first of the three phases of flowering. It is an *initial chemical reaction* that begins the flowering process; it occurs before any physical changes take place in the plant. The factors which trigger blooming are well known for some species but unknown for others. Generally as a plant becomes older and larger, its likelihood of flowering increases.

Causes of Flower Induction. Induction of flowering is sometimes caused by environmental factors such as temperature or day/night length. In nature these factors program flowering into a seasonal pattern and will cause it to occur at the same time each year. Flower-promoting conditions can also be simulated to induce flowering at a preselected time. Controlling environmental conditions is a common practice among commercial greenhouses in growing such plants as poinsettias and Easter lilies for holiday sales.

Cool temperature is a main environmental factor necessary for the flowering of many plants. Some plants which grow and flower quite profusely in northern climates either die or grow only vegetatively in warmer locales. They must be *vernalized* (subjected to cold temperatures) before they can be successfully grown. It is because of insufficient vernalization that fruits such as apples, cherries, and pears are not grown in most parts of Florida.

In addition to certain fruit trees, biennial plants may also require vernalization for flowering. The cold received during the winter following the first season of growth induces flowering and completes the life cycle of the plant. Cabbages, carrots, and beets are all biennial plants which require vernalization. However, since they are raised for their roots and leaves and harvested the first year, the vernalization requirement is not important.

Figure 3-4 An oak tree retaining its dead leaves in winter, a characteristic of juvenility.

The vernalization needs of bulbs such as tulips, hyacinths, and crocuses are important since they are grown for their flowers. In warm climates the bulbs of these plants are sold "prechilled"; that is, they have already been vernalized by refrigerated storage. They will flower the following spring regardless of the climate they are grown in. However, they will not bloom more than 1 year unless they are growing in a climate where they will receive yearly natural vernalization.

Daily light duration is a second environmental effect that can control flowering in plants. It is common knowledge that the nights are longest in winter, shortest in summer, and intermediate in spring and fall. The further north or south of the equator, the more extreme is the difference between the shortest night of summer and the longest of winter. Plants which respond to these changes in the day/night ratio are called *photoperiodic* and are common in all but tropical climates near the equator.

Two main types of photoperiodic plants have been studied intensively; *short-day* and *long-day* plants.[3] Short-day plants flower only when the nightly dark period is *longer* than their "critical photoperiod" of, for example, 12 hours. Long-day plants will flower only if the dark period is *less* than their critical photoperiod. They can also flower under continuous light or during a long dark period interrupted at some point with a short light period. In an otherwise long night, a short period of light has the effect of cutting the dark period into two shorter periods. Provided each half is less than the critical period, the plant will respond as if it was exposed to only short nights, and flowering will take place (Figure 3-5).

Exactly how night length triggers flowering is only partially understood. A chemical called *phytochrome,* which alternates between two different forms, each with differing wavelengths of light in the red to far-red range, is known to be associated with the process (see Chapter 16 for a discussion of the wavelengths of light).

Light intensity can also control flowering. Plants which are "day neutral" and not affected by dark/light ratio frequently respond to intensity. Greater light intensity is often required to induce flowering that is sufficient to keep a plant growing vegetatively. Consequently, many houseplants bought for their blooms never flower again after being moved from a greenhouse to indoors. The light intensity is simply too low to induce flowering.

Stress, such as from drought or the crowding of a plant's roots, can also stimulate flowering. One possible explanation is that the natural production of chemical compounds which trigger flowering of a mature plant is intensified when the plant is dying, and the remainder of the stored carbohydrates is used for flowering and continuation of the species through seed.

FLOWER INITIATION

The second phase of flowering is flower initiation. Like induction, it is a phase of flowering invisible to the naked eye, but changes are taking place in the microscopic parts of the plant. It is during this phase that the vegetative meristems found at the stem tips or leaf axils change to flower meristems capable of developing into blossoms. This radical changeover from a vegetative to a reproductive structure takes place over several days or weeks. The center of the vegetative meristem, which was formerly mounded, becomes flattened. Small knobs of cells emerge in a spiral around the center (Figure 3-6); these are the beginnings of petals, stamens, and other flower parts. The meristem stops lengthening at this time but development continues.

[3]The confusing emphasis in naming of day length stems from the first research in photoperiod. At that time it was mistakenly believed that the length of light rather than darkness caused flowering.

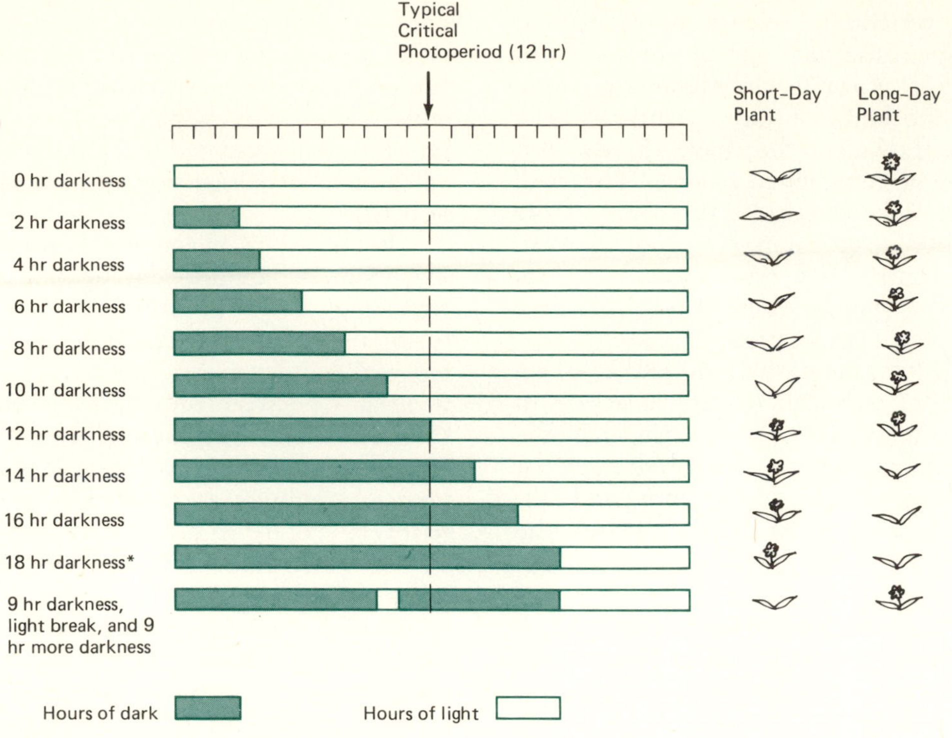

Figure 3-5 The flowering response of long-day and short-day plants to varying dark/light ratios.

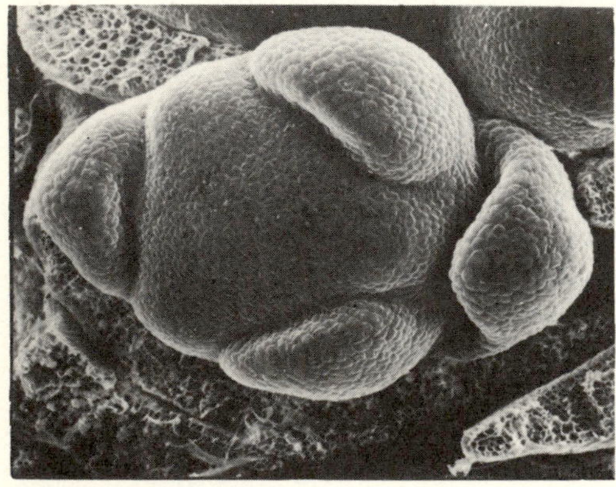

Figure 3-6 Developing flower meristem of an Easter lily (*Lilium longiflorum* 'Ace').

FLOWER DEVELOPMENT

The length of time from induction to bloom may vary from several days to 6 or 8 months. Cold-climate woody plants usually have long developmental periods. The flower buds of these plants are partially formed in late summer, are vernalized over the winter, and finish developing and open in early spring. The opening of the flower is its final stage of development.

POLLINATION AND FERTILIZATION

After a flower opens, it is generally receptive to pollination. Whether it pollinates itself

(self-pollination) or is cross-pollinated by another plant will affect the resulting seeds.

During pollination dust-size pollen grains found on the anthers are deposited by wind or insects on the tip of the stigma. Like seeds, they germinate and grow downward toward the ovary where the eggs are located. After reaching the ovary, the uniting of sperm (contained in the pollen) and eggs (in the ovary) occurs in the process of *fertilization* (Figure 3-7).

In some instances pollen produced by a flower may be unable to fertilize the eggs produced by that same flower. This *self-incompatibility* is found in about 40 percent of cultivated plants (see Chapter 7). In some cases the pollen grain simply never germinates. In others it germinates but grows poorly and never reaches the egg.

Cross-pollination is thus quite common. Separate fruiting and pollinating tree varieties are frequently planted in fruit orchards to overcome self-incompatibility problems, since flowers which are not pollinated and fertilized usually will not develop fruit.

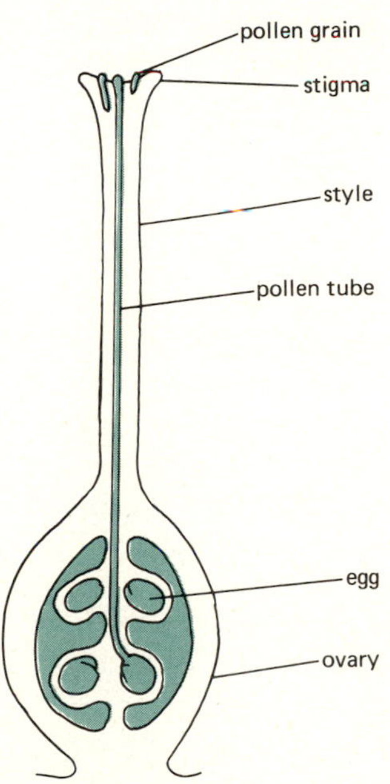

Figure 3-7 A germinating pollen grain growing toward the ovary.

Cross-pollination can be a hindrance as well as an advantage. Sweet corn and field corn will cross-pollinate if planted too close together, making the resulting sweet corn less tasty. Sweet and hot peppers will also cross-pollinate. The fruit, which develops from the ovary, will still be edible. But the seeds are the sexual union of two parents, and a mild pepper with hot-tasting seeds could be expected.

Although fertilization of the egg by pollen usually must occur before a fruit will develop, there are cases in which it is not necessary. For example, pollen does not fertilize the egg but the fruit still develops, resulting in a seedless fruit.

Another way in which seedless fruits form is through the spontaneous death of the seed embryo in its early stages of development. Fertilization is needed to start fruit growth, but after the death of the embryo, the fruit will continue to mature to full size. Seedless fruits formed in these two ways are called *parthenocarpic.*

For the majority of cultivated fruits, living seeds are necessary for the fruit to develop normally. For example, if only half the eggs inside an apple are fertilized, only one side will grow; a misshapen fruit is thus produced. Cucumbers sometimes have this problem, with a portion staying small while the remainder develops normally (Figure 3-8). In cases where the fruit contains only one seed, the fruits will drop prematurely if the embryo dies. This happens in the case of apricots, peaches, and cherries.

FRUIT DEVELOPMENT AND ENLARGEMENT

During fruit development sugars photosynthesized in the leaves constantly flow into the

Figure 3-8 A greenhouse-grown cucumber that was not fully pollinated and failed to develop normally.

fruit. They supply energy for the developing ovary which will become the fruit.

RIPENING

When a fruit reaches the end of its enlargement period, ripening begins. The fruit may become soft or change color, and its flavor can change from sour to sweet.

Softening of fruits is the result of the breaking down of compounds called *pectic substances* which strengthen the cell walls and cement cells together. In overripening, mushiness occurs because too much of the pectic substance has been lost. Mealiness in apples results from such overripening. The cells are so easily separated from each other that they are chewed without breaking and releasing the flavor of the apple.

Color changes result from the depletion of chlorophyll and the accumulation of other pigments. The chlorophyll continually decreases while the other pigments increase in intensity until coloration is complete. Carotene is a pigment which gives orange color to fruits such as oranges and persimmons. Anthocyanin is another pigment, responsible for the red color of ripe strawberries and apples.

Senescence

Senescence is the aging of a plant or any of its parts. It occurs as a part of the natural life cycle of the plant or as a result of environmental factors.

The natural life span of the plant often determines when senescence begins. Annual plants begin senescence at the flowering stage and will die soon after seed formation is complete. Senescence after reproduction is also the rule for monocarpic plants which live several years before flowering takes place.

Senescence of perennial plants is frequently called *decline.* Asparagus is a perennial vegetable that suffers from decline. The productive life of asparagus plants is generally 20 to 25 years. Once the plants reach this age, the yield drops off considerably, and the bed is replanted.

An entire plant need not die for the term senescence to be applied. The top portions of herbaceous perennials and bulbs senesce annually, but the root system remains alive. The lower branches of trees and leaves of houseplants such as red-margin dracaena (*Dracaena marginata*) and bamboo palm (*Chamaedorea erumpens*) frequently die while the top continues growing rapidly.

A colorful show of leaf senescence occurs during the fall when trees turn color. The coloring is due to the same pigment changes which cause the ripening of fruit. However, environmental conditions trigger this senescence, in this case the shortening of day length in fall and the cooler temperature.

ABSCISSION

The dropping of leaves, flowers, fruits, or other plant parts is called *abscission.* This is a process that involves the manufacture of plant hormones (see p. 56) and the formation of a "zone of abscission." In simple leaves, this zone is formed at the point where the petiole connects to the stem (Figure 3-9). In compound leaves, the main abscission zone is still at the point where the petiole reaches the stem, although individual leaflets may also form abscission layers and drop.

Flower abscission is often connected with pollination, and again a hormone (auxin, p. 56) is involved. The pollen contains auxin which it carries to the pistil, triggering abscission of the flower parts. Since the petals are no longer needed to attract pollinating insects, their function is complete and their abscission funnels more carbohydrate to the developing fruit.

Fruit abscission can occur at any point in development, but is most common after ripening. Dropping of ripe fruit is a type of seed dispersal. As with leaf drop, an abscission layer generally forms before a fruit drops.

Dormancy

Dormancy is a stage of plant development in which growth slows or stops. It affects all phases of growth from seed to mature plant and plays an important part in adapting plants to their environment and ensuring their survival.

Dormancy occurs during periods not suitable for plant growth and is usually seasonally related. Winter dormancy is found among plants in many parts of the United States. Dry-season dormancy is found in other areas of the country which experience distinct wet and dry seasons.

Leaf abscission is a common indicator of dormancy, signaling the onset of winter dormancy in many deciduous (leaf-dropping) trees. Although photosynthesis can take place at temperatures below freezing (such as in evergreens like spruce and pine), the leaves of most plants are not adapted to withstand subfreezing temperatures. Instead, they abscise in autumn, and new leaves survive as buds.

The changeover from dormancy to active growth is called *breaking dormancy.* It normally results from changing environmental conditions—from those which induced the

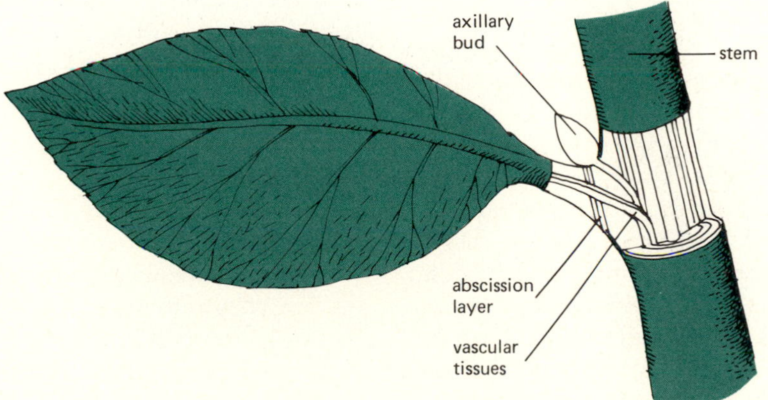

Figure 3-9 A leaf just prior to abscission.

dormancy to those which encourage photosynthesis and growth.

BREAKING SEED DORMANCY

Seeds usually enter dormancy just prior to abscission of the fruit or senescence of the parent plant. The conditions required to break dormancy and start germination are specifically geared to ensure the survival of the seedling.

For example, many seeds require winter vernalization before breaking dormancy and sprouting. This assures that germination will only take place in the spring, when the plant will have adequate time and suitable temperature to establish itself. If germination occurred as soon as the seed touched the ground in the fall, the young plant would be unlikely to survive the approaching winter. Simulated vernalization is sometimes used on seeds to break their dormancy in commercial nursery crop production. This process, called *stratification,* involves storing the seeds at temperatures near freezing for 1 or more months. Peach and apple seeds are stratified before planting.

Another environmental condition that breaks dormancy in some seeds is heat. Forest fires destroy existing trees but also trigger the germination of dormant pine seeds. Because the seed coats of some pine seeds are woody and thick, only the intense heat from a forest fire will weaken them enough to allow water to be absorbed and the embryos to emerge.

Most seeds with hard seed coats require less drastic measures to break dormancy. Usually the seed coat erodes enough by weather to permit germination. It can, however, be weathered artifically by a process called *scarification.* This involves cutting, scraping, or otherwise injuring the seed coat enough to allow water absorption and subsequent germination.

Cold, heat, and weathering are all ways used to break seed dormancy but are seldom normal practices for the home gardener. The application of water is usually all that is required to start the germination process.

PLANT HORMONES AND GROWTH-REGULATING CHEMICALS

Plant hormones are chemicals made within the plant body which produce changes in growth. They are effective in extremely small quantities and cause such responses as root formation, seedless fruit formation, leaf drop, and stem lengthening. Plant growth regulators are manufactured chemicals like plant hormones. They have the same effects and may be chemically quite similar to hormones, but they do not occur naturally.

At least nine groups of plant hormones have been identified, and it is likely that others will eventually be discovered. Scientists researching plant hormones encounter a number of difficulties. First, the effects of hormones are not the same from species to species. Second, slight changes in hormone concentrations can completely alter their effects. Third, two or more hormones frequently are found together, and it is difficult to determine which chemical is responsible for an effect.

Auxins

Auxins were the first hormones discovered. In varying concentrations auxins have such diverse effects as promotion of rooting, formation of underground tubers and bulbs, prevention of fruit formation, defoliation, and prevention of abscission of leaves and fruits.

Synthetic auxins, which can be purchased in garden centers, have a number of practical home uses. Powders sold to encourage rooting of cuttings are composed of manufactured auxins in a talcum powder base. 2,4-D sold to control lawn weeds is also an auxin. Applied at high concentrations, it kills many plants, but at very low concentrations it is a growth enhancer.

High auxin concentrations can also be used to prevent fruit on ornamental trees when the fruit is an undesirable feature. Ginkgo and horse chestnut trees both have large bothersome fruits which can be prevented from forming by using an auxin spray in spring. In commercial agriculture, synthetic auxins are used to defoliate plants before harvest, prevent sprouting of potatoes in storage, and prevent premature fruit drop in orchards.

Gibberellins

Gibberellins are a second group of hormones. The first plant activity that was associated with gibberellins is stem elongation. Additional experiments showed other effects including breaking dormancy of seeds, buds, and tubers and inducing flowering in plants which normally require vernalization or a correct photoperiod. Increases in flower, leaf, and fruit size were also found to be caused by gibberellins, leading to speculation that they would be a wonder chemical for agriculture, but this did not occur. However, gibberellins are used in greenhouses to form tall tree-form fuchsias and geraniums from cuttings because of their stem-elongation ability. They are also used to increase the size of grapes and to substitute for vernalizing azaleas and fruit trees in the South.

Cytokinins

Cytokinins are the third plant hormones, which have less pronounced effects on growth than auxins or gibberellins. Cytokinins are believed to work in conjunction with light to increase cell division and enlargement. This leads to stimulation of leaf growth, lengthening of stems, and formation of buds. They have also been shown to prevent chlorophyll degeneration and to break axillary bud dormancy.

Synthetic Growth Inhibitors

Synthetic and natural growth inhibitors make up the fourth classification of growth-regulating chemicals. These synthetics are sold under the trade names Phosphon, B-Nine, and Cyocel for use on florist crops such as poinsettia and chrysanthemum. They slow elongation of the stems, making the plant sturdier and fuller. On fruit crops they are used to improve color and firmness and prolong storage life. One growth inhibitor, maleic hydrazide, is occasionally used on hedges and lawns to slow growth and decrease maintenance.

Abscissic acid is a growth inhibitor that induces abscission and dormancy and inhibits seed germination. However, its action can be counteracted by growth-inducing chemicals like auxins, gibberellins, and cytokinins.

Ethylene gas is another inhibitor, whose use dates back to the Chinese practice of ripening fruits in incense-filled rooms. It was later determined that it was the ethylene gas given off in the fumes which caused the accelerated ripening. Ethylene is also produced by the ripening fruits themselves and by cut flowers, or it can be manufactured.

In addition to ripening fruits, ethylene also ages plant parts such as flowers and induces flowering in a limited number of species. Pineapples and other members of the bromeliad family (many of which are used as houseplants) will bloom when treated with ethylene. The home practice is to enclose a ripening apple (a source of ethylene) in a plastic bag with the plant for about a week. If the plant is mature enough, flowers will appear in 1 or 2 months.

Vitamins

Vitamins, particularly B vitamins, are occasionally sold as stimulants for plant growth and for use after transplanting. They are the same vitamins sold for human use but act more as hormones in plants.

The effectiveness of vitamins on improving plant growth has not been fully determined. When vitamins are used on bean seeds, limited experiments have shown that they improve germination rates and decrease the time from seed sowing to harvest. Increases in yield have also been reported. When vitamins are used with auxins, improved rooting of cuttings has been found, although it is questionable whether this is due to the vitamin or the auxin. Overall, it is very possible that vitamins do improve plant growth. However, they should never be used in place of fertilizers or proven effective chemicals sold for use on plants.

Selected References for Additional Reading

Galson, A. W., and P. J. Davis. *Control Mechanisms in Plant Development.* Englewood, N J: Prentice-Hall, 1970.

Leopold, A. C., and P. E. Kriedemann. *Plant Growth and Development.* 2d ed. New York: McGraw-Hill, 1975.

Wareing, P. F., and I. D. J. Phillips. *The Control of Growth and Differentiation in Plants.* New York: Pergamon Press, 1970.

Wilkins, M. B. *Physiology of Plant Growth and Development.* New York: McGraw-Hill, 1969.

Zimmerman, M. H., and C. L. Brown. *Trees, Structure and Function.* New York: Springer-Verlag, 1974.

Chapter 4

Plant Propagation

SEXUAL AND VEGETATIVE PROPAGATION

Plants are reproduced, or propagated, either sexually or vegetatively. Sexual propagation involves seeds or spores which result from the union of male and female cells. Vegetative propagation occurs by using a part of an existing plant to generate new plants which will have traits identical to those of the plants from which they were derived.

Propagation by seed occurs in almost all cultivated plants, and is used extensively in propagation. However, vegetative propagation is equally, if not more, important for several reasons. It is usually faster than seed propagation. Most important, exact genetic duplicates of the parent plant can be created. This is important for the preservation of desirable characteristics (such as fast growth or high yield) which could be lost in sexual reproduction. Seedlings may resemble the parent plant on which they were produced, but they will seldom be exactly like it. This is a result of genetic recombination of male- and female-carried genes. Just as brothers and sisters produced by the same parents are not all alike, seeds produced by a plant will not grow into identical plants. The gene mixing will result in the appearance of a variety of different traits.

Whether a plant is reproduced sexually or vegetatively depends on many factors, among them: the ease of germinating the seed, the number of plants which must be grown, and the importance of preserving traits possessed by one parent plant. Table 4-1 groups plants by their uses and lists the ways in which they are normally propagated.

Seed Formation

Seeds can be formed as a result of a plant fertilizing itself (self-pollination) or fertilization by another plant (cross-pollination). Offspring resulting from cross-pollination are called *hybrids* and carry traits of both parents. Although the term "hybrid" can apply to any seed resulting from cross-pollination, a seed package bearing the word "hybrid" indicates that the included seeds are the result of special breeding and are generally superior to other varieties. Selected hybrid seeds produce healthier, faster-growing plants, a phenomenon called *hybrid vigor.*

Hybrid seed results from the controlled crossing of two groups of plants of known genetic makeup. The crossing of the two "genetically pure lines" produces seed which will have the best traits of both parents.

Table 4-1 Primary Propagation Methods for Cultivated Plants

Plant category	Primary method of propagation
Annual flowers	Seed
Annual vegetables	Seed
Fruit trees	Seed for rootstock, followed by grafting of vegetatively propagated scion
Groundcovers	Vegetatively by cuttings
Houseplants	Vegetatively by cuttings
Ornamental trees	Seed
Perennial flowers and bulbs	Vegetatively by cuttings, underground storage organs, or division
Perennial vegetables	Vegetatively by division
Turfgrass	Seed
Vines	Vegetatively by cuttings

The example shown at the bottom of the page uses two pure lines of cucumbers to produce hybrid cucumber seed.

Seed saved from hybrid plants will not grow good plants the following year. Only the original hybrid seed bears the desirable traits. The seeds included in its fruit result from random cross- or self-pollination. They will not produce the same superior plants.

Plant Breeding

Home plant breeding is an interesting activity, but it seldom yields very spectacular plants. Desirable characteristics are the result of luck in most cases. For example, a person who crosses red and white petunias may assume that pink offspring will be produced due to the mixing of red and white. But plant genetics is a complex science, so the result is more likely to be a combination of reds, whites, and shades between.

When breeding plants, a few basic rules of genetics must be followed. First, the two plants to be crossed nearly always must belong to the same genus and often to the same species. Varieties within a species will cross-pollinate in most cases.

Only in rare instances do plants cross between genera. One example was the cross

Inbred parents	Traits of parents	Traits of resulting hybrid offspring
Cucumber A line	Long, thin shape (desirable trait)	Long, thin shape
	Pale color (undesirable trait)	Deep green color
	Matures early in season (desirable trait)	Matures early
	Large, tough seeds (undesirable trait)	Small, tender seeds
Cucumber B line	Short, thick shape (undesirable trait)	
	Deep green color (desirable trait)	
	Matures late in season (undesirable trait)	
	Small, tender seeds (desirable trait)	

of English ivy (*Hedera helix*) and Japanese fatsia (*Fatsia japonica*) to produce the popular plant *Fatshedera,* a semivining shrub with ivylike leaves intermediate in size between the parents.

Under natural conditions, seeds from cross-pollination make up only 4 percent of all seeds produced. Understandably so, since the pollen-bearing anthers within a flower are close to the female style, making self-pollination most likely.

Among plants which never self-pollinate are species which produce male and female flowers on separate plants (holly and bittersweet, for example). Clearly, self-pollination of these species is impossible, and seeds produced are the result of pollination between separate plants.

The mechanics of plant breeding, either self- or cross-pollination, are not difficult. Self-pollination can be achieved by sealing a flower in a paper bag just before it opens (usually in the early morning). The flower will open within the bag and, having no other source of pollen, will self-pollinate.

For cross-pollination, flowers which are ready to open should be selected from two parent plants. The petals and anthers of the female parent should then be removed with small scissors and the blossom enclosed in a paper bag. Proceeding to the male parent, the flower should be opened by hand and a small paintbrush inserted to pick up some of the pollen. The pollen should then be taken to the female parent, brushed lightly over the style (Figure 4-1), and the bag replaced.

In several days the pollen will fertilize the eggs, and the pistil will shrivel and fall. The bag can then be removed and the flower tagged with the names of the parents.

Seed Harvesting

When seed is mature and ready to be harvested, usually either the seed pod splits or the fruit drops to the ground. The color of the mature seeds will darken as they ripen.

Cleaning removes the ovary tissues which surround the seed. Dry fruits such as the pods of beans or fruits of many trees usually split open when mature, and the seeds can be shelled out with the fingers. With fleshy fruits such as peach or squash, the fruit should be cut apart and the seeds extracted and dried on a paper towel for several days.

Storing Seed

The best conditions for storing seed depend on the species. However, for general storage of vegetable and flower seeds, a sealed container kept in the refrigerator is sufficient. Glass jars, plastic containers, and freezer bags all are suitable because they maintain constant humidity. The average refrigerator temperature of about 40°F (4.4°C) slows respiration and keeps the seed viable for the greatest possible time.

The life span of seeds in storage varies from days to years among species. Excess vegetable seeds are commonly stored for use the following year. Chapter 6 gives the maximum time seeds of common vegetables can be stored under recommended storage conditions and still produce satisfactory plants.

Even under ideal conditions, the longer the seeds are stored, the fewer seeds will germinate. The seedlings which do emerge will be weaker. Therefore when planting seed which has been stored, it is advisable to sow the seed thickly to compensate for decreased germination.

GROWING PLANTS FROM SEED

Seed Growing Outdoors

Sowing seed outdoors is called *direct seeding* and is used for many vegetables and a few flowers and herbs. Since the plants are seeded where they will mature, it is important to

Figure 4-1 Pollination of a hibiscus bloom by brushing the pollen on with a paintbrush.

choose the site carefully. It should provide the proper light for the species and should also be well drained (water should not stand after a rain). The soil must be loose and crumble easily in the hand. Soils which crust or pack should be conditioned as described in Chapter 5 and spaded and worked to a depth of about 1 foot (30 centimeters) to assure suitable soil conditions for root growth. Clods should be broken and any rocks or debris removed.

The seeds should be sown at a depth and spacing recommended on the package. If sowing directions are not given, the general rule is that seeds should be planted one and one-half times as deep as their diameter. Spacing between seeds can be estimated from the mature size of the plant. Thicker sowing is permissible and even recommended, with the understanding that weaker seedlings will be removed to achieve proper spacing for mature plants.

Soil used to cover seeds should be free of even small clods to enable the emerging shoots to reach the soil surface easily. Rubbing the soil between the palms of the hands will pulverize it to the proper degree.

Timing is important when planting seeds outdoors. If planted too early in spring, the seeds will fail to germinate or germinate slowly because of cold weather. Planted too late, they may not mature within the growing season. Package directions are the best source of planting information.

Seeds planted in moist soil do not require

immediate watering. They are able to absorb moisture from the soil to start germination. In dry soil, watering will supply moisture and settle the soil around the seed so that the emerging root will have immediate contact with a source of water and nutrients.

Maintaining adequate soil moisture during the germination and establishment parts of a plant's life is essential. A drying of the soil at these times will be fatal to the seedlings.

Soil crusting can occur after rain or watering; it can be recognized by the smooth, glazed appearance of the soil surface. Air and water entry into the soil is inhibited and germination reduced. If soil begins to crust, the surface should be broken carefully with a light rake or hand tool so emerging seedlings are not injured.

A liquid fertilizer "starter solution" (Chapter 5) can be applied to seedlings after germination is complete. Although this is not essential, research has shown that a starter fertilizer will increase the growth rate of seedlings substantially. Seedlings are considered "established" after adult, or "true," leaves (Figure 4-2) appear. True leaves indicate that the root system is absorbing water and nutrients from the soil and the plant is no longer dependent on carbohydrate stored in the seed.

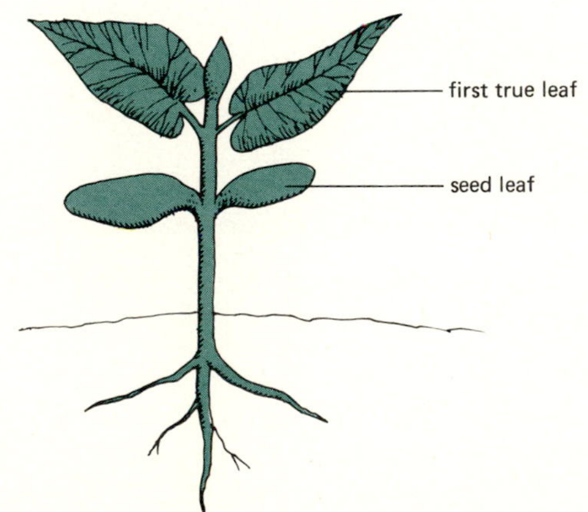

Figure 4-2 Seed leaves and true leaves of a typical seedling.

Natural Reseeding of Outdoor Plants

A number of cultivated plants produce seed which germinates the following season without the aid of humans. Among these plants are tomatoes, corn, and petunias. Seedlings which grow untended are called *volunteers* and generally do not produce plants of the same quality as the parents.

Seed Growing Indoors and in the Greenhouse

Seedlings of vegetables, flowers, and herbs can be raised indoors or in a greenhouse. They should be started 6 to 8 weeks ahead of when they will be required for the garden.

A container for raising seedlings need not be elaborate. The bottom of a milk carton or plastic dairy product container is sufficient, provided holes are made at the base for draining of excess water. For large numbers of seedlings, plastic, styrofoam, or wood nursery flats can be purchased (Figure 4-3). Compressed peat moss disks (Figure 4-4) which swell with water and form a plantable container for one seedling are also available in garden centers. Cells of plastic egg cartons can also be used for individual seedlings.

The material in which seeds are planted is called the *growing medium,* which must be fast draining and not prone to packing. Commercial potting soils sold for houseplants fit these requirements and are convenient for raising seedlings. Pure garden soil should not be used, since it packs when restricted in a container and becomes unsuitable for seedling growth.

Sterilization of the growing medium is an important precaution when growing seed indoors because it kills fungi in the medium which cause plant diseases. To sterilize the medium, it should be dampened and placed in a baking dish and covered with foil or a lid. The oven temperature should be set at

Figure 4-3 A wooden nursery flat used for growing large numbers of seedlings. The plastic is used to retain moisture around the germinating seeds.

180°F (85°C) and the medium baked for about an hour or until it is heated throughout. If all the growing medium is not used, any unused soil should be kept in a sealed plastic bag to prevent recontamination.

Houseplant potting soil seldom requires any further preparation than sterilization. After it has cooled, it can be poured into a clean seeding container and the seeds planted immediately. Spacing between seeds can be close provided the young plants will be transplanted soon after they emerge.

Water can be supplied to container-grown seedlings by sprinkling or misting whenever

Figure 4-4 A compressed peat moss disk (Jiffy 7 peat pellet) before and after water is added to it.

the growing medium begins to dry. An easier way is to water the seeds after planting and then enclose the seed container in a clear plastic bag (Figure 4-3). The plastic prevents moisture from evaporating. No further watering will be required until after the seeds have germinated and the plastic is removed. If the container is made of plastic or other waterproof material, it is not necessary to enclose the entire container. A sheet of plastic food wrap can be laid across the top of the container to hold in moisture.

For most rapid germination, seeded containers should be placed at a temperature of 70 to 80°F (20–27°C). A sunny area is not required for germination and, in fact, should not be used for containers covered or enclosed in plastic. The heat from the sunlight will build up under the plastic and kill the seeds. Instead, the container should be placed where it will receive bright filtered light, such as behind a sheer curtain.

Though not essential for germination, bright light or direct sunlight is needed after seedling emergence. The limited light entering through a window, even a bright south window, is often not sufficient to raise healthy transplants. If the seedlings become pale and "stringy" (Figure 4-5) instead of compact, poor garden plants will result. For raising most transplants, artificial light (Chapter 16), a greenhouse, or a greenhouse window is advisable. Exception can be made for seedlings of houseplants (such as palms and Norfolk Island pines) and for bright-light climates listed in Chapter 1.

Figure 4-5 "Stringy" tomato seedlings raised in insufficient light.

DAMPING OFF

The period from seedling emergence until several true leaves have formed is the primary time that seedlings are susceptible to damping-off disease. If a fungus is present in unsterilized growing media, the organism rots the seedling stem, which discolors and shrivels at soil level. Within a few days the plant collapses and dies (Figure 4-6).

Once damping-off disease is detected, it can spread rapidly and kill the majority of the seedlings. Fungicidal soil drenches (see Chapter 18) are moderately effective in stopping the spread, but the best prevention is sterilizing the growing medium before planting. Use of an inorganic material such as vermiculite or perlite to cover the seeds is also helpful.

TRANSPLANTING SEEDLINGS

It is advisable to transplant seedlings to larger growing containers as they begin to crowd each other. As leaves of neighboring plants

Figure 4-6 A damping-off infection in seedlings.

overlap, less light will be available to each seedling, and growth will slow accordingly. Roots will compete for nutrients and also cause stunting. At this point, usually corresponding with emergence of true leaves, the seedlings should be transplanted into separate growing containers. Seedlings are very delicate, and transplanting must be done carefully to avoid injury. A pencil or knife can be used to lift up groups of seedlings (Figure 4-7), after which they can be gently separated with the fingers. Seedlings should always be picked up by a leaf rather than by the stem, since any injury to the latter will kill the seedling.

When transplanting, the growing medium should be firmed around the roots with the fingertips, and each plant should be watered to assure contact between the growing medium and the roots. The seedlings should be kept in a shaded location for 2 to 3 days to minimize transpiration water loss while the roots damaged during transplanting resume water absorption.

After this period of restablishment, the seedlings can be replaced in bright light and fertilized weekly with dilute liquid fertilizer (Chapter 5).

HARDENING OFF

Seedlings raised on a windowsill, in a greenhouse, or under artificial light grow under less strenuous conditions than they will be exposed to in a garden. To acclimatize them to outdoor conditions before transplanting, they should be subjected to a *hardening-off* period. When the weather is suitable, the transplants should be moved outdoors for several hours each day. Gradually, the time can be increased until they are brought indoors only if night temperatures are predicted to drop below the range acceptable for the species. After about a week the transplants will be "hardened" and less succulent appearing. Hardened plants are less likely to wilt or "shock" at transplanting.

TRANSPLANTING OUTDOORS

If possible, an overcast day should be chosen for transplanting outdoors to minimize wilting. The roots should be disturbed as little as

Figure 4-7 Lifting out seedlings for transplanting with a knife.

possible and the plant watered in afterward. Some wilting should be expected, but if it is severe or lasts more than 1 to 2 days, the plant will need protection during reestablishment. Shade should be provided by a cardboard tent or newspapers. The soil should be moist at all times so the roots can easily replace water lost through transpiration.

GROWING PLANTS FROM SPORES

Of all the commonly cultivated garden and indoor plants, only one type, ferns, bears spores. On seed plants, the egg and sperm (pollen) unite and mature into seeds on the parent plant. In ferns, a spore drops on the ground and grows into a flat plant called a *prothallus.* The prothallus then develops the sperm and egg, which unite and form the new plant. The spore is only the beginning of sexual reproduction in the fern and not the end as a seed is.

Mature, healthy ferns (with the exception of highly ruffled types which are sterile) develop hundreds of organs called *spore cases* on the undersides of their fronds (Figure 4-8). The spore cases resemble small brown dots or lines, and protect and release the spores.

Several months are required for a fern to develop and ripen spores. When the spores are mature, the spore cases open and release them into the air. The spores themselves look very much like a brown powder.

To capture spores before they are released, the maturity of the spore cases can be tested by lightly tapping the frond over a piece of white paper. When specks appear on the paper, the frond should be cut off and enclosed in an envelope for several days. The spore cases will then open in the envelope, yielding a tiny amount of spores in the bottom.

Spores have the same growing requirements as seeds: warmth, moisture, air, and a growing medium. A small pot or flat should be filled with damp houseplant soil, then misted, and the spores scattered over the surface. The pot should be enclosed in plastic and set in a shaded location. In several weeks the spores will begin to germinate and grow into flat, green structures called *prothalli* (plural of prothallus; Figure 4-9). As they grow, the prothalli should be misted daily because the egg and sperm they form can only unite in water. Fronds will then grow through the center of the prothalli, which will slowly die. The young ferns can then be transplanted to individual pots as soon as they are large enough to handle.

Figure 4-8 Spore cases on the underside of a holly fern leaf.

VEGETATIVE PROPAGATION

Propagating plants vegetatively involves the use of nonsexual plant organs such as leaves, stems, and roots. Whereas a plant propagated from seed may not develop identically to its parent, vegetatively propagated plants almost always do. Vegetative propagation thereby preserves the work of plant breeders and natural mutations which have yielded valuable plants.

A second reason is to reproduce plants

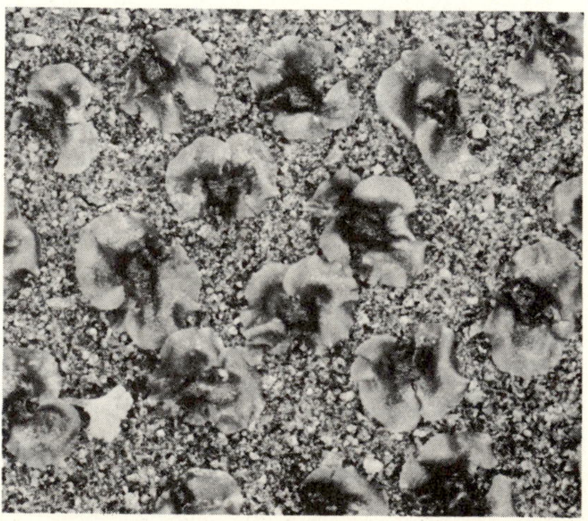

Figure 4-9 Fern prothalli.

which seldom flower or produce only sterile flowers. Foliage plants and navel oranges fall in this group.

A third reason is the relatively short time it requires to produce a mature plant. A vegetatively propagated cutting can be rooted in 2 weeks. To raise the same size plant from seed would require a month or more.

Cuttings

Cuttings or "slips" are the most widespread vegetative propagation method. Cuttings are vegetative plant parts such as leaves, stems, and roots which regenerate their missing parts to form new plants. The environment required for growing cuttings is the same as for germinating seeds: warmth, moisture, and a growing medium.

ROOTING MEDIA

Most parts used for vegetative propagation are taken from aboveground portions of the plant and must regenerate roots to become independent, growing plants. The growing medium that supports and surrounds the rooting area will determine whether roots will form and their quality. It should drain quickly to admit air to the rooting area yet should retain some moisture. There is not one superior rooting medium. Many combinations of materials are used, including the following five: two parts (by volume) sand and one part peat moss, one part perlite and one part vermiculite, pure vermiculite, pure perlite, and pure sand. The ingredients listed in these media are discussed in more detail in Chapter 15. As with seed media, media used for rooting should always be sterilized.

CUTTINGS FROM OUTDOOR PLANTS

Shrubs, vines, herbs, groundcovers, and a few perennial flowers are propagated by cuttings taken at different times of the year. Cuttings from woody outdoor plants are classified as hardwood, semihardwood, or softwood, depending on the degree of woodiness of the cutting. A fourth classification, herbaceous, includes cuttings from nonwoody plants such as many groundcovers.

Hardwood Cuttings. Hardwood cuttings (Figure 4-10) should be taken in late fall through early spring from the matured wood of the past season's growth. They are generally about 6 to 10 inches (15–25 centimeters) long and may come from either deciduous or evergreen plants such as juniper, viburnum, and apple.

The specific procedures for rooting cuttings vary among species, but generally the cuttings are wrapped in plastic and stored over winter in the refrigerator. In spring they are unwrapped, and the lower 1 to 2 inches (2.5–5 centimeters) is pushed into damp rooting medium. The cuttings should be enclosed in a plastic bag to prevent moisture loss and placed in a 70 to 80°F (21–25°C) location.

Roots should form in less than 6 weeks, and dormant buds should begin growing shortly thereafter. The cuttings can be hardened off and transplanted several weeks later.

Semihardwood Cuttings. Semihardwood cuttings are taken in summer from the par-

Figure 4-10 Hardwood cuttings of juniper (left), sycamore (middle), and boxwood (right).

tially matured new growth of woody plants. The cuttings are made 3 to 6 inches (8–15 centimeters) long and, since they lose water through their leaves very rapidly, should be placed as they are cut in a premoistened plastic bag.

The procedure for rooting is the same as for hardwood cuttings, but rooting is often more rapid.

Softwood Cuttings. Softwood cuttings are taken in late spring from succulent new growth produced that season. They generally root more consistently and quickly than either hardwood or semihardwood cuttings. Picking the cuttings at the right stage of maturity is crucial to success. Shoots which are growing rapidly are too tender and prone to rotting. Ideally, the shoots should be slightly flexible but should snap when bent to a 90° angle. They should be harvested into wet plastic bags and rooted, using the same directions as for hard- and softwood cuttings.

Because the younger parts of plants generally root more easily than more mature parts, softwood cuttings are the most reliable way to vegetatively propagate most woody outdoor plants.

Herbaceous Cuttings. Herbaceous cuttings are the equivalent of softwood cuttings but are taken from herbaceous plants such as coleus and impatiens. Most houseplants are also herbaceous; the types of cuttings used to propagate them are discussed on pages 72–74.

Herbaceous cuttings can be taken and rooted at any time during the growing season. They are generally 2 to 4 inches (2.5–5 centimeters) long and taken from the tip of a stem. As with all the types of cuttings previously discussed, the bases should be planted in damp rooting medium and the cuttings enclosed in a plastic bag to increase humidity. Keeping the cuttings wet from the time they are cut until they are stuck in the medium will reduce wilting.

Root Cuttings. Root cuttings are sections of thickened roots without leaves or stems attached. Whereas most outdoor plants can be propagated by one or more types of top cuttings, only a few will grow from root cuttings. These plants have the ability to generate an adventitious bud on the root section.

Root cuttings of most outdoor plants are best taken in early spring just before new growth begins. At this time carbohydrate reserves are at the highest point, and they are ready to emerge from dormancy. Like other cuttings, root cuttings should be protected from drying by being enclosed in a plastic bag and can be harvested by simply digging out medium-sized roots from around the base of an established plant. The sections can be cut into 3-inch (7- to 8-centimeter) lengths and planted horizontally ½ inch (1–2 centimeters) deep in rooting medium. The medium should be kept damp, and shoots will appear in 4 to 6 weeks.

INDOOR PLANT CUTTINGS

Cuttings of indoor plants can be of the following four types, depending on the part of the plant from which they originate. The procedure for rooting the cuttings is the same as for cuttings of outdoor plants.

Stem Tip Cuttings. Stem tip cuttings are the most common type (Figure 4-11) and consist of the top 2 to 4 inches (5–10 centimeters) of a growing stem.

Leaf Bud Cuttings. Leaf bud cuttings (Figures 4-11 and 4-12) are taken from a plant after the stem tip has been used. Each cutting includes a short section of stem, one leaf, and its corresponding axillary bud. When the base of the cutting roots, the axillary bud breaks dormancy and forms the new stem of the plant. Any plant that can be propagated by stem tip cuttings can be propagated by leaf bud cuttings, although the latter method is slower.

Stem Section Cuttings. Stem sections or cane cuttings (Figures 4-11 and 4-12) are short pieces of thickened, leafless stem. They are used primarily for obtaining large numbers of

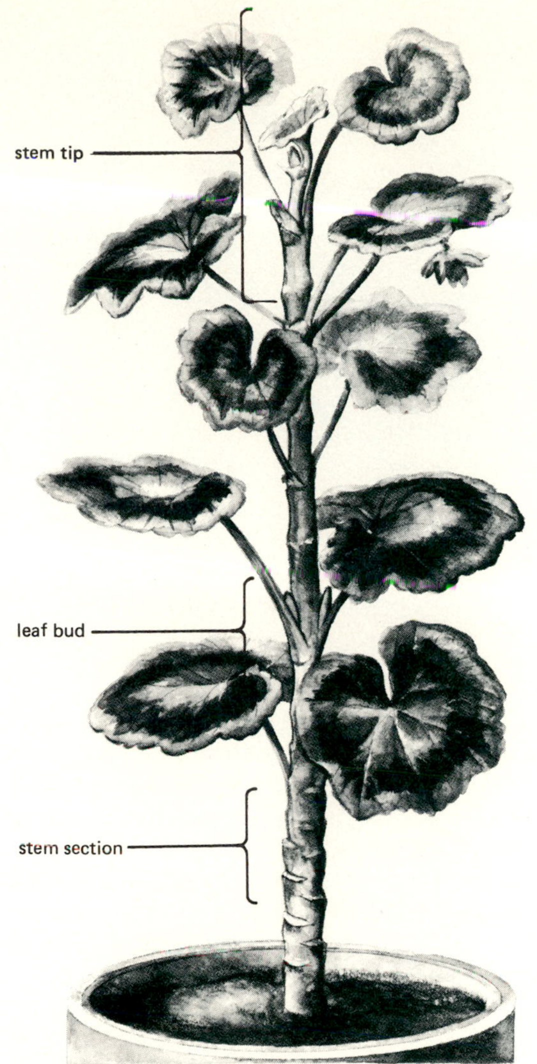

Figure 4-11 A typical plant, showing where stem tip, leaf bud, and stem section cuttings would be found.

offspring from cane-forming plants such as dumbcanes (*Dieffenbachia*) and Hawaiian ti plants (*Cordyline*). First, the top of the plant is removed for a stem tip cutting. The remaining stem is then cut into pieces containing two to three nodes each. The sections are laid horizontally half-buried in the rooting medium. A plastic covering is not essential provided the medium remains moist, and roots and shoots will be generated in 1 to 3 months.

Leaf Cuttings. Leaf cuttings are composed of a single leaf and sometimes its petiole. Most indoor plants cannot be propagated using only a leaf; there are only certain species which generate buds where the leaf blade joins the petiole or along the leaf veins or margins (Table 4-2). The propagation procedures vary among genera. Leaves of African violet and peperomia are picked with the petiole attached. The leaf is buried in the rooting medium up to the blade, and new plants form at the soil line (Figure 4-13).

With bryophyllum and other succulents such as jade plant, burro tail, and echeveria, leaves are picked directly off the plant and laid on the surface of the medium (Figure 4-13). Since they are succulents, a medium largely composed of sand is best; enclosing them in a plastic bag is not advisable. With bryophyllum, up to 20 new plants will form around the leaf margin. With others, single plants will appear where the leaf was broken from the plant.

Fibrous-rooted begonias such as Rex begonia produce new plants along the veins of mature leaves and also where the petiole connects to the leaf blade. Whole leaves or pieces of leaves should be laid on a damp rooting medium so that the veins make contact. Cutting the veins with a knife stimulates bud formation, and three to five cuts can be made on each leaf.

Another houseplant grown from a leaf is the snake plant (*Sansevieria* spp.). A mature leaf can be cut crosswise in 1-inch (2.5-centimeter)-long sections and the base of each stuck in rooting medium (Figure 4-13). Remember that the orientation of the cuttings is critical. The end of the section which was originally on top must be up, and the part which was lower then planted in the rooting medium. Sections inserted upside down do not root. After the leaf section roots, a new

Figure 4-12 Rooting stem tip (left), stem section (middle), and leaf bud (right) cuttings.

plant will be produced at the base and grow up beside the leaf section.

It should be noted that variegated or yellow-banded sanseverias propagated by this method produce green offspring. To maintain variegation, the variety must be propagated by division.

In all cases propagation from leaf cuttings is slow. The generation of an adventitious bud and roots is an energy-consuming process. Several months will be required before the young plants are large enough to be transplanted.

Attempts to propagate plants by cuttings meet with varying degrees of success, depending on the care which one takes and the readiness with which the species generates new roots. Following the suggestions given here will maximize your chances of success with most species. But for plants which are difficult to propagate, consulting one of the references listed at the end of the chapter may be necessary.

First, select cuttings from shoots without

Table 4-2 Selected Plants Propagated by Leaf Cuttings

Common name	Botanical name
Rex begonia	*Begonia* × *rex-cultorum*
Jade plant	*Crassula argentea*
Echeveria	*Echeveria* spp.
Bryophyllum	*Kalanchoe beauverdii, laxiflora,* and *pinnata*
Peperomia	*Peperomia* spp.
African violet	*Saintpaulia ionantha*
Snake plant	*Sansevieria* spp.
Burro tail	*Sedum morganianum*
Christmas cheer	*Sedum* × *rubrotinctum*

Figure 4-13 Leaf cuttings of a snake plant (*Sansevieria*) (left), begonia (middle), and peperomia (right). Note the locations of the new plants which form.

flowers or flower buds, or if unavoidable, the flowers should be picked off. A cutting with blossoms will channel energy into the reproductive phase and will be less likely or slower to root.

Also note the positions of nodes on the stem of the cutting. Roots are usually generated first at these sites, so cuttings should be made with at least one node near the base.

Also, leafy cuttings wilt easily and, once severely wilted, are less likely to root. Therefore foliage should always be kept moist after cutting and before the cuttings are stuck in the medium to slow water loss from the leaves.

Any leaves on a cutting which will be covered after the cutting is stuck into the rooting medium should be removed. If left on, they will rot and provide a breeding ground for disease organisms.

Likewise, leaves which die and drop from the cuttings should be removed along with whole cuttings which appear dead.

Finally, use of a rooting hormone can increase rooting speed and success. The auxins discussed in Chapter 2 are available in powder form in a talc base, as dissolvable powders for mixing with water, or as liquids, and are applied to the stems of cuttings before insertion into the rooting medium.

Gentle tugging of rooting cuttings about once per week is the standard test for determining whether rooting has occurred. If the cutting slips out easily, no anchoring roots have formed. The cutting should be inspected for signs of rotting and, if still healthy, can be reinserted into the medium. If the cutting does not pull out with gentle tugging, it already has roots. The plastic bag can then be opened to accustom the plants to normal humidity and be removed entirely after several days. After 1 week the cuttings can be transplanted to pots.

ROOTING CUTTINGS IN WATER

A few outdoor plants, such as willow and pussy willow, and many tropical indoor plants (Table 4-3) can be rooted and grown in water. Stem tip cuttings are usually used, with the lower leaves removed and the cutting absorbing water through the cut end until roots are formed.

One of the easiest methods is to fill a glass with water and cover the mouth with foil. Holes can then be punched in the foil and the cuttings inserted. To minimize transpiration the cuttings should be kept out of direct sun.

Crown Division

Crown division is probably the most common and reliable propagation method. It is used for many herbaceous perennials, a few shrubs, and many houseplants such as ferns, asparagus ferns, African violets, and spider plants.

In division, one plant is separated into two or more pieces, each containing a portion of roots and crown (Figure 4-14). With herbaceous perennials, the entire plant is dug up and cut apart. The sections are then replanted and watered.

The division can be done at any time in the growing season, however, plants in active growth must be treated carefully to minimize water loss through the leaves. Therefore division of dormant plants in early spring is common in cold-winter areas of the country.

Shrubs to be divided must be multistemmed, that is, have a number of stems growing out of the ground instead of branches arising from one single short trunk. They should be divided, when dormant if possible, by cutting through the crown with a spade (Figure 4-14) so that it is broken down into several sections. Each section is then dug, keeping as many roots as possible, transplanted to the new location, and pruned to lessen shock. A part of the parent crown can be left to rejuvenate the shrub.

Figure 4-14 Division of a plantain lily into several smaller crowns.

When dividing houseplants, the parent can be removed from its pot, cut or pulled apart, and the sections repotted. Pruning is normally unnecessary. If wilting occurs, the newly potted sections can be left in plastic bags for several weeks until the roots are reestablished.

Layering

Layering is another method for propagating vines and also shrubs with a trailing growth habit or flexible branches. A low, flexible shoot is bent to the ground during any time the plant is in active growth and held in place by a bent wire or heavy stone. It is then

Table 4-3 Selected Cuttings for Rooting in Water

Common name	Botanical name
Chinese evergreen	*Aglaonema* spp.
Wax begonia	*Begonia semperflorens*
Angel wing begonia	*B.* spp.
Grape ivy	*Cissus rhombifolia*
Coleus	*Coleus* × *hybridus*
Dumbcane	*Dieffenbachia* spp.
Dracaena	*Dracaena* spp.
Pothos	*Epipremnum aureum*
Botanical wonder	*Fatshedera lizei*
Rubber plant	*Ficus elastica*
Nerve plant	*Fittonia vershaffeltii*
Tahitian bridal veil	*Gibasis geniculata*
English ivy	*Hedera helix*
Polka dot plant	*Hypoestes phyllostachya*
Impatiens	*Impatiens wallerana*
Bloodleaf	*Iresine herbstii*
Prayer plant	*Maranta leuconeura*
Philodendron	*Philodendron* spp.
Swedish ivy	*Plectranthus* spp.
African violet	*Saintpaulia ionantha*
German ivy	*Senicio macroglossus*
Arrowhead	*Syngonium podophyllum*
Wandering Jew	*Zebrina* spp. and *Tradescantia* spp.

mounded with soil a short distance back from the tip. Within a few months, roots should have formed on the covered portion of stem, and the layer can be cut from the parent plant and transplanted.

Runners, Rhizomes, and Stolons

These three structures are types of horizontal stems produced as a natural means of vegetative reproduction. Runners (Figure 4-15) grow aboveground and are found on such plants as strawberries, ferns, spider plants, and strawberry begonias. Rhizomes and stolons grow at ground level or below ground and are typically found on bamboo grasses and some irises. There are subtle botanical differences among the three, but basically they are handled the same in propagation.

As modified stems, these organs have buds, nodes, and internodes. A runner will have only one bud at the tip, whereas stolons and rhizomes may have several. The nodes are the sites where new plants will form. With a runner on an outdoor plant, the stem can be positioned as it begins to form the new plant. This is a common practice to avoid overcrowding in strawberries (Chapter 8). The runner will root without additional attention and can be severed from the parent and moved afterward if desired.

Because houseplants are grown in pots, the runner plants may never make contact with a rooting medium without help and can live attached to the parent indefinitely. When a new plant is wanted, a small pot of growing medium can be placed under the runner plant. Roots will form in less than a month, and the runner can then be detached. When this method is impractical (such as with hanging plants), the runner can be detached and treated as an unrooted cutting.

Figure 4-15 Rooting a runner of a spider plant (*Chlorophytum*).

Propagating plants from rhizomes and stolons is done much the same way. The stem connecting the young plant to its parent is cut after a moderate amount of roots has formed, and the young plant is transplanted to the new location.

Suckers and offsets (Figure 4-16) are young shoots which grow from the roots or stems of mature plants. They are functionally similar to rhizomes and stolons and are found in many shrubs and houseplants such as bromeliads, succulents, and cacti. Offsets on cacti (Figure 4-17) are frequently produced on top of the plant and can be broken off and rooted without difficulty. Suckers from the bases of plants may or may not have developed root systems independent from the parent. If so, they can be transplanted directly. If not, they are treated as cuttings.

These underground organs are produced by some herbaceous perennials such as lilies and amaryllis as an underground repository of stored carbohydrate. Botanically, these are modified stems with nodes, buds, and modified leaves. Their natural means of vegetative reproduction is the formation of smaller storage organs around the base of the parent (Figure 4-18). These can be broken off (preferably while the plant is dormant) and planted in new locations. Blooming of young storage organs may take 2 to 3 years, since a minimum size must be reached before flowering will occur.

Air layering is a technique used primarily on such houseplants as rubber plants to encourage stem rooting of a plant that is still growing

Figure 4-16 A snake plant (*Sansevieria*) with two young offsets.

Figure 4-17 A pincushion cactus (*Mammillaria*) with offsets.

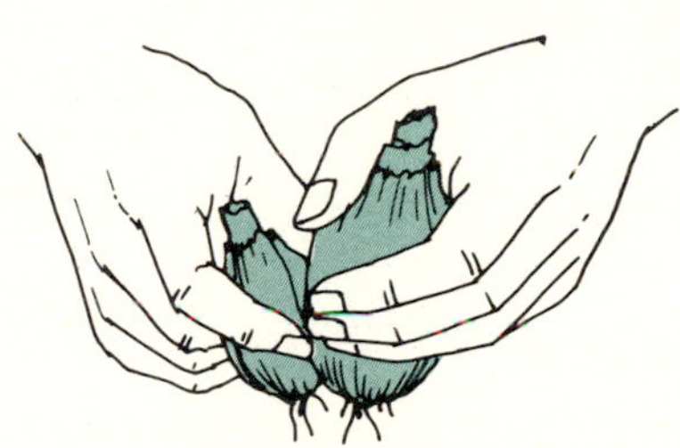

Figure 4-18 Removing a daughter bulb from the parent.

in its pot. It can be used to obtain new plants from a large branched specimen, to shorten a plant which has grown too tall, or to make a plant that has become tall and leafless short and bushy again. New plants formed by air layering can be very large, up to 3 feet (1 meter) tall if desired.

Many plants having woody stems can be air-layered (Figures 4-19, 4-20, and 4-21). As the first step, choose the section of the stem where the new root system of the plant is desired. Then a strip of bark about 1 inch (2 centimeters) wide should be cut around the stem and the bark pulled off. This "girdling" removes the phloem and cambium but not the xylem, which will still translocate water to the top of the plant.

Next, place a handful or two of damp sphagnum moss over the girdled area (Figure 4-19) and wrap with plastic (Figure 4-20).

Figure 4-19 Placing sphagnum moss around a branch to be air-layered.

Twist-ties or tape secure the moss at both ends and seal in moisture (Figure 4-21).

Within 2 to 3 months, roots should penetrate the sphagnum. During the waiting period the sphagnum should be checked every 1 to 2 weeks for moisture since roots will not form if the moss is dry. When a good supply of roots has formed, the air layer can be cut from the parent and transplanted to its own pot. If it wilts severely, humid conditions should be provided for about a week.

Budding and Grafting

Budding and grafting are fairly complex methods of propagation used for reproducing valuable fruit and ornamental plants in nurseries; they are discussed only briefly here. A gardener who wishes to try either budding or grafting should plan the project ahead of time and consult several of the reference books listed at the end of the chapter for more in-depth information.

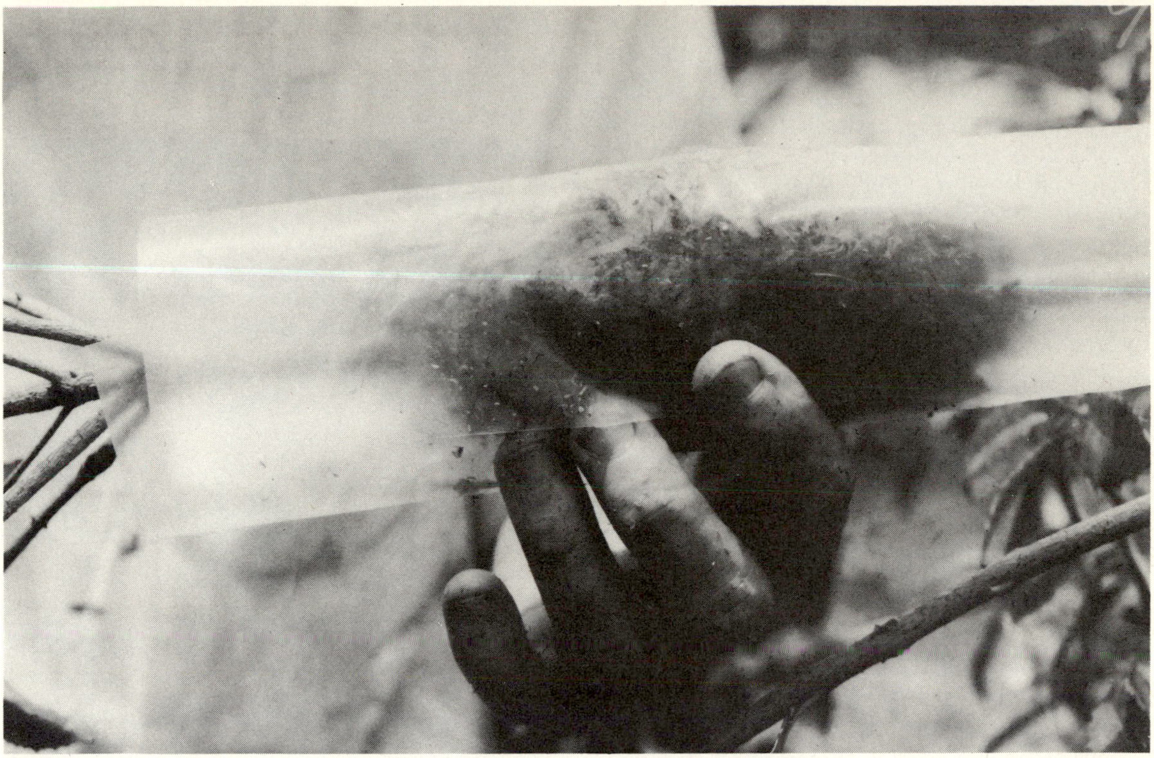

Figure 4-20 Wrapping plastic around the sphagnum.

Essentially, budding and grafting unite parts of plants so that they heal together and function as a single unit. Budding transfers a bud of one plant to another, whereas grafting attaches a small branch to another plant.

Most frequently, budding or grafting combines two varieties of a species into one plant that exhibits the best features of each. For example, a plant with a vigorous root system but little ornamental value (called the *rootstock*) could be grafted onto a plant having good ornamental qualities but weak roots (called the *scion*).

Grafting can also serve other purposes: (1) repair of girdled trees which would otherwise die; (2) creation of unusual plant forms such as tree roses or trees with weeping heads atop strong, straight trunks; and (3) changeover of fruit trees in an old orchard to a new variety.

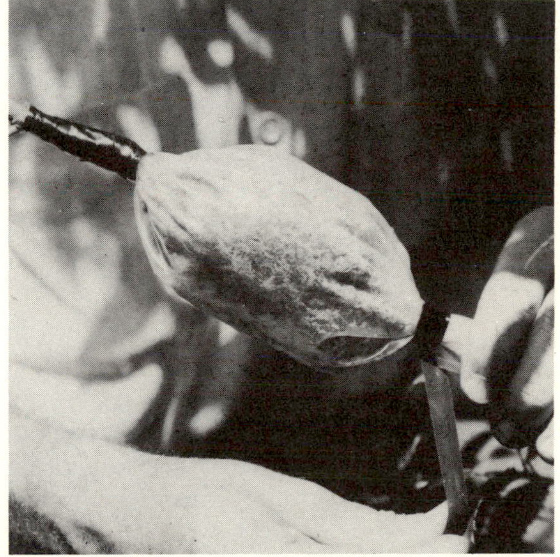

Figure 4-21 Sealing the plastic in place with tape.

Grafting and budding rely on the activity of cambium cells and must both be done when these cells are actively dividing and will heal the grafted area quickly. To determine the correct stage for grafting, bark is examined for *slippage.* If it can be separated easily from the underlying wood, the cambium cells are active and the grafting or budding is feasible. This is usually in early spring before new growth starts.

GRAFTING

There are many styles of grafting. The style used will depend on the species and the reason the graft is being made. General rules for grafting are:

1. The rootstock diameter must be equal to or larger than that of the scion, but the scion wood is usually a pencil thickness or slightly larger.
2. The cambium of the stock and scion must be in contact, and preferably over as great an area as possible. If cambium contact is not made, the graft will not heal, and the scion will die.
3. The stock and scion must fit tightly together, and the joint must be protected from drying. A tight graft union is achieved by wrapping the area with special rubber ties or waxed string. Drying is then prevented by a coating of wax.

Figure 4-22 shows the basic steps involved in a style of graft called *whip grafting.*

BUDDING

Budding (Figure 4-23) involves removing a patch of bark with a bud from a scion and laying it directly against the cambium of the stock. It is a less risky process than grafting because the stock is only slightly damaged if the union does not heal. As with grafting, the area of the union must be protected from drying with rubber ties.

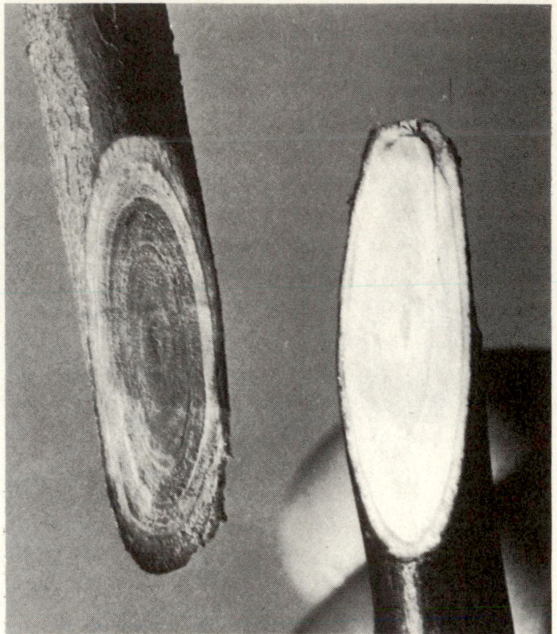

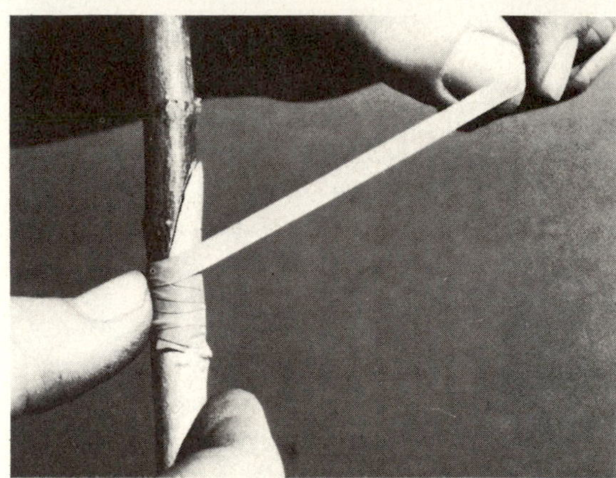

Figure 4-22 Steps in performing a whip graft. (a) Stock and scion ready for grafting. (b) Stock and scion joined and secured with a rubber strip.

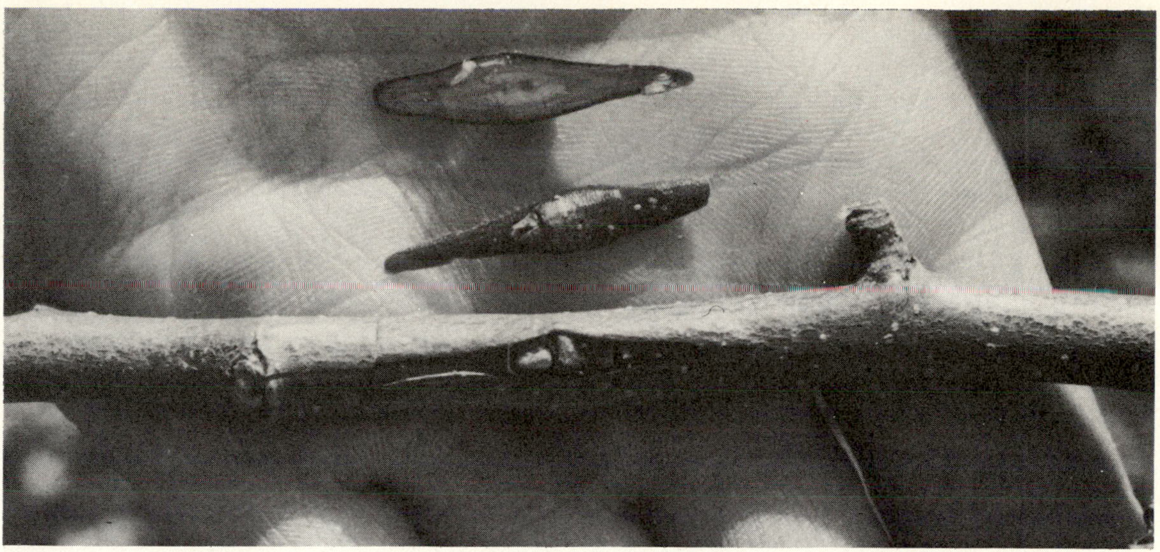

Figure 4-23 Bud patches to be used in budding and a bud patch in place on a branch.

Selected References for Additional Reading

Beaty, J. Y. *Plant Breeding for Everyone, How to Find and Develop New Plant Varieties.* Boston: Charles T. Branford, 1954.

Free, M. *Plant Propagation in Pictures.* New York: Doubleday, 1978.

Kester, D. E., and H. T. Hartman. *Plant Propagation: Principles and Practices.* 3d ed. Englewood Cliffs, N J: Prentice-Hall, 1975.

United States Department of Agriculture, Forest Service. *Seeds of Woody Plants in the United States.* Handbook 450. Washington, D.C.: U.S. Government Printing Office, 1974.

PART II

GROWING PLANTS OUTDOORS

Chapter 5

Outdoor Soils and Fertility

The soil is a blanket of loose material which covers the majority of the Earth's land surface. It is a dynamic mixture of minerals, decaying plants and animals (organic matter), microorganisms, water, and air. Relative to plants, the soil serves two primary purposes: first, it provides mechanical support for the plant by permitting roots to grow through it; second, it provides a reservoir for water, air, and nutrients essential for healthy plant growth. How well a particular soil performs each of these functions largely determines the fertility of the soil, that is, how well plants will grow in it.

CLASSIFICATION

Mineral soils are classified primarily by the size and quantity of mineral particles present in the soil. Soil particles range from clay particles, which are so small that they require an electron microscope to be seen, to silt, and then to sand, which can be seen readily by the unaided eye. Each type of soil particle possesses unique properties which influence plant growth. However, the properties of silt are in many respects intermediate between those of clay and sand, and silt is present in relatively small quantities in many soils. Therefore, a basic understanding of soils can be obtained by considering mainly the properties of clay and sand particles. Awareness of these properties will enable the gardener to adapt cultural practices so that the soil environment will be the most conducive to healthy plant growth.

Organic soils (including muck and peat) contain predominately decaying organic matter with smaller amounts of mineral components. They are generally found in small pockets in the north central and northeastern sections of the United States and Canada as well as in the Everglades region of Florida. Organic soils, while very fertile, require careful management; they do not, in many respects, respond in the same manner as mineral soils. Since organic soils are not important as home garden soils, they are not discussed in this text. Persons gardening in organic soils should consult a soils text or their county extension agent for further information.

Soils seldom contain only one type of particle. Instead they contain several types in

Table 5-1 Sample Classifications of Soil by Particle Types

Soil	Average particle size	Suitability as garden soil
Clay	Smallest	Fair
Loamy clay	Small	Fair
Clay loam	Medium	Good
Loam	Medium	Good
Sandy loam	Medium	Good
Loamy sand	Large	Good
Sand	Largest	Fair

varying percentages, and the soil is classified according to the type of particle which predominates (Table 5-1). For example, if a soil contains mostly clay particles but also a small amount of sand, it will be classified as a clay soil. When the percentage of sand increases to a point where it nearly equals the percentage of clay, the soil is classified as a loam. When the percentage of sand further increases to the point where it greatly surpasses the percentage of clay, the soil is designated as a sandy soil. Thus the term "loam," which is often used on seed packages to describe the ideal soil in which seeds should be planted, refers to a natural soil having nearly equal percentages of sand and clay particles.

As mentioned previously, clay and sand particles each possess unique characteristics. If a soil is classified as a loam, its properties will be intermediate between those of clay and sand particles. The higher the percentage of one type of particle, the more closely the soil properties will resemble those of that particle group. A loamy clay soil, for example, mostly possesses the characteristics of clay particles but is modified somewhat by the sand it contains. A clay loam, on the other hand, is basically a loam with a slight extra amount of clay.

Characteristics of Clay Soils

Clay soils (Table 5-2) are often referred to as "heavy" soils, composed primarily of platelike clay particles. These particles are flat and disk-shaped, and their small size and flattened shape are responsible to a large extent for the properties of clay soils. Because the particles are flat, they are able to lie in close contact with one another. This closeness makes the spaces between individual clay particles relatively small compared to the spaces between more rounded particles such as sand. This might be likened to pennies (clay particles)

Table 5-2 Properties of Sand and Clay Soils

Clay	Sand
Ability to hold a large amount of water	Inability to hold water
Slow movement of air and water in the soil (poor drainage)	Rapid movement of air and water in the soil (good drainage)
Small, numerous pore spaces	Large and fewer pore spaces
Tendency to expand when wet and contract and crack when dry	Does not expand or contract as a result of water
Good ability to attract and hold nutrients for plant growth (high cation-exchange capacity)	Poor ability to attract and hold nutrients for plant growth (low cation-exchange capacity)
Tendency to compact under pressure when wet	No tendency toward compaction
Plastic when wet and highly moldable	Not plastic when wet

being stacked in a jar as compared to marbles (sand particles). The spaces between the pennies are much smaller than the spaces between the marbles (Figure 5-1). Conversely, the jar filled with marbles, which is likened to a sandy soil, contains fewer particles but larger spaces. These small spaces between clay particles (called *pore spaces*) account for one of the main properties of clay soils: slow water and air movement into and out of the soil.

Water movement through pore spaces causes a soil to be either fast- or slow-draining and determines whether its drainage is good (fast) or poor (slow). Thus soils composed mostly of clay are considered poor-draining, hold water for long periods of time, and can be detrimental to plant growth.

Another property of clay soils attributable to the flattened shape of the particles is *plasticity.* This is the property of clay to be sticky and moldable when wet (Figure 5-2) and hard and cloddy when dry. Since clays are highly plastic when wet, they are also subject to compaction, a packing down of the soil particles which practically closes up all the pore spaces (Figure 5-3). Compaction is

Figure 5-2 The plasticity of clay allows it to be formed into ribbons when wet.

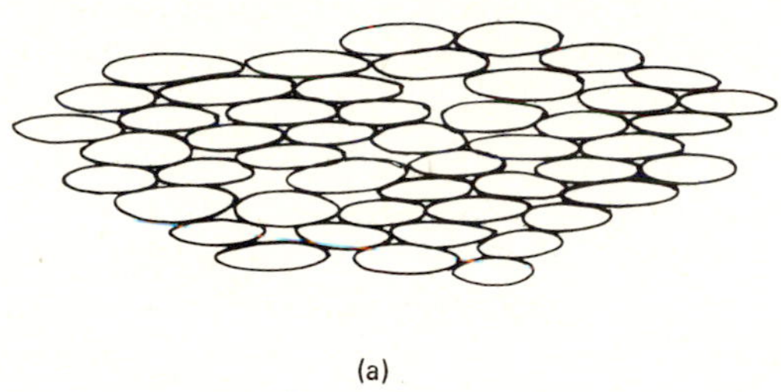

(a)

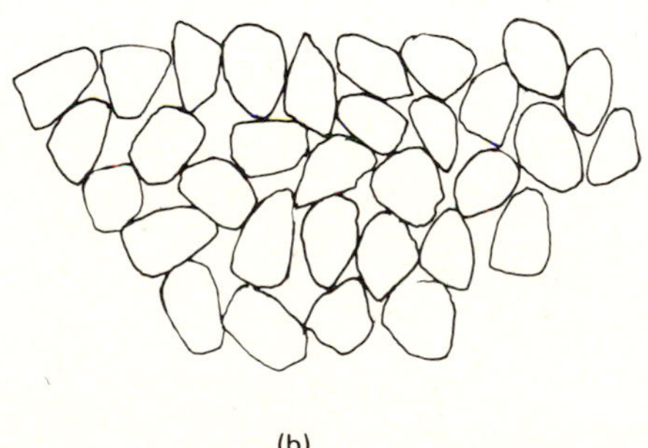

(b)

Figure 5-1 Clay particles (a) lay flat and have smaller pore spaces between them than sand particles (b).

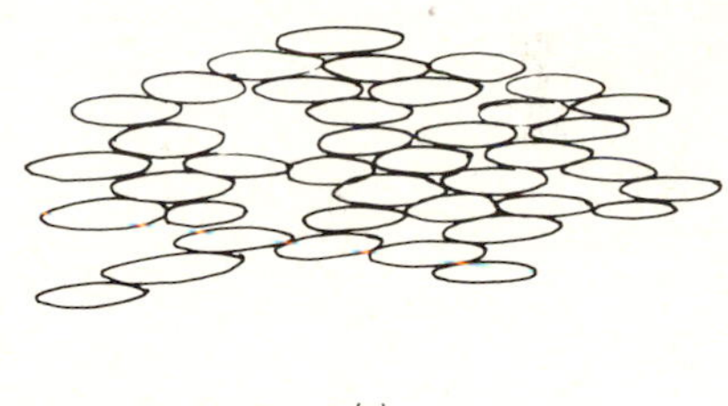

(a)

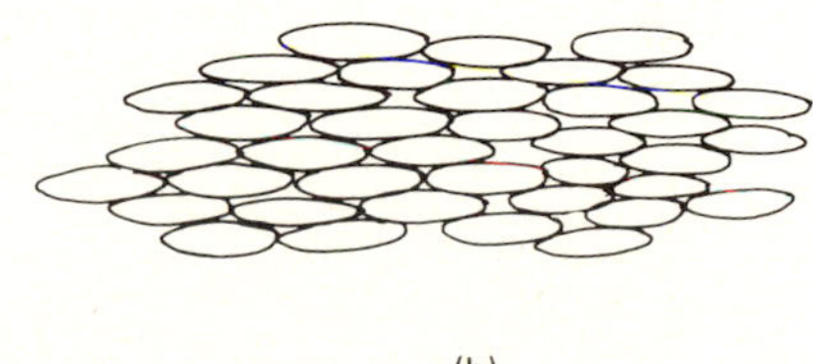

(b)

Figure 5-3 (a) Aggregated clay soil. (b) Compacted clay soil. Note the decreased pore space in the compacted soil.

the result of pressure such as by walking or driving on the soil when it is wet. It makes water penetration and drainage poor, resulting in poor plant growth due to a lack of air in the soil. Another result of the plastic properties of clay is that it expands when water is added and contracts when it drys out. This tendency to expand and contract results in the surface cracking of the soil that is often evident when clay soils become very dry.

Clay soils also have a high capacity to attract and hold nutrients. This property (called *cation-exchange capacity*) will be discussed in more detail later. Basically, it means that nutrients tend to be stored in clay soils rather than washed away by rainfall or irrigation.

Characteristics of Sandy Soils

Sandy soils (Table 5-2), or "light" soils, can be thought of as possessing the opposite properties of clay soils. Whereas water moves through clay soils slowly via the small pore spaces, in sandy soils it moves freely and rapidly through large pore spaces. Sandy soils do not retain moisture well; nor do they expand and contract, become sticky or cloddy, or compact readily. The cation-exchange capacity of sandy soils is quite low, so nutrients wash away readily when water drains through.

The difference between the properties of clay and sand particles is due mainly to the larger size of sand particles and their rounded or irregular shape. Being irregularly shaped, they do not pack tightly (compact). Nor do they slide easily past one another when wet, as clay particles do. Consequently they lack the plastic properties inherent in sandy soils.

Characteristics of Loam Soils

Loam soils, possess qualities of both sand and clay proportional to the relative amount of clay or sand particles in the loam soil. They are considered to be ideal for most plants because they do not have the extreme properties of either sands or clays. For example, clay soils may become waterlogged and unworkable for many days after rain, and sandy soils hold so little water that irrigation must be applied frequently. However, a loam soil will absorb a large quantity of water and not require frequent irrigation; it will dry enough to be workable in a much shorter time than will a clay soil. Loams also hold more nutrients than do sandy soils; however, they are not as prone to becoming cloddy or compacting as clay soils.

ORGANIC MATTER IN THE SOIL

In addition to the mineral components of soil such as sand and clay, organic matter is also a valuable component. In fact plants can be grown in pure organic matter quite satisfactorily without any inclusion of sand or clay. Organic matter in the soil is the result of vegetative matter such as leaves and roots and animal products such as manure returning to the soil and decomposing. Though organic matter is present in most soils in relatively small quantities, it greatly influences the properties of the soil. It commonly, for example, is responsible for half of the cation-exchange capacity of a soil. In sandy soils organic matter increases water- and nutrient-holding capacity; in clay soils it improves drainage and air movement and decreases compaction and plasticity. Its net effect, then, is improvement of any soil to which it is added.

One of the most important effects of organic matter on clay soils is to cause aggregation of the clay particles, which have a tendency to lay very close together and exclude air. When organic matter decomposes, the resulting humic acid acts as a glue and causes the clay particles to clump together loosely (Figure 5-3). Since the small clay particles are then clumped together, they form much larger particles which function more like sand. The effect of this aggregation is twofold: the soil has larger pore spaces, and water and air movement is improved, resulting in better plant growth. When clay soils are worked while wet, these aggregates are broken apart, resulting in a poorer soil. For this reason, clay soils should never be disturbed until they are dry enough to let the soil crumble easily in the hand.

CATION-EXCHANGE CAPACITY

The cation-exchange capacity (CEC) of a soil is the relative capacity of the soil to attract and hold nutrients (cations) on the surface of the soil particles. Soil and organic matter particles have negative charges; since particles with opposite charges attract each other (like a magnet), they are able to attract positively charged ions from the soil water and hold them against the surface of the soil particle (Figure 5-4). In soils with a low CEC such as sand, the number of cations each sand particle can attract is quite low, due largely to the small surface area of each particle. Thus, when nutrients are added to the soil, only a small amount can be held by the soil particles. The rest remain dissolved in the soil water and are washed away readily with rainfall or irrigation.

In soils with a high cation-exchange capacity such as clays or those high in organic matter, a very large quantity of nutrients can be held by each soil particle. Thus, when a fertilizer such as potassium is added, a large amount of the nutrient will be held in the soil. These absorbed nutrients cannot be washed away by watering; instead, they are stored in the soil for future use by plants.

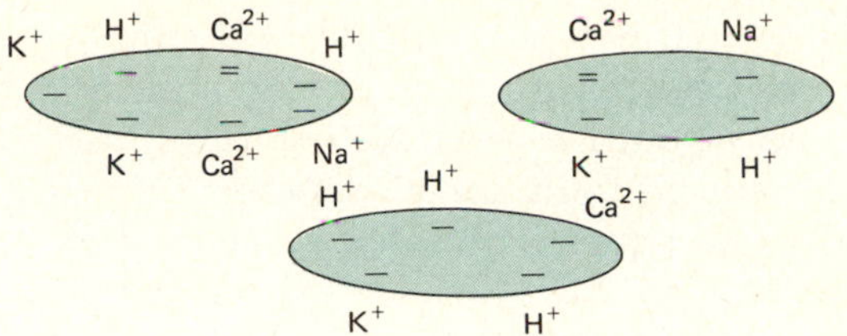

Figure 5-4 Negatively charged clay particles attract and hold cations, providing a reservoir of nutrients.

Understanding cation-exchange capacity is important in fertilizer application. Fertilizers need to be applied less frequently to soils high in cation-exchange capacity because of their nutrient storage capabilities. A high cation-exchange capacity is therefore a desirable soil quality.

AIR AND WATER IN THE SOIL

Varying amounts of air and water are contained in any soil. Most higher plants require that plant roots be in contact with adequate amounts of both water and air. Thus a good soil will provide a balance between water and air. Whenever either water or air is lacking in the soil, the plant will die. In a sandy soil that tends to provide abundant quantities of air but does not retain water well, lack of water commonly restricts plant growth. On the other hand, in a heavy clay soil where water drainage is poor, plant growth may be restricted by a lack of adequate air for root respiration.

In any soil the pore spaces are filled with either air or water. At any given time the percentage of pores filled by water or air is determined by the size of these pores and the relative abundance of either air or water. Therefore the balance between air and water is partially determined by the soil itself (from pore size and number) and partially determined by the quantity of water which has been supplied to the soil.

Pore size affects the air/water ratio in the soil because small pores hold water better than do large pores. When water is added to the soil, it has a tendency to move downward in response to gravity. It is a force called *adhesion,* the attraction of solid materials (soil particles) to liquids, which holds water in the soil rather than allowing it to flow through. Adhesion can be demonstrated by dipping a marble in water and observing the thin film of water which adheres to it. At the same time that water is adhering to soil particles, it is also *cohering* to other water molecules. *Cohesion* is illustrated by the droplet of water that forms at the bottom of the marble after it is removed from the water (Figure 5-5). This droplet is composed of a layer of water molecules adhering directly to the marble and another layer cohering to the first layer of molecules. When too many layers of water molecules cohere to the first layer, the force holding them together becomes weaker, and the droplet will fall from the marble. The same phenomenon occurs in soil. If a pore space is small, a greater quantity of water will remain, due to the combined forces of adhesion and cohesion.

Using this information and knowledge of the size of pore spaces found in clay and sand, it is easily seen why clay has a higher water-holding capacity than sand. The presence of a large number of small pores in clay, as compared to the fewer but larger pores in sandy soil, explains the difference.

Water, in addition to moving downward in the soil, may also move sideways or upward. Water moving in these directions is due to the combined forces of adhesion and cohesion and is called *capillary water.* Capillary water is analogous to water that moves upward in a straw which has been placed in a glass of water (Figure 5-6). In this case the water adheres to the sides of the straw, and then the water molecules cohere to each other, thus creating a column of water that extends above the level of water in the glass. In the soil, water moves upward in a column through the pores to a root through which the plant absorbs it or to the surface, where it evaporates. Through the process of capillary water movement, water which has not been utilized by the plant or has evaporated is constantly replaced by water from below. Capillary action

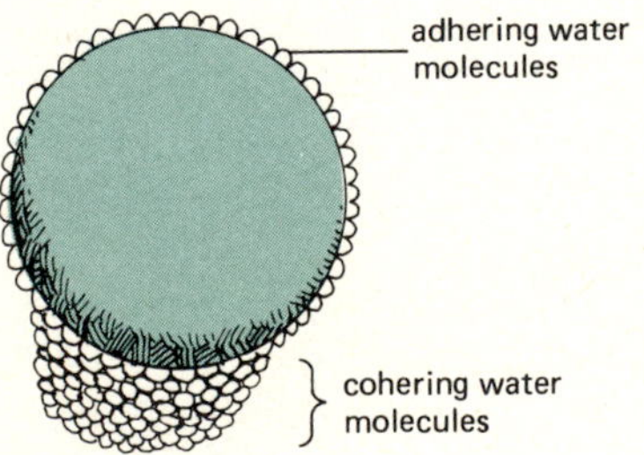

Figure 5-5 Adhesion and cohesion of water molecules around a marble show how water is held in the soil.

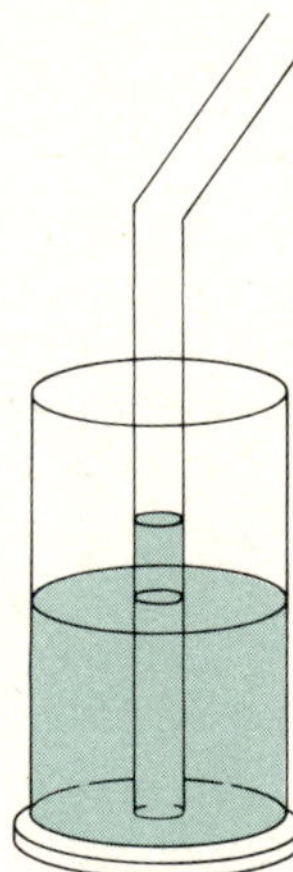

Figure 5-6 Capillary action moves water upward in this straw in the same way that water moves upward in the soil.

illustrates the importance of watering deeply rather than applying only a small quantity of water to the soil surface.

SOIL LAYERS

Most soils possess distinct layers within them which affect plant growth. For the gardener, the most important layers are the *topsoil* and the *subsoil* (Figure 5-7). The topsoil is a layer located at the soil surface above the subsoil and may be from less than an inch to as much as 30 inches thick. This is the layer of soil that is usually involved in soil preparation activities. Topsoil is characteristically higher in organic matter and nutrients than subsoil and is thus more suitable for plant growth.

In soils which have not been disturbed,

Figure 5-7 Topsoil and subsoil layers.

the topsoil is usually quite deep due to the organic matter that has accumulated from the growth and death of plants on the surface. This accounts for the usual high fertility of newly cultivated soils. Unfortunately, during the construction of houses much or all of the topsoil is removed by grading, leaving only the relatively infertile subsoil in which to garden. In order to improve the fertility of subsoil, a soil improvement program using soil amendments must be followed.

In addition to the natural topsoil and subsoil layers, another layer called a *hardpan* (Figure 5-8) may occasionally be present. A hardpan is a layer of compacted soil that slows or stops the movement of water and air and the growth of roots into lower layers. This results in the total displacement of air in both large and small pores, with consequent damage to plant roots due to the absence of air. When air is lacking, roots become susceptible to several disease organisms present in the soil and will be weakened or killed.

Hardpans may occur naturally or as a result of heavy equipment moving over the soil. If a hardpan is present, the best solution is to dig through it with a posthole digger or, if this fails, to construct raised planters. The presence of a hardpan and accompanying poor drainage can be identified by filling the hole with water and waiting for it to drain away. If the water remains after a half-hour, the hardpan is still present.

SOIL ORGANISMS

The soil is inhabited by a vast and ever-changing population of small, often microscopic, plants and animals. These organisms affect the soil and the higher plants growing in it in a number of ways. They improve drainage and aeration, assist in the decomposition of organic matter, aid in plant nutrient and water uptake, cause diseases, or feed directly upon plants. Although some are harmful, causing plant injuries or diseases, most are beneficial and form an essential part of the soil complex. Included among soil-inhabiting organisms are such animals as earthworms, insects, and nematodes and such microscopic plants as mycorrhizae, algae, fungi, and bacteria.

Perhaps to the home gardener the most interesting, beneficial animal in the garden soil is the common earthworm. These animals feed by passing soil containing organic matter through their bodies, where it is subjected to

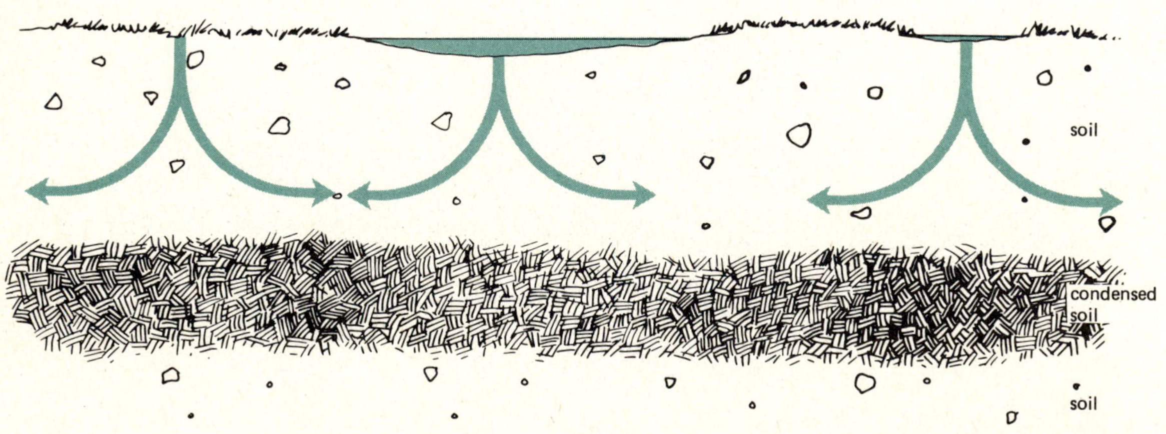

Figure 5-8 A soil hardpan.

digestive enzymes and a grinding action. The material that has passed through the earthworm's body is called a *casting.* Castings are much higher in organic matter and nutrients than the surrounding soil. In addition, the earthworm burrows increase soil aeration and drainage.

Earthworms are most abundant in moist clay or loam soils with a high organic matter content. In cold climates they are least abundant in the spring, with populations increasing all summer. Since earthworms are killed by early fall freezes in bare soil, they should be protected by surface mulches to maintain high populations in the home garden. After the first cold snap of the year, earthworms begin to acquire cold tolerance and migrate deeper into the soil where the temperature is more favorable.

SOIL pH

The pH of a soil is a measure of its acidity or alkalinity. It is measured on a logarithmic scale that runs from 0 to 14, although a soil pH below 4 or above 9 is uncommon (Figure 5-9). A pH of 7 is neutral, whereas a pH above 7 is alkaline and below 7 is acid. The optimum range for the majority of plants is 6.5 to 7.0. Many plants adapted to a lower pH are called "acid-loving." Other plants are tolerant of alkaline soils.

A soil pH that is not optimum may be corrected by adding materials to the soil to change the pH. Since the pH scale is a logarithmic one, a pH of 6.0 is 10 times more acid than a pH of 7.0; a pH of 5.0 is therefore 100 times more acid. Thus relatively large quantities of acid- or alkaline-forming materials must be added to cause even a small change in pH, especially in clay soils.

Lime is one material utilized to raise soil pH. It is available in several different grades and types, all of which will make the soil more alkaline. Ground limestone or calcitic limestone ($CaCO_3$) is the most common and the least expensive type available. Dolomitic limestone ($Ca \cdot MgCO_3$), although slightly more expensive than calcitic limestone, performs the added function of providing magnesium, a necessary nutrient lacking in some soils. Whichever type of lime is used, it will take between 3 and 6 months to accomplish the desired pH change. This is due to the slow speed with which lime dissolves. To speed up the pH change as much as possible, buy finely ground lime, and mix it thoroughly into the soil.

Acidification of the soil is generally accomplished through the use of the chemical sulfur or aluminum sulfate or by using organic

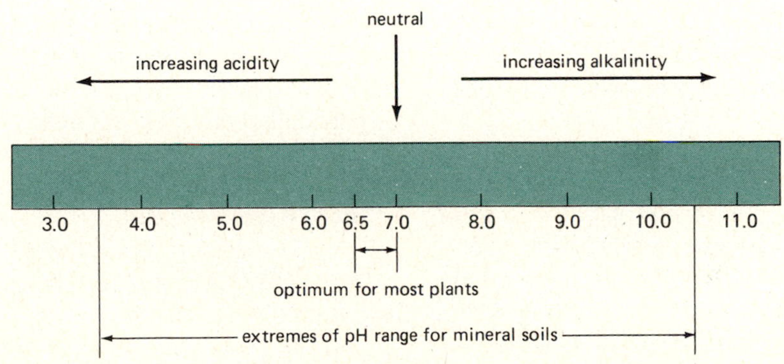

Figure 5-9 The pH scale of soil.

amendments such as oak leaves, pine needles, and sphagnum moss which are acid-forming. When using organic amendments, large quantities are needed, and the change is very slow. Sulfur or aluminum sulfate changes soil pH slowly, and only after the use of relatively large quantities. Due to the high cost of acidifying materials as compared to lime, large-scale soil acidification is generally not done. Instead the soil immediately surrounding an acid-loving plant is adjusted. Agricultural sulfur is the most commonly used acidifying agent and is less costly than aluminum sulfate; however, a greater quantity is required than of the latter, and the effect is slower. Aluminum sulfate is more acid-forming that sulfur but must be used carefully since, if overused, the aluminum can become toxic to plants.

SOIL IMPROVEMENT

Soil Testing

The best way to determine the pH and nutrient content of a soil is to have it professionally tested. The results can then be used as a guide in the soil improvement program. In most of the United States a soil test will be performed for a small charge by the Cooperative Extension Service. The usual procedure is to submit a soil sample to the county agent, who will forward it to the soil testing laboratory located at the land grant university. In other areas commercial soil testing laboratories must be used.

The most important step in the soil testing procedure is taking the sample. The sample must be representative of the area being tested or it will be meaningless. In order to obtain a representative sample, a combined sample is taken from the area being tested. The best technique is to establish a grid pattern over the area being sampled (Figure 5-10) and then take a small sample from each grid intersection. Each individual sample can best be taken by digging a small hole 6 to 8 inches deep and then skimming off a small slice of soil (Figure 5-11) from the top to the bottom of the hole. At least a dozen samples should be combined in a plastic bucket and mixed thoroughly, and then a 1-pint sample should be taken from this. This composite sample can then be submitted for analysis.

Soil Improvement through Addition of Amendments

A soil amendment is a material that is added to a soil to improve physical properties such as drainage, CEC, aeration, and water retention. It may or may not supply nutrients, depending on the choice of amendment. The addition of soil amendments for soil improvement is a fundamental gardening practice and will greatly influence gardening success.

The best amendment is chosen by considering several factors including the problem to be corrected and its seriousness, the properties of the various amendments available, and the cost of the amendment. Often the best amendment will not be practical because of cost, and a less expensive amendment will have to be used.

The most important rule to remember about the use of any soil amendment is that amendments must be used in large quantities to achieve the desired benefits. The addition of a small amount will not make any appreciable difference in soil properties. The amount of amendment required will depend on the particular amendment chosen and the severity of the soil problem. As a general rule, an amount of amendment equal to 25 percent of the volume of the soil should be applied.

The easiest way to apply an amendment is to spread it over the surface of the soil to a depth of several inches (8–10 centimeters) and then spade or rototill it into the soil. The thickness of the amendment layer needed can be determined by comparing the soil depth

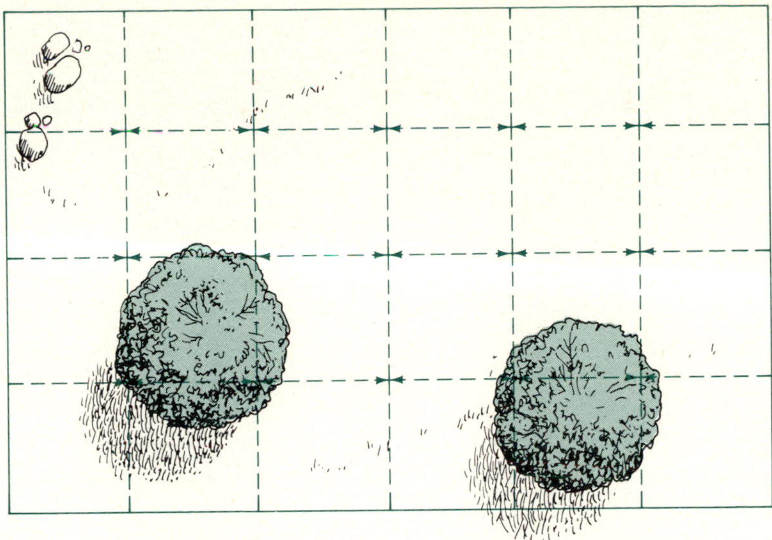

Figure 5-10 Following a grid pattern for soil testing will assure a representative soil sample.

to be spaded or rototilled to the amendment depth. For example, if a 2-inch (5-centimeter) amendment layer is used and the soil is spaded 8 inches (20 centimeters) deep, the amendment added equals approximately 25 percent of the soil volume. If large quantities of amendments are added, the process is best repeated several times, with about a 2-inch (5-centimeter) layer of amendment incorporated each time.

Figure 5-11 Slicing off an undisturbed section of soil for testing.

Each type of soil amendment has properties which must be evaluated when determining its suitability. An important consideration is the quantity of nitrogen it contains and how rapidly it decomposes. If the amendment decomposes rapidly and contains little nitrogen, nitrogen may actually be taken from the soil to supply the nitrogen needed for the process of decomposing the amendment. This nitrogen will be released when decomposition is completed, but in the meantime nitrogen deficiency symptoms will often occur in plants growing in the area. This effect will be intensified if the amendment is finely ground and the soil is warm and moist, since these conditions are conducive to the most rapid decomposition of the amendment and thus the greatest nitrogen deficit. To counteract this, a fertilizer containing nitrogen should be

Table 5-3 Soil Amendments

Amendment	Uses	Nitrogen requirement	Comments
Inorganic			
Perlite	Work into heavy soils to improve drainage and aeration and reduce compaction	None	Expensive; lasts a long time; best for planters; lightweight
Sand (coarse)	Same as perlite	None	Less expensive than perlite but required in large quantities for substantial improvement of clay soils; heavy; do not use beach sand because of salts
Topsoil (loam)	Work into very sandy soils to improve water and nutrient holding	None	Inexpensive but heavy; may be contaminated with weed seed, nematodes, fungi, and other undesirable organisms
Vermiculite	Improves water and nutrient holding in sandy soils	None	Expensive; lightweight; if soil is worked frequently, the structure of the vermiculite will be damaged, reducing its effectiveness
Scoria (lava rock, cinder rock)	Same as perlite	None	Less expensive than perlite; medium weight
Organic			
Bark	Mix with heavy soils to improve drainage and aeration and decrease compaction; some improvement of water- and nutrient-holding capacity in sandy soils	Add about 1 lb of actual nitrogen per cubic yard	Usually inexpensive; breaks down fairly slowly; finely ground bark is best
Compost (also called humus)	Improves both clay and sandy soils; contains some nutrients	None	Variable; may have a high ash content
Leaf mold	Similar to compost but different in that it is derived primarily from leaves	None	May acidify soil slightly
Manure (composted)	Improves both clay and sandy soils; contains some nutrients	None	May be high in both salts and ash; may have a strong odor; inexpensive

Amendment	Uses	Nitrogen requirement	Comments
Michigan peat	Improves both clay and sandy soils if used in large quantities	None	Inexpensive, but large quantities are needed; high ash content; not readily available west of the Rockies.
Redwood compost	Best for clay soils, with some improvement of sandy soils	Add about ¼ lb of actual nitrogen per cubic yard	Inexpensive, however, does not hold water or nutrients well
Sawdust	Improves clay soils	Add about 1 lb of actual nitrogen per cubic yard	Very inexpensive, however, breaks down so rapidly that nitrogen deficiencies may develop despite the addition of nitrogen; improves soil for a relatively short time; if used, mix with a better amendment; some toxicity may occur if used fresh, depending on source; avoid walnut
Sphagnum peat moss	Excellent for clay or sandy soils; high water and nutrient capacity	None	Expensive, but less is required; hard to wet when fully dry; acidifies soil
Sludge	Improves both sandy and clay soils; adds some nutrients	None	Inexpensive; may have a high ash content and a strong odor; do not use in the vegetable garden if root vegetables are grown

added with the amendment. Amendments which take up nitrogen excessively include sawdust, wheat straw, and bark.

Other considerations are salinity and ash content. Certain amendments are high in chemicals such as sodium and ammonium salts. This may not be a serious problem in the eastern United States and Canada where abundant rainfall occurs; however, in the drier western parts of the hemisphere, rainfall is often insufficient to wash away the salts, resulting in salt toxicity to plants.

Ash content refers to the amounts of mineral matter present in the amendment. In organic amendments the presence of large quantities of ash is usually undesirable, as ash does not improve the soil. Leaf molds, manures, and compost often contain large amounts of ash.

Inorganic amendments such as sand, vermiculite, and perlite are usually too expensive to be used as amendments in garden soils, although they are commonly used in potting soils. When they are used, the effects are often quite permanent, whereas organic amendments must be added periodically. Table 5-3 contains a list of soil amendments, their characteristics, and uses.

Composting

Compost (Figure 5-12) is one soil amendment that can be made rather than bought and which is excellent for soil improvement. It is composed of partially decomposed plants and is made by piling garden refuse such as weeds, grass clippings, fallen leaves, and plant clippings in a pile to rot. The piling helps keep in moisture and heat generated by the decomposition process and speeds the normal rotting process.

A compost pile can be relatively simple or quite complicated. In its most basic form, the plant refuse can be simply piled in an out-of-the-way spot and allowed to decompose slowly without any further attention. With this method, compost will be made very slowly, and 3 to 4 months may pass before appreciable compost forms.

To form compost at a faster rate, a compost heap must be built. A wire or wooden bin should be constructed as shown in Figure 5-13. A layer of refuse should then be spread in the bottom 6 to 9 inches (15–23 centimeters) deep, sprinkled with a handful of high-nitrogen fertilizer, covered with a thin layer of soil to supply microorganisms, and wet slightly. The layering and wetting process should be repeated until the bin is full. The composting material should be kept damp by sprinkling with water as necessary and turned and mixed with a pitchfork every 4 weeks to ensure even decomposition of the entire pile.

Figure 5-12 Compost.

If not turned, the pile will generate heat and decompose mainly in the center, with the outer layers rotting only very slowly. Compost can be considered ready for use when the material is evenly dark-colored and decomposed to the point where the materials are not recognizable. It is then used for soil amending or mulching.

Among the other materials which can be composted are vegetable peelings and rotten vegetables, egg shells, and other kitchen refuse other than meat scraps. Plants which are diseased should not be composted since there is a chance the disease-causing organisms will survive despite the heat of decomposition (often up to 150°F) (66°C). Woody materials such as twigs and small branches can be used but will break down very slowly. Their decomposition can be hastened by chopping them into small pieces before adding them to the pile.

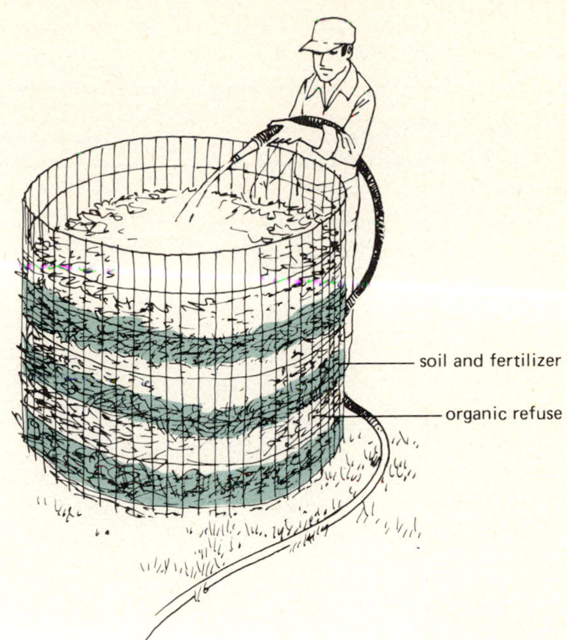

Figure 5-13 Cross section of a compost pile built in layers for rapid decomposition. Keeping the pile moist speeds the process.

PLANT NUTRITION

All plants obtain a large number of mineral nutrients from the soil. Since many nutrients are known to be utilized by plants, they are grouped for ease of discussion into two groups based on the relative quantities utilized by the plant. The first group, called *macronutrients*, includes those nutrients which are used in fairly large quantities by plants. The macronutrients include nitrogen (N), phosphorus (P), potassium (K), calcium (Ca), magnesium (Mg), and sulfur (S). The second group, called *micronutrients*, or trace elements, is equally important but is used by plants in small quantities. The most important micronutrients are iron (Fe), copper (Cu), zinc (Z), boron (B), molybdenum (Mb), chlorine (Cl), and cobalt (Co). An example of the difference in quantity of nutrients used by a plant is that an acre (hectare) of corn might easily remove 150 pounds (60 kilograms) of nitrogen from the soil while removing less than an ounce (28 grams) of cobalt.

For optimum plant growth to occur, all of the nutrients needed by that plant not only must be present in the soil but must be present in a form which a plant can utilize.

Macronutrients

The macronutrients are those nutrients used by plants in the largest quantities and thus are the primary ingredients in most garden fertilizers. Of the six macronutrients plants obtain from the soil, calcium, magnesium, and sulfur are generally present in large enough quantities to be adequate for many years and seldom need to be added when fertilizing. For cases where these nutrients are lacking, a list of fertilizer sources is given in Table 5-4. Phosphorus, potassium, and particularly nitrogen are frequently not present in sufficient quan-

Table 5-4 Chemical Soil Amendments for pH Adjustment and Micronutrient Addition

Name	Chemical formula	Speed of effectiveness	Use
Limestone	$CaCO_3$	Slow	Raise soil pH
Hydrated lime	$Ca(OH)_2$	Rapid	Raise soil pH
Gypsum or calcium sulfate	$CaSO_4$	Medium	Add sulfur, which displaces sodium in soils and improves drainage
Sulfur	S_2	Slow	Add sulfur and lower soil pH
Epsom salts	$MgSO_4{\cdot}7H_2O$	Rapid	Add magnesium
Aluminum sulfate	$Al_2(SO_4)_3$	Rapid	Add sulfur and lower soil pH
Calcium nitrate	$Ca(NO_3)_2{\cdot}2H_2O$ (analysis 15-0-0)	Rapid	Add calcium and nitrogen and raise soil pH
Ammonium sulfate	$(NH_4)\ 2SO_4$ (analysis 20-0-0)	Rapid	Lower pH sharply, add sulfur and nitrogen
Magnesium sulfate	$MnSO_4$		Lower pH and add magnesium
Iron sulfate	$FeSO_4$		Lower pH and add iron
Chelated iron	9–12% Iron		Add iron in absorbable form
Borax	$Na_2B_4O_7{\cdot}10H_2O$		Add boron
Copper sulfate	$CuSO_4$		Add copper and lower pH
General micronutrient fertilizers	Contain a combination of most essential micronutrients		Add deficient micronutrients

tities for optimum growth; they are therefore the main macronutrients supplied in fertilizers.

NITROGEN

Nitrogen is used in large quantities by plants. When nitrogen is lacking, plants grow slowly and lose their deep green color. The application of nitrogen will often result in a flush of leafy growth and the return of a deep green color within a day or two. Anyone who has applied nitrogen fertilizers to a lawn can vouch for the resulting fast growth and increased frequency of mowing.

Since nitrogen stimulates rapid vegetative growth, it is desirable to apply it frequently to plants grown for their foliage such as turf, leafy vegetables, and indoor foliage plants. High levels of nitrogen fertilization do, however, have the disadvantage of causing leafy growth instead of flowers and fruit. A common example is the vegetable gardener who uses high-nitrogen lawn fertilizer on tomatoes. The result is a tomato plant of astounding size with healthy, dark green leaves but which never produces many flowers or fruit. Another disadvantage of high nitrogen fertility levels is that succulent new growth of shrubs and trees not only is more susceptible to disease but may continue too late in the summer, resulting in cold injury when winter arrives. Thus, when nitrogen fertilizers are used, the consequences must be assessed and the quantity and timing of each application carefully regulated to maintain adequate, but not excessive, levels of nitrogen in the soil.

A characteristic of nitrogen that greatly affects fertilizing rate and frequency is that nitrogen is not bound to soil colloids as are many other nutrients. Virtually all the nitrate nitrogen added to the soil will remain dissolved in the soil water. As a consequence, when nitrogen is in the nitrate form usable to plants, it readily leaches from the soil. This means that whenever heavy rainfall follows nitrogen applications, much of the nitrate nitrogen will be leached away and lost to the plant. The drainage characteristics of the soil should thus be assessed when deciding how often to fertilize with nitrogen.

In a sandy soil nitrogen will be quickly leached out when rainfall occurs, so slow-release fertilizers or frequent small applications of quick-release (nitrate) fertilizers rather than a single large application are best. Slow- and fast-release fertilizers and the various forms of nitrogen contained in fertilizers and used by plants are discussed under "Fertilizers."

In slower-draining clay soil the loss of nitrogen is slower and the benefits obtained from slow-release fertilizers are less valuable. Another consequence of the lack of nitrate adsorption is that virtually all of the nitrates will remain in the soil solution. If too much fertilizer is applied, severe root damage to plants may result. With nutrients which are adsorbed, burning is less likely since some of the nutrients will be removed from the soil solution by adsorption to soil particles.

Nitrogen can be added to the soil in many different forms, each having characteristics which determine its suitability for a particular usage. Consult Table 5-5 for characteristics of a number of common nitrogen sources.

PHOSPHORUS

Phosphorus, the second macronutrient, influences many plant functions including flowering and fruiting, root development, disease resistance, and maturation. To the home gardener, phosphorus may be thought of as a root and flower stimulator and thus of particular importance to plants grown for their flowers and fruit or to newly transplanted plants undergoing establishment.

Phosphorus behaves very differently from nitrogen in the soil. It is insoluble in the soil water and, when added to the soil, reacts readily with aluminum, iron, calcium, and other elements, forming various compounds unusable by plants. Consequently, relatively little phosphorus is ever in the soil water at any one time, and there is negligible leaching and little danger of plant damage from overapplication. The problem, then, with phosphorus is to maintain enough of it in the soil water for plant use. This can best be done by maintaining the pH between 6.0 and 7.0, maintaining high amounts of organic matter, and wherever possible, placing the fertilizer in bands next to plants to minimize soil contact.

Since phosphorus is insoluble, it will remain where it is placed and will not leach through the soil. However, it must be supplied in the root zone of the plant. Surface applications of phosphorus will have little effect on deeply rooted plants.

An important, rather specialized use of phosphorus is its use as a transplant or starter fertilizer. (Starter fertilizers are characteristically low in nitrogen and potassium but high in phosphorus.) Starter fertilizers are applied as a liquid at the time of transplanting and

Table 5-5 Garden Fertilizers and Amendments and Their Characteristics

Name	Analysis	Speed of nutrient release	Effect on pH
Inorganic and chemically manufactured			
Ammonium sulfate	20-0-0	Fast	Very acid
Sodium nitrate	15-0-0	Fast	Alkaline
Potassium nitrate	13-0-44	Fast	Neutral
Ammonium nitrate	33-0-0	Fast	Acid
Urea CO	46-0-0	Fast	Slightly acid
Monoammonium phosphate	11-48-0	Fast	Acid
Diammonium phosphate	18-46-0	Fast	Acid
Treble superphosphate	0-40-0	Medium	Neutral
Superphosphate	0-20-0	Medium	Neutral
Potassium chloride	0-0-60	Fast	Neutral
General balanced granular	10-10-10, 5-10-10, and so on	Various	Various
Organic			
Bone meal, steamed	2-27-0	Slow	Alkaline
Bone meal, raw	4-22-0	Slow	Alkaline
Hoof and horn meal	13-0-0	Slow	
Dried blood	12-0-0	Medium	Acid
Sewage sludge	2-1-1	Medium	Acid
Activated sewage sludge (microorganisms added)	6-5-0	Medium	Slightly acid
Cattle manure (fresh)	0.5-0.3-0.5	Slow	Slightly acid
Chicken manure (fresh)	0.9-0.5-0.8	Slow	Slightly acid
Horse manure (fresh)	0.6-0.3-0.6	Slow	Slightly acid
Sheep manure (fresh)	0.9-0.5-0.8	Slow	Slightly acid
Swine manure (fresh)	0.6-0.5-0.4	Slow	Slightly acid
Wood ashes	0-2-6	Medium	Alkaline
Rabbit manure (dried)	2.25-1-1	Slow (unless steamed)	Slightly acid
Cattle manure (dried)	2-3-3	Slow (unless steamed)	Slightly acid
Swine manure (dried)[a]	2.25-2-1	Slow (unless steamed)	Slightly acid
Seaweed	1.7-0.75-5	Slow	

[a]The nutrient of the dried, processed manures not listed can be approximated as four times that of the fresh manure.

can bring about less transplanting shock and a resulting quick resumption of growth. It is not uncommon for tomatoes which received a starter fertilizer to produce fruit several weeks sooner than a similar plant which did not receive starter fertilizer.

POTASSIUM

The third main fertilizer element is potassium, essential for starch formation, movement of sugars in the plant, formation of chlorophyll, and flower and fruit coloring.

Potassium (or potash) is abundant in many soils. However, most of the potassium is not available for plant use, necessitating the addition of potassium in usable forms. Potassium, unlike phosphorus, leaches fairly readily and may be wasted if heavy rainfall occurs after application. The problem of leaching is less severe in clay soils since potassium will be held to soil particles and thus prevented from leaching. This potassium can then be drawn upon and used by plants at a later time.

Heavy applications of this nutrient are undesirable because the plants will absorb more potassium than they need if it is available. This results in fertilizer waste with no benefits in plant growth or appearance, or what is called *luxury consumption.* Moreover, overdoses of potassium will injure or "burn" plants (overdoses of nitrogen are more likely to do so).

Potassium fertilizers can be used most efficiently by being added frequently in small amounts rather than in a single, larger application.

MICRONUTRIENTS

Although micronutrients are just as important to plant growth as macronutrients, they are needed in such small quantities that soils generally contain enough to supply the needs of most plants for many years. Occasionally micronutrients are lacking or unavailable. Frequently, however, a micronutrient deficiency occurs even though large quantities of the nutrient are present in the soil. At certain pH levels, nutrients will react chemically with other elements in the soil, forming chemical compounds which cannot be absorbed or utilized by the plant. Thus each nutrient has a definite pH range within which it is most available to plants. In most soils, as can be seen in Figure 5-14, the majority of nutrients is readily available between pH 6.5 and 7.0. If the soil pH shifts out of this range, certain elements are likely to become unavailable. If micronutrient deficiencies occur consistently, a check on soil pH is desirable.

Correcting micronutrient deficiencies involves correcting the soil pH, if practical, and applying the deficient nutrient to the foliage and roots of the plant. When applying micronutrients to the soil, it is best to purchase the nutrient in a "chelated" form. Although a chelated nutrient is more expensive, it is formulated to be unreactive with other elements in the soil and will remain available to the plant. Foliage applications of micronutrients (as discussed under "Fertilizer Application") will speed the elimination of deficiency symptoms.

FERTILIZERS

Fertilizers are compounds which, when properly used, provide the necessary nutrients for plant growth. In common usage, they are divided into organic and inorganic types.

Organic fertilizers are derived from the decomposition of plant and animal products and include blood meal, bone meal, manure, and sewage sludge. Organic fertilizers have the advantage of acting as a soil amendment in addition to supplying nutrients. Since nutrients from organic fertilizers are released slowly during decomposition, burning of plants caused by too much fertilizer dissolved in the soil solution at one time is less likely. In general, organic fertilizers are much more expensive than chemical fertilizers on the basis of the amount of nutrients supplied.

Inorganic fertilizers are manufactured using such raw materials as natural gas and phosphate rock. They usually are much more concentrated than organic fertilizers. Some inorganic fertilizers are formulated to release nutrients rapidly so that a rapid plant response to the fertilizer occurs. Others release nutrients slowly over a longer period of time. These two different types of fertilizers are described as "fast release" and "slow release," respectively. Synthetic organic fertilizers such as urea generally react in the same manner as inorganic fertilizers.

Chemical fertilizers have been much maligned by organic gardening enthusiasts; however, they are a beneficial tool to the gardener if used properly and in conjunction with sound soil-management practices, such as returning plant residues to the soil.

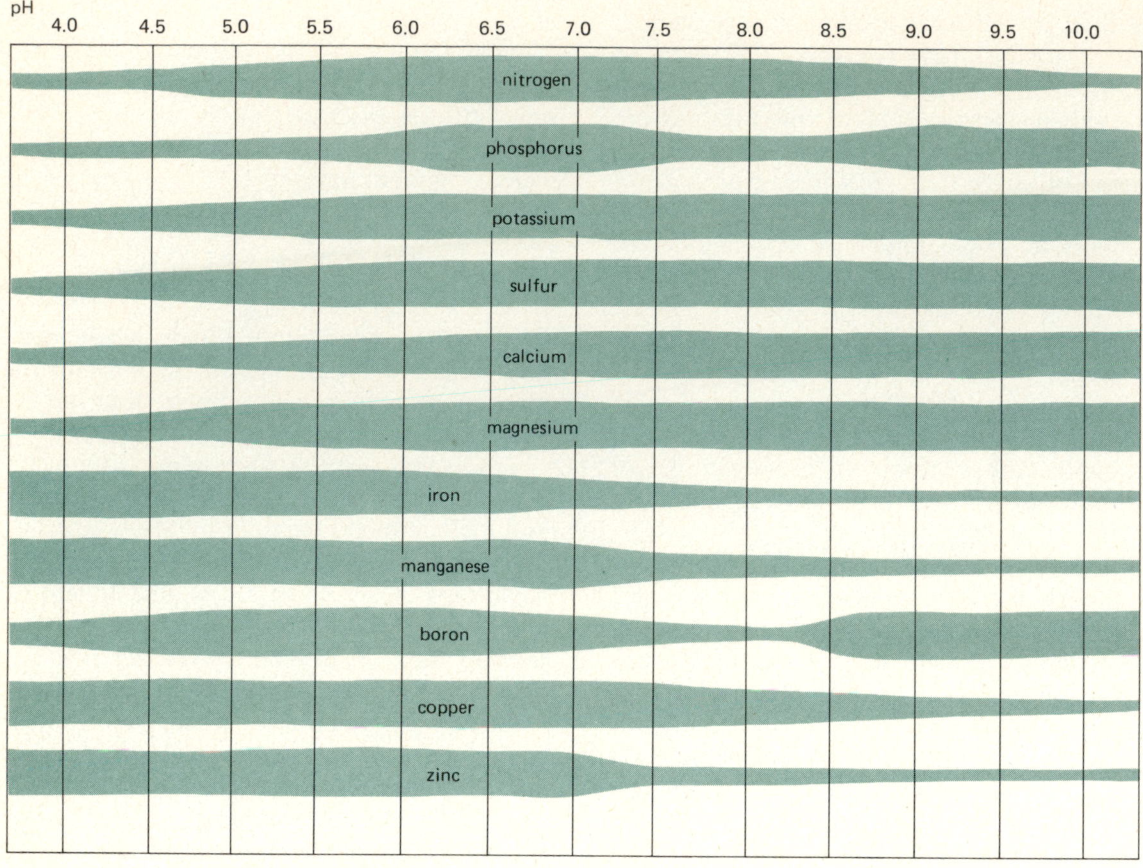

Figure 5-14 The general relation of pH to the availability of plant nutrients in the soil.

To understand the difference between organic and chemical fertilizers, it is helpful to understand the nitrogen cycle, defined as the series of transformations which nitrogen undergoes in the environment (Figure 5-15).

Nitrogen comprises about 70 percent of the air. However, plants, with the exception of those in the legume family and a few grasses, cannot use this atmospheric nitrogen. Legumes (beans, peas, peanuts, and others), with the aid of bacteria which live on their roots, are able to enrich the soil by "fixing" atmospheric nitrogen found in the air spaces in the soil.

Most plants can use nitrogen only when it is present in the soil in either the nitrate (NO_3^-) or the ammonium (NH_4^+) chemical form. Though some plants use nitrogen in the ammonium form effectively, most garden plants use nitrogen primarily in the nitrate form. This means that a considerable amount of decomposition must occur before nitrogen (which is added in the form of leaves or blood meal, for example) moves through the decomposition cycle to the point that nitrates are available. This accounts for the slow-release characteristic of organic fertilizers.

If a chemical fertilizer such as ammonium nitrate is added to the soil, part of the nitrogen is already in the nitrate form and part is in the ammonium form. The portion in the nitrate form can be used immediately

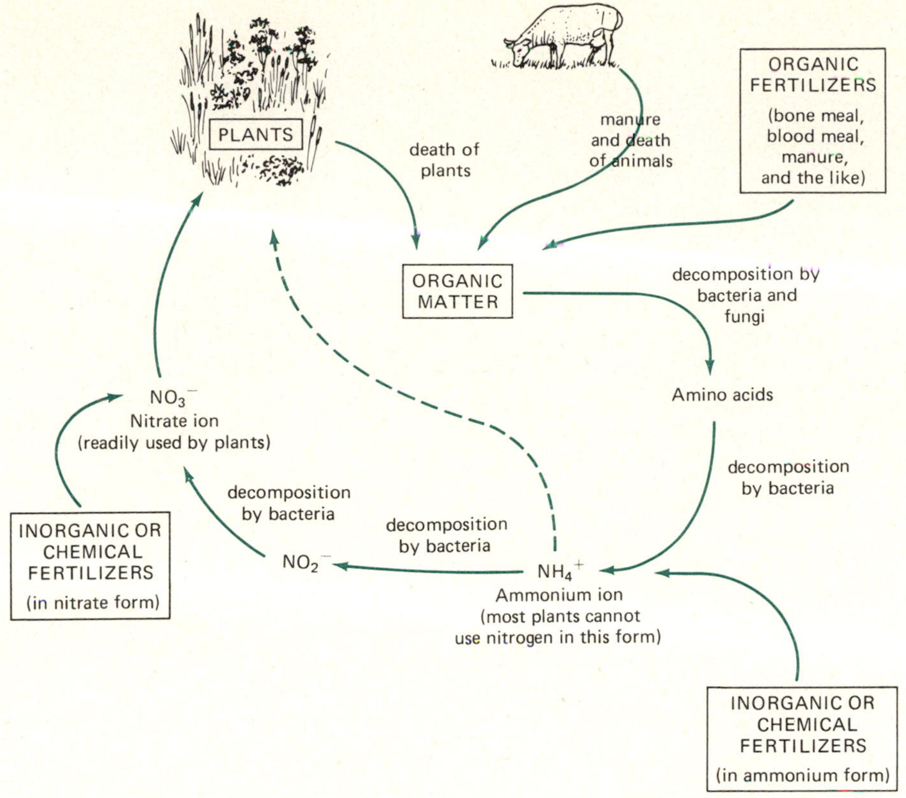

Figure 5-15 The nitrogen cycle, showing the breakdown of nitrogen-containing compounds from organic matter so that nitrogen can be utilized by plants.

by the plant. Depending on the plant, the ammonium portion can be either used immediately or converted into nitrates through the decomposition process. This accounts for the rapid plant response to ammonium nitrate and the possibility of burning plants due to high concentrations of usable nitrogen.

It is important to realize that whether the fertilizer is added in an organic form, such as animal manure, or in the form of a chemical fertilizer, such as ammonium nitrate, the plant absorbs the nitrogen primarily as nitrates. The only difference is that the nitrogen is added at a different point in the nitrogen cycle. The main disadvantage of chemical fertilizers is that if organic matter is not added to the soil (by compost, natural plant residues, or other means), the amount of organic matter in the soil will decline, causing a worsening of physical characteristics of the soil.

Fertilizer Forms

Fertilizers are usually available in several different forms—liquids, powders, and granules.

Liquids are popular primarily because of their ease of application (usually with a hose-end sprayer) and the quick response obtained. Since all the nutrients are in a water-soluble form, it is easy to burn plants if too much fertilizer is used. In addition, much of the nitrogen will leach away if heavy rainfall or irrigation occurs after fertilization, and the benefits of the fertilizer will be short-lived.

Powdered fertilizers are occasionally used by home gardeners because of their low cost. These fertilizers are formulated as very fine powders and are difficult to handle because of their light weight and their tendency to blow away. If designed to be applied in powder form, they are a poor buy because the dust will cling to leaves, resulting in fertilizer burn. Others are designed to be mixed in water and applied as liquids, and these can be very useful for applying both macro- and micronutrients.

The most common and widely used form of fertilizer is the granular form. During the manufacture of a granular fertilizer, the nutrients are compressed into bead-sized particles which are heavy enough to eliminate problems of blowing. Moreover, since the nutrients need not be immediately dissolvable in water (as in liquid fertilizers), slow-release formulations prolong the release of nutrients and decrease the risk of burning.

Purchasing Fertilizers

Several factors are involved in determining which fertilizer is the best buy: analysis, form, organic or inorganic, slow or fast release, and price.

The analysis of a fertilizer is a statement of the quantity and type of nutrients contained in the fertilizer and is required by law to be printed on every bag of fertilizer. The standard form for the statement of fertilizer analysis is a set of three numbers separated by hyphens, for example, 5-10-20. Each number refers to the percentage of a particular nutrient contained in the fertilizer (Figure 5-16). The first number always refers to the percentage of nitrogen, the second number to the percentage of phosphorus in a chemical form called *phosphorus pentoxide* (P_2O_5), and the third number to the percentage of potassium in a chemical form called *potash* (K_2O). Thus, for the sample analysis above (5-10-20), the fertilizer contains 5 percent nitrogen compound, 10 percent phosphorus compound, and 20 percent potassium compound. Another way of stating this is that the fertilizer contains 5 pounds of nitrogen compound, 10 pounds of phosphorus compound, and 20 pounds of potassium compound in every 100 pounds of fertilizer.

Fertilizers are often grouped into three categories based on the relative amounts of nitrogen, phosphorus, and potassium they contain. The first group is called *balanced fertilizers* since all the numbers in the analysis are identical (8-8-8). The second group, called *complete fertilizers,* contains nitrogen, phosphorus, and potassium but in varying amounts (5-10-15). The last category, called *single element fertilizers,* contains only one of the three major nutrients such as urea (46-0-0).

The analysis is the first item to look at in buying any fertilizer since nutrients are the important part of the fertilizer. Other ingredients may be present which make the volume of fertilizer larger and make it appear to be a better bargain. These "filler materials" may be useful as soil amendments but cannot be considered a good buy when purchased at the price of fertilizer.

Consider both the total quantity of nutrients (by adding the numbers) and the relative amounts of nitrogen, phosphorus, and potassium. If you are buying a lawn fertilizer where the primary nutrient needed is nitrogen, the nitrogen analysis is a prime consideration. For most, a balanced fertilizer which contains about equal proportions of nitrogen, phosphorus, and potassium should be chosen.

Next look at the form of the fertilizer. It is difficult to compare liquid pricewise with either granular or insoluble powder forms due to the weight of the water; however, the latter two can be compared. Powdered fertilizers should be distinctly less expensive than gran-

Figure 5-16 Three bags of fertilizer showing the analysis in large numbers.

ular forms. If a substantial difference in price does not exist, there is little justification for purchasing a powdered fertilizer.

Whether a fertilizer is organically derived may be an important consideration if you are committed to the exclusive use of organic materials. If an organic fertilizer is selected, you should be prepared to pay a much higher price for the same quantity of nutrients and should not expect a high fertilizer analysis. This is because, by definition, an organic fertilizer will contain bulky organic material in which the nutrients are contained, and shipping costs will often be a substantial component of the price. Either fast- or slow-release fertilizers can be purchased in inorganic form but the slow-release types will be more expensive. Whether or not they are worth the added expense depends on how they are to be used.

Price often becomes the primary consideration in the purchase of fertilizers for home use. In order to determine the best buy, all of the factors discussed above should be weighed relative to the price.

Fertilizer Application

TOPDRESSING

The method of fertilizer application will vary according to the type of plant being fertilized, but the basic rule is that the fertilizer must reach the root zone. For shallow-rooted plants such as turfgrasses, groundcovers, annual flowers, and some shrubs, surface applications, called *topdressings*, followed by rainfall or irrigation will be sufficient to supply nutrients to the root zone. When surface applications are used, the fertilizer should be broadcast around the plant, being careful to remove fertilizer that collects around the stem or upon the leaves.

SIDEDRESSING

Sidedressing is a form of surface fertilizer application useful on rows of plants such as are found in vegetable gardens and some landscape plantings. Using this method, a narrow band of fertilizer is applied along one or both sides of the row (Figure 5-17) over the root area of the plants but away from the crown, where it could cause burning injury by contact. This banding makes fertilizing long rows of plants relatively easy and avoids the sloppiness and fertilizer waste that can occur with topdressing. It also minimizes contact of the fertilizer granules with the soil and thus helps prevent susceptible nutrients such as phosphorous from reacting with soil components.

For more deeply rooted plants such as large shrubs and trees, surface applications are not effective since the roots are located 18 inches (45 centimeters) or more below the soil surface; most of the phosphorus and a large amount of the potassium will never leach down to the roots. In addition, if surface fertilizer applications are used routinely, surface roots will be encouraged and reduce the plant's ability to survive drought.

NEEDLE FEEDING

Two methods are commonly employed to place fertilizer in the root zone of trees. The first is the use of a feeding needle, often called a *root feeder.* This apparatus (Figure 5-18) attaches to a garden hose and contains a fertil-

Figure 5-17 Sidedressing a row of peas.

Figure 5-18 A liquid feed needle.

izer chamber and a pipe with a pointed tip. When the water is turned on, it passes through the fertilizer chamber, dissolving the fertilizer, through the pipe, and out the tip. The water passing through the tip aids the operator in pushing the pipe to the desired depth in the soil. This method is easy and provides both water and available nutrients to the root zone when injected at several points over the root zone of the tree.

DRILL HOLE

Another effective technique that does not involve the purchase of any special equipment is the drill hole method, whereby holes are pounded 12–18 inches (30–45 centimeters) deep into the ground throughout the root zone of the tree (Figure 5-19). These holes are then filled with a granular fertilizer according to the recommended rate and covered with peat moss or topsoil. To be effective, 10–20 holes for each inch (2.5 centimeters) of trunk diameter must be made throughout the root zone of the tree. Holes may be made with a crow bar, reinforcing bar, pipe, or auger.

Fertilizer spikes are a modification of the drill hole method. The same results can be obtained using the drill hole method and at less expense.

FOLIAR FEEDING

Another method of supplying small quantities of nutrients to plants is by means of foliar feeding. Diluted water-soluble fertilizers are sprayed on plant foliage, and the plant then absorbs small quantities through the leaves. Since leaves absorb only small quantities of nutrients, this method is best reserved for micronutrients. The advantage of foliar fertilization is that if deficiency symptoms are obvious, improvement will be evident in a matter of days. In addition, where the micronutrient is likely to become unavailable in the soil, soil application can be supplemented. As a general rule, foliar fertilization does not supply enough nutrients to completely satisfy requirements and should be used in addition to regular fertilizer that is added to the soil.

Figure 5-19 The drill hole method for fertilizing trees.

Selected References for Additional Reading

Buckman, H. O., and N. C. Brady. *The Nature and Properties of Soils.* New York: Macmillan, 1969.

Bush-Brown, J., and Bush-Brown, L. *America's Garden Book.* rev. ed. New York: Charles Scribner's Sons, 1966.

Donahue, R. L., J. C. Shickluna, and L. S. Robertson. *Soils, An Introduction to Soils and Plant Growth.* 4th ed. Englewood Cliffs, N J: Prentice-Hall, 1977.

Hausenbuiller, R. L. *Soil Science Principles and Practices.* Dubuque, IA: W.C. Brown, 1972.

Janick, J. *Horticultural Science.* 2d ed. San Francisco: W.H. Freeman, 1972.

Organic Gardening and Farming Editors. *Organic Fertilizers: Which Ones and How to Use Them.* Emmaus, PA: Rodale Press, 1973.

Chapter 6

Vegetable Gardening

Vegetable gardening is one of the most rewarding forms of horticulture. It combines the enjoyment of cultivating plants and watching them grow with the pleasure of harvesting fresh produce at the peak of taste and maturity. Few people will argue that "home grown" tastes better than "store bought," and there are several reasons why. Vegetable varieties grown on a large scale must have certain characteristics including the property of shipping well without bruising or becoming overripe. Therefore many commercial vegetable varieties are bred to be less succulent and must frequently be picked and shipped while not completely ripe. The qualities of flavor are sometimes considered second to these characteristics of shipping ability. Second, from the moment a ripe vegetable is picked, its quality will normally begin to slowly deteriorate. In most varieties of sweet corn, for example, sugar begins to turn to starch almost immediately and the flavor of the vegetable will be completely altered in less than 2 hours. Although not always the case, the general rule of "the fresher, the better" is fairly reliable.

The economics of vegetable gardening for saving on food bills have been widely acclaimed. Studies quote that a vegetable plot of a given size will supply all the vegetable needs of a family for a year. Although this is true with intensive, systematic cultivation and conscientious use of all vegetables as they become available, for most families it will not apply. It is more reasonable to regard a vegetable garden as a type of money saver; at the same time its recreational value can be enjoyed as well as the superior vegetables harvested and eaten at their prime.

TYPES OF VEGETABLES

Just like all other plants, vegetables are either annual, biennial, or perennial. Most fall in the annual group including peppers, squash, and beans. Others are technically biennials but are treated as annuals because only the vegetative parts are eaten. Still fewer are perennials, including asparagus and rhubarb. Surprising to many people is the fact that the tomato is a perennial, but it is not frost-tolerant and is therefore treated as an annual. Table 6-1 lists the common garden vegetables and their classifications in this regard.

More important than whether a vegetable is annual or perennial is its classification as either warm season or cool season (Table 6-1). Warm-season vegetables thrive at daytime temperatures ranging from 65 to 90°F (18–32°C) and at nighttime lows not less than about 55°F (13°C). At lower temperatures they grow slowly. Some, like tomatoes, fail to develop fruit, whereas others, like peppers, develop fruit but produce only small

Table 6-1 Garden Vegetable Cultivation

Vegetable	Cool/warm season	Grown as annual, biennial, or perennial	Edible portion	Moderate planting for family of four	Distance between plants	Distance between rows
Artichoke, globe	Cool	Perennial in zones 9–10, annual in others	Flower buds	3 or 4 plants	4′ (1.2 m)	5′ (1.5 m)
Artichoke, Jerusalem	N/A	Perennial	Tubers	10–15 plants	1′ (0.3 m)	5′ (1.5 m)
Asparagus	N/A	Perennial	New shoots	30–40 plants	1′ (0.3 m)	5′ (1.5 m)
Bean, lima	Warm	Annual	Fruit	15–25′ (4.5–7.6 m) row	6″ (15 cm) (bush) 2′ (0.6 m) (pole)	30″ (0.76 m)
Bean, snap or green	Warm	Annual	Fruit	15–25′ (4.5–7.6 m) row	3″ (8 cm) (bush) 2′ (0.6 m) (pole)	2′ (0.6 m)
Beet	Cool	Annual	Leaves, root	10–15′ (3–4.5 m) row	2″ (5 cm)	18″ (0.45 m)
Broccoli	Cool	Annual	Flower buds	7–10 plants	2′ (0.6 m)	3′ (1 m)
Brussels sprout	Cool	Annual	Axillary buds	7–10 plants	2′ (0.6 m)	3′ (1 m)
Cabbage	Cool	Annual	Leaves	10–15 plants	2′ (0.6 m)	3′ (1 m)
Carrot	Cool	Annual	Root	20–30′ (6–9 m) row	2″ (5 cm)	18″ (0.45 m)
Cauliflower	Cool	Annual	Flower head	10–15 plants	2′ (0.6 m)	3′ (1 m)
Celery	Cool	Annual	Petioles	20–30′ (6–9 m) row	5″ (13 cm)	18″ (0.45 m)
Chinese cabbage	Cool	Annual	Leaves	10–15′ (3–4.5 m) row	1′ (0.3 m)	30″ (0.75 m)
Collard	Cool	Annual/perennial	Leaves	10–15′ (3–4.5 m) row	18″ (4.5 m)	3′ (1 m)
Corn	Warm	Annual	Seeds	4 rows of 20–30′ (6–9 m)	6″ (15 cm)	3′ (1 m)
Cucumber	Warm	Annual	Fruit	6 plants	2′ (0.6 m)	4′ (1.2 m)
Eggplant	Warm	Annual	Fruit	4–6 plants	2′ (0.6 m)	3′ (1 m)
Endive	Cool	Annual	Leaves	10–15′ (3–4.5 m) row	10″ (25 cm)	18″ (0.45 m)
Kale	Cool	Annual	Leaves	10–15′ (3–4.5 m) row	10″ (25 cm)	2′ (0.6 m)
Kohlrabi	Cool	Annual	Enlarged stem base	10–15′ (3–4.5 m) row	3″ (8 cm)	2′ (0.6 m)
Leek	Cool	Annual/perennial	Enlarged stem base	10′ (3 m) row	2″ (5 cm)	2′ (0.6 m)
Lettuce	Cool	Annual	Leaves	10–15′ (3–4.5 m) row	12″ (0.3 m) (head) 6″ (15 cm) (leaf)	2′ (0.6 m)
Melon	Warm	Annual	Fruit	5–6 hills	6″ (15 cm)	6′ (1.8 m)
Mustard green	Cool	Annual	Leaves	10′ (3 m) row	8″ (20 cm)	18″ (0.45 m)
Okra	Warm	Annual	Fruit	10–20′ (3–6 m) row	18″ (0.45 m)	3′ (1 m)
Onion, green	Cool	Annual	Leaves	10–15′ (3–4.5 m) row	1″ (2–3 cm)	18″ (0.45 m)
Onion, bulb	Cool	Annual	Bulb	30–40′ (9–12 m) row	3″ (8 cm)	18″ (0.45 m)
Parsnip	Cool	Annual	Root	10–15′ (3–4.5 m) row	3″ (8 cm)	18″ (0.45 m)
Pea, English	Cool	Annual	Seeds	30–40′ (9–12 m) row	2″ (5 cm)	3′ (1 m)
Pea, edible pod or Chinese	Cool	Annual	Fruit	30–40′ (9–12 m) row	2″ (5 cm)	4′ (1.2 m)
Pepper	Warm	Annual	Fruit	5–10 plants	2′ (0.6 m)	3′ (1 m)
Potato	Warm	Annual	Tubers	100′ (30 m) row	1′ (0.3 m)	30″ (0.75 m)
Pumpkin	Warm	Annual	Fruit	1–3 plants	4′ (1.2 m)	6′ (1.8 m)
Radish	Cool	Annual	Root	4′ (1.2 m) row	1″ (2–3 cm)	18″ (0.45 m)
Rhubarb	N/A	Perennial	Petioles	2–3 plants	3′ (1 m)	4′ (1.2 m)
Rutabaga	Cool	Annual	Root	10–15′ (3–4.5 m) row	3″ (8 cm)	18″ (0.45 m)
Spinach	Cool	Annual	Leaves	10–20′ (3–6 m) row	3″ (8 cm)	18″ (0.45 m)
Squash, summer and zucchini	Warm	Annual	Immature fruit	2–4 plants	2′ (0.6 m)	4′ (1.2 m)
Squash, winter	Warm	Annual	Mature fruit	2–4 plants	4′ (1.2 m)	6′ (1.8 m)
Sweet potato	Warm	Annual	Root	50′ (15 m) row	1′ (0.3 m)	3′ (1 m)
Swiss chard	Cool	Annual	Leaves	3–4 plants	1′ (0.3 m)	30″ (0.75 m)
Tomato	Warm	Annual	Fruit	10–20 plants	2′ (0.6 m) (bush) 1′ (0.3 m) (staked)	3′ (1 m)
Turnip	Cool	Annual	Root	10–15′ (3–4.5 m) row	2″ (5 cm)	18″ (0.45 m)

fruits which are not fleshy or full. These vegetables will not live through a frost.

The opposite is true with cool-season vegetables. They will tolerate light frost and grow best at daytime temperatures from 50 to 65°F (10–18°C). In warmer weather, quality is often poor, for example, lettuce may become bitter. Some cool-season vegetables will flower when leaves alone are wanted, and this too is undesirable. The flowering, called *bolting,* is a response to the shortening of the nights and the warmer weather during the summer.

Another factor affecting the culture of vegetables is whether they are grown for roots, leaves, or fruit (Table 6-1). First, fertilizer selection must be considered. Leaf vegetables produce best with high-nitrogen fertilizer, which encourages vegetative growth. Root and fruit vegetables, on the other hand, are best with a fertilizer formula containing less nitrogen. In fact an excess of nitrogen will prevent some vegetables from bearing at all. Second, leaf and many root vegetables can be grown successfully in semishady areas. Fruit-producing vegetables generally require full sun since they are unable to generate the necessary energy for flowering and fruiting without bright sunlight and may produce only foliage in shady situations. This understanding is important for city dwellers whose gardens are shaded by tall buildings or for other gardenders who find themselves restricted to a shady area. Undue time and effort need not be wasted on fruit crops which will produce no reward. The space can be reallocated to vegetables which will grow more successfully.

CLIMATIC FACTORS

Vegetable gardening is considered a spring and summer activity by most people, with seed sowing in spring and harvest through summer, ending at the first killing frost. But in mild-winter areas of the country, vegetable gardening continues all year. In most other areas it can be extended far beyond the traditional season by choosing the right crops and cultural practices.

Vegetable Gardening in Mild-Winter Areas

The mild-winter areas of the country (see Figure 6-1) are the vegetable bowl of the nation. Here most of the fresh vegetables are grown for winter consumption. Gardeners located in one of these areas are fortunate, for they can grow fresh vegetables the entire year and can easily supply all their vegetables from the garden.

The climate characteristics of the locale will determine which vegetables can be raised in each season. In the deep South, for example, winter will be cool and nearly frostless and summer will be very hot. The cool-season vegetables can be raised in winter and the warm-season vegetables in summer when the temperatures exceed the acceptable range for the former. In southern Florida, the only tropical area of the country, winter temperatures are moderate (50–75°F, 10–24°C), high enough for warm-season vegetables but not too high for cool-season vegetables. In the summer only warm-season crops could be grown. Along much of the coast of California temperatures are moderated by the Pacific Ocean; there is a minimal difference between winter and summer temperatures. They often fall just short of the acceptable range for warm-season vegetables, and gardeners in this area grow primarily cool-season crops all year.

Vegetable Gardening in Temperate Climates

Most of the United States has widely varying winter and summer temperatures. Temperatures rise slowly in spring, reach a maximum in summer, and cool slowly as winter approaches. In these climates spring and fall are

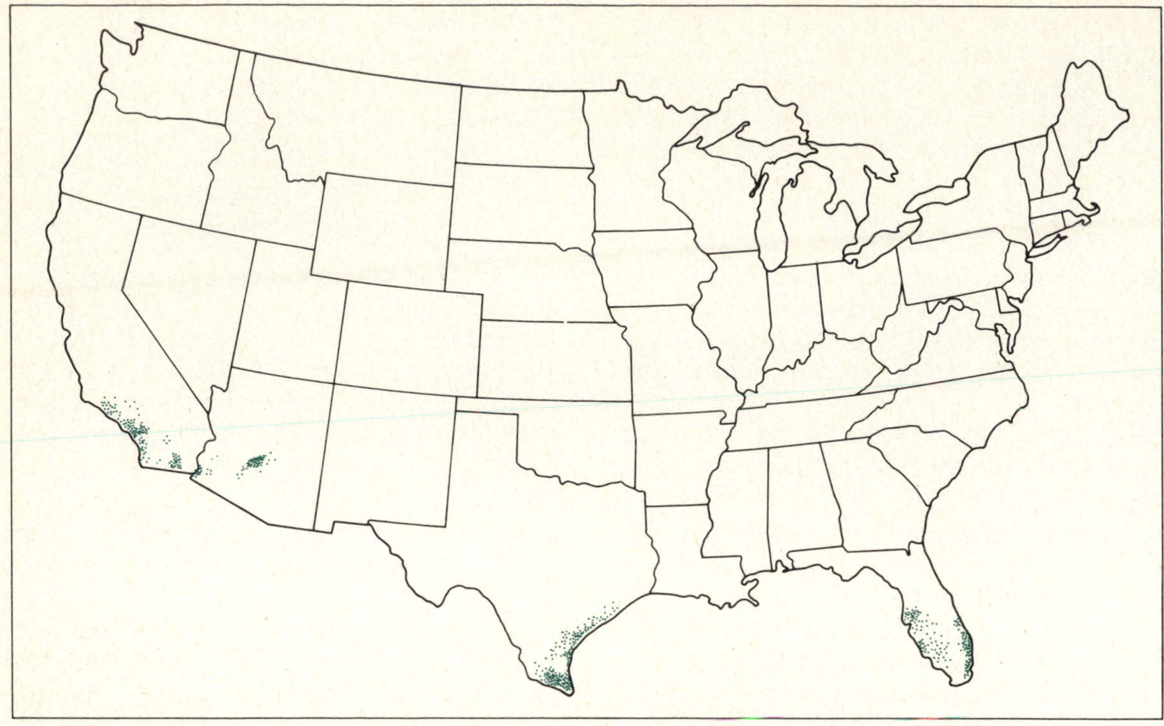

Figure 6-1 Areas of the continental United States where vegetables can be grown throughout the winter.

the best seasons for cultivating cool-season crops, whereas summer is most conducive for raising warm-season crops.

The number of days from the last spring frost until the earliest fall frost defines the period of growing warm-season crops. This "frost-free period," an important concept in vegetable gardening, varies widely with latitude, ranging from 250 days in the deep South to 60 days in parts of North Dakota. Figure 6-2 shows maps of the United States divided into zones by the date of the last frost in spring and the first frost in fall.

The importance of knowing the number of frost-free days lies in the variety selection of warm-season crops. Modern plant breeding offers the home gardener a wide selection of vegetable varieties bred for taste, canning suitability, and rate of maturity. The rate of maturity is the number of days from either seed or transplant (depending on the vegetable) to the time when the crop is ready for harvest. For example, there are tomato varieties which mature in as little as 54 days and others that require 80 or more frost-free days.

The benefits of fast-maturing vegetables to gardeners in the far North are obvious. Even gardeners in more moderate climates value fast-maturing vegetables because they are ready for harvest earlier in the season.

Cool-season vegetables are not restricted to the frost-free period. Not only are they tolerant of light frost, but their flavor may actually be improved by it. Extending the vegetable gardening season in temperate climates is largely a matter of taking advantage of the spring and fall growing periods, which have occasional frosts but are generally favorable for the growth of cool-season vegetables.

Early spring, for example, is fine for sow-

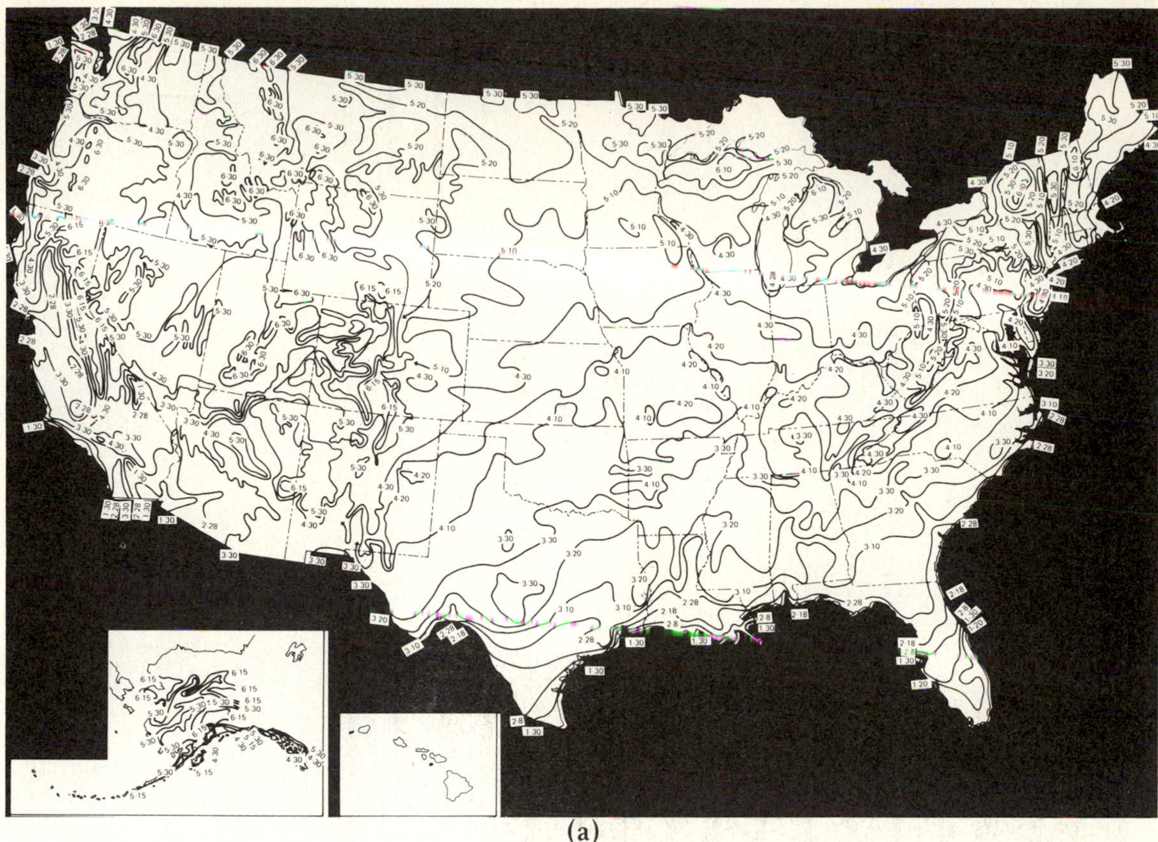

(a)

Figure 6-2 (a) Average dates of the last frost in spring throughout the United States. [(b) is on the next page.]

ing peas, lettuce, carrots, and other cool-season vegetables grown from seed. It is also ideal for setting out transplants of broccoli, cabbage, or cauliflower. Although they may appear to grow slowly at first, they will establish their roots and grow faster later in the season. Planting time for cool-season crops starts as soon as the ground is thawed and dry enough to be workable.

FALL VEGETABLE GARDENING

Fall vegetable gardening is a second way to extend the growing season. It involves delaying the planting of cool-season vegetables so that they mature after the rest of the crops have been harvested. Although many of the cool-season vegetables planted in spring are started as transplants, they are usually seeded when grown for a fall crop. This adds 6 to 8 weeks to the "days to maturity" designation of the variety and must be factored in when deciding the date to sow seed. Crops such as leaf lettuce and radishes may be started in late summer or even fall, since they mature quickly, but slower crops should be seeded in early summer. The following vegetables are suitable for summer sowing and fall harvest:

Beets	Kohlrabi
Broccoli	Leeks (sow in spring or overwinter seedlings)
Cabbage	Lettuce (leaf and semihead)
Carrots	Mustard greens

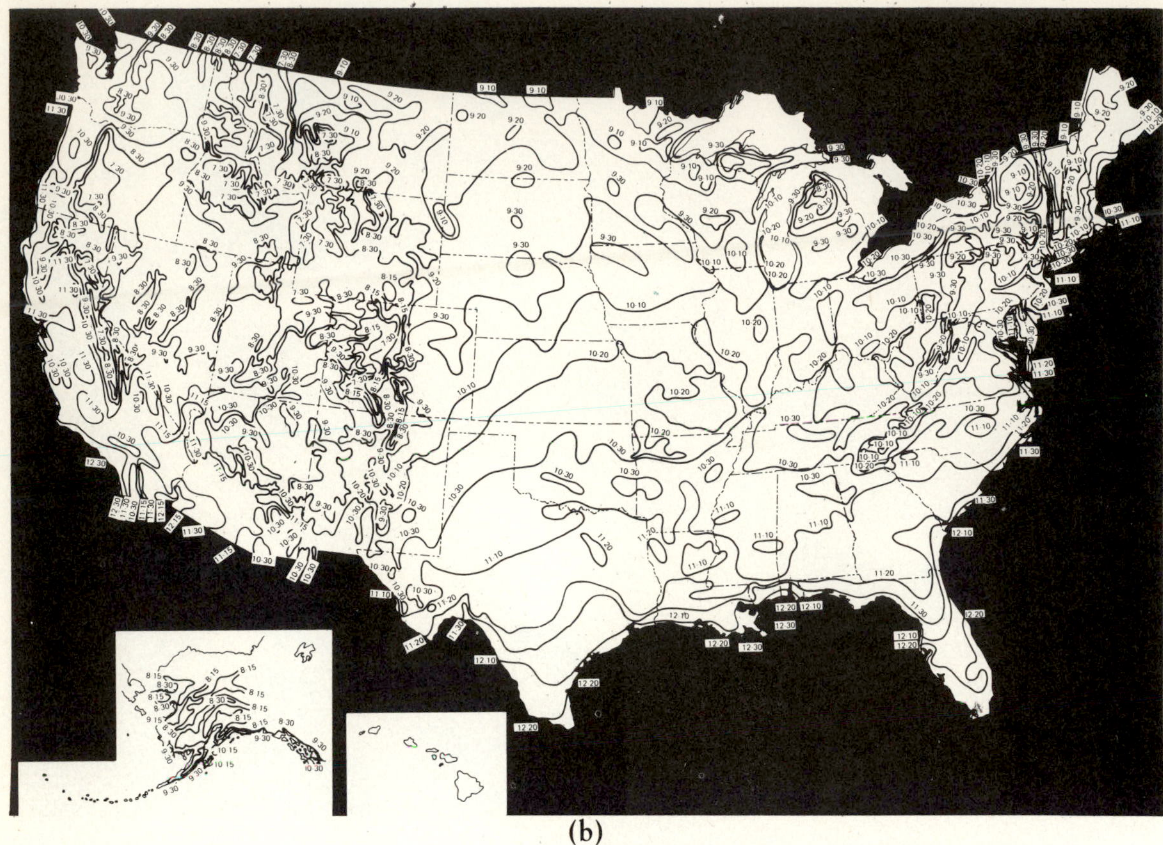

(b)

Figure 6-2 (b) Average dates of the first frost in fall throughout the United States.

Cauliflower
Chard
Chinese cabbage
Collards
Endive
Kale
Parsnips (sow in spring)
Peas, English and edible pod
Radishes
Spinach
Turnips

OVERWINTERING VEGETABLES IN THE GROUND FOR CONTINUOUS HARVEST

The cold tolerance of some root vegetables comes as a surprise to many gardeners. It is not unusual to discover these vegetables growing through the snow in spring and still good to eat. Occasionally even leaf vegetables such as lettuce and parsley will remain green under the snow cover and renew growth in the spring.

This overwintering phenomenon is due not only to the cold hardiness of the vegetable itself but also to the insulating effects of snow cover and the latent heat of earth (Chapter 20). When these two conditions combine, the soil may remain unfrozen or just at freezing all winter, despite air temperatures reaching far below 32°F (0°C).

It is both possible and practical to make use of these conditions to "store" root vegetables in the ground and extend their harvest through winter. The technique is fairly simple. A section of garden should be set aside for the winter harvest area and planted with root vegetables timed to mature in late summer or fall. These include beets, carrots, Jerusalem artichokes, leeks, parsnips, radishes (Chinese or icicle type), rutabagas, and turnips. In fall the area should be mulched with compost, newspapers, leaves, or any other suitable material. The depth of the mulch needed to prevent freezing will depend on the expected winter temperatures. If winter temperatures only reach 20–25°F (−7–4°C), a layer of mulch 6–8 inches (15–20 centimeters) thick would be sufficient. The depth should be increased, ranging up to 2 feet (60 centimeters), if the temperature will drop below 0°F (−18°C). Continual snow cover on the mulch will add additional insulation. As a precaution, the winter garden can be located near the house in a protected area to preserve the ground heat.

Harvest of the vegetables can begin any time. The mulch should be removed, the roots harvested, and the mulch replaced each time.

PLANNING A VEGETABLE GARDEN

Choosing the Area

A potential vegetable area should have both fast-draining soil for vigorous root growth and full sunlight to encourage maximum growth rate and flowering. Accordingly, the garden should be situated away from trees and buildings. A previously gardened area will probably possess better soil than one which has been planted in grass, but any soil could be improved to a suitable quality by adding amendments, as discussed in Chapter 5, or by green manuring, as described later in this chapter.

Determining the Size of the Garden

For the inexperienced gardener, a small plot is preferable to a large one. After the first gardening season, the amount of work involved and the number of vegetables produced will be known. The gardener can expand or decrease the size accordingly. A 25 × 25-foot (8 × 8-meter) plot will enable a first-time gardener to grow most popular vegetables except for large space consumers like melons and potatoes.

Garden Layout

The first tasks are to decide which vegetables will be grown and to lay out on paper where they will be located in the garden. Rows should be planted running east and west, if possible, to prevent shading. Taller or trellised crops should be placed north with shorter crops to the south. Figure 6-3 shows a typical layout for a plot producing many different kinds of vegetables.

It is often difficult to estimate how much of each vegetable to plant; Table 6-1 gives an average planting for a family of four for each vegetable. When using these estimates, you should divide the amount to be planted into two or three parts, and stagger the sowing of the seed so that the harvest is scattered throughout the season. This practice, called *succession cropping,* will assure a steady supply of vegetables instead of an overabundance at one time.

For example, two leaf lettuce sowings made 2 weeks apart in spring will provide lettuce into summer. A third sowing in late summer will yield a fall crop, giving a total of three successive crops of lettuce. Succession cropping requires careful management of the garden space. It entails more work than the straight "plant and harvest" method, but the crop rewards are greater throughout the season. Almost all vegetables with the exception of the slow-maturing warm-season crops can be succession-cropped.

Vegetable Variety Selection and Seed Purchasing

The Cooperative Extension Service of most states yearly evaluates new and existing veg-

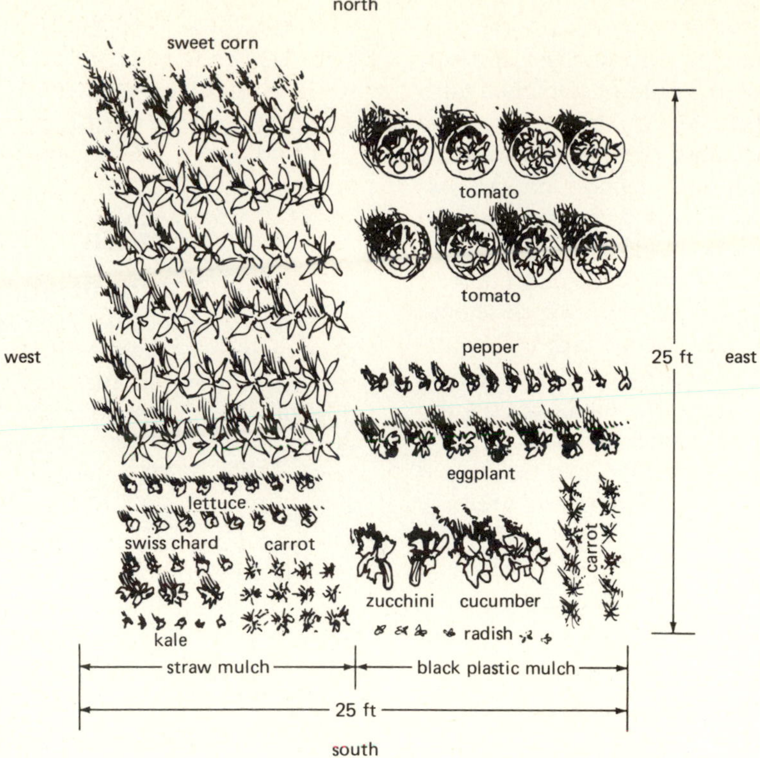

Figure 6-3 A typical vegetable garden plan.

etable varieties to determine the best ones for the climate of that state. A list of the recommended varieties is then compiled, published, and made available to the public. This list is probably the single most important information source on vegetable varieties a gardener can have.

Seed catalogs list numerous varieties and their merits but say little about their drawbacks. The gardener must evaluate the suitability of each variety to his needs and climate, noting days to maturity, resistance to disease, and other characteristics. It is wise to use catalogs from companies located in a climate similar to one's own so that the varieties that are offered are geared to similar climates. For example, southern gardeners should avoid most catalogs from northern companies. It would be unfortunate to pick a variety by the picture, only to find it a poor performer under local growing conditions.

Many people buy their seeds from racks in groceries or nurseries. If so, the packages should be examined for the year in which they were packaged for use. If the stamped date is not the current year, the seed is old and may germinate poorly. Moreover, a seed rack may have a limited selection of varieties, all or none of which may be recommended for the area.

"Seed tape" is an expensive way to purchase seeds but will save labor. The seeds are sandwiched between layers of a plasticlike material that dissolves when the row is irrigated.

Some small seeds are available in pelleted form, having been coated to make them larger and more uniform in size for use with mechanical seeders. Pelleted seeds are more expensive and may germinate more slowly than normal seeds, so they are not generally used for home gardens.

Overbuying vegetable seed is a common mistake and results in leftover seed at the end of the growing season. Most vegetable seeds can be kept for the following season by storing them in the refrigerator (see "Seed Storage," Chapter 4). Table 6-2 lists the relative keeping quality of various vegetable seeds.

Table 6-2 Approximate Life of Vegetable Seeds Stored under Cool Conditions

Vegetable	Years
Asparagus	3
Bean	3
Beet	4
Broccoli	5
Brussels sprout	5
Cabbage	5
Carrot	3
Cauliflower	5
Celery	5
Chard, Swiss	4
Chinese cabbage	5
Collard	5
Corn	1–2
Cucumber	5
Eggplant	5
Endive	5
Kale	5
Kohlrabi	5
Leek	3
Lettuce	5
Muskmelon	5
Mustard	4
Okra	2
Onion	1–2
Parsley	2
Parsnip	1–2
Pea	3
Pepper	4
Pumpkin	4
Radish	5
Rutabaga	5
Spinach	5
Squash	5
Tomato	4
Turnip	5
Watermelon	5

INTENSIVE OR BLOCK GARDENING

In intensive or block gardening (Figure 6-4), vegetables are grown in blocks rather than rows. It was the original form of gardening practiced before the plow was invented and is still popular in countries where land is scarce. In the United States block gardening is useful in cities because it gives maximum yield from minimum space.

For block planting, a garden should consist of squares or rectangles no more than 4 feet (1.2 meters) across to keep the plants within reach of the path. Seed is sown by scattering it evenly over the area rather than in row fashion.

A variation of the original block garden involves growing vegetables in raised beds. The soil in these beds is then thoroughly conditioned and fertilized to grow superior-quality crops.

USING VEGETABLES AS ORNAMENTALS

Vegetables can be used not only for eating but also as ornamentals (Figure 6-5). Leaf lettuce, for example, makes an attractive flower border. Varieties of vegetables selected for their colorful flowers, foliage, or fruits are available through seed catalogs. A few are listed in Table 6-3.

Patio pot gardens are another ornamental use of vegetables. Ornamental peppers and chard grow well in containers. Cherry tomatoes can be planted in hanging baskets, and runner beans trained on strings to screen a porch or balcony. By using vegetables as ornamental plants, the gardener with space

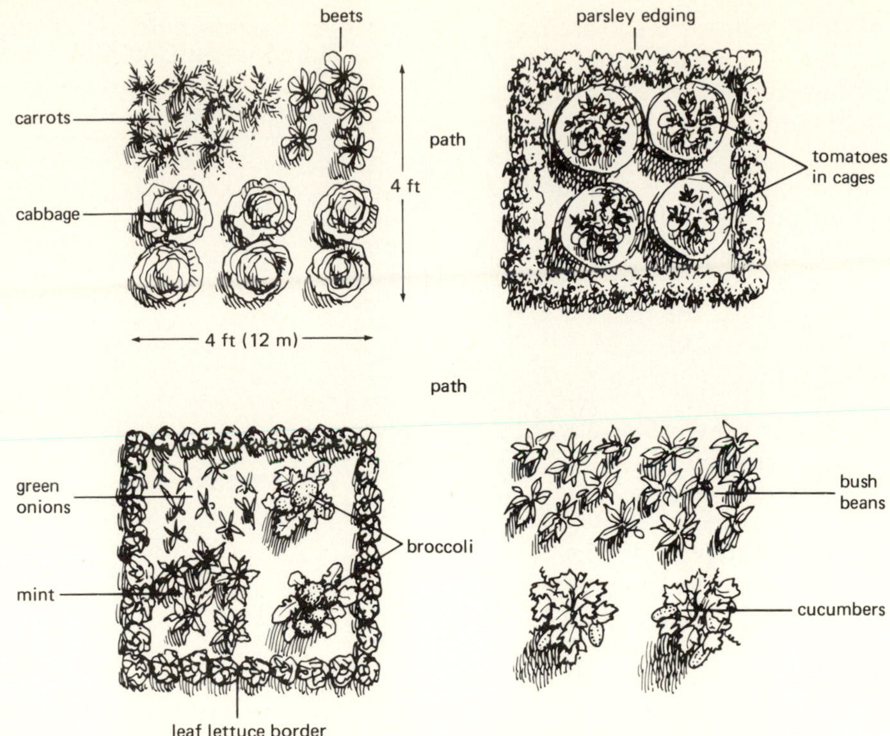

Figure 6-4 A block garden made up of four blocks.

restrictions can also make better use of his limited area of land.

PLANTING THE GARDEN

Sowing Seed

With the exception of block gardening, vegetables grown from seed are planted in rows or hills. When sowing in rows, each row should be marked with a labeled stake and the seed sown thickly and in accordance with package directions. Heavy sowing assures a sufficient stand of plants in case of poor germination.

With slow-germinating seeds such as carrots, a fast-germinating seed is sometimes planted in the same row to mark its location until the other seedlings emerge. In soils which tend to crust, the rapidly germinating seeds also break the surface to ease the emergence of less vigorous crops. Radishes germinate in 3 days and are often used for row marking. They can be pulled as soon as the main crop emerges but are usually left in place. Since radishes mature in as little as 21 days, they can be harvested in time to prevent crowding of the main crop. This technique, called *interplanting,* saves space and is frequently done with small, fast-maturing crops such as radishes and leaf lettuce. A full crop of leaf lettuce, for example, can be raised by seeding between transplants of broccoli or cauliflower (Figure 6-6).

Directions on packages of squash and vining vegetables like cucumbers normally specify planting in "hills." Hill planting refers

Figure 6-5 Dark green oakleaf lettuce used as a border in a perennial flower bed.

to grouping several plants together, not planting seeds in a mound of soil.

After seeding it is important to keep the surface of the soil moist. Drying the soil surrounding the germinating seeds will kill the emerging seedlings and necessitate resowing.

Buying Transplants

Selecting well-grown transplants contributes to the success of the garden, and quality should not be sacrificed in favor of price. The plants should be short, sturdy, and with foliage to the base. Those with yellowed foliage or bare stems should be avoided because they have been growing too long in the transplant container and are likely to be root-bound and slow to begin growth.

Setting Out Transplants

If possible, an overcast day or early morning should be selected for transplanting to lessen shock. Watering the plants before removing them from the pack is advisable.

Figure 6-6 A crop of leaf lettuce planted between broccoli plants.

If the plants were grown in individual cells of a plastic pack, there will be very little root shock during transplanting. However, if after removing a transplant, it is found to be root-bound, the root ball should be cut shallowly on each side (Figure 6-7). The cutting will force branching of the roots into the soil. If uncut, the roots may continue growing within the original soil ball, and establishment will be slow.

Transplants grown in posts of compressed peat moss can be transplanted with the pot intact, but the top of the pot should be broken off down to the soil. If the rim is not broken and is exposed to the air, it will wick water from the peat moss, restricting the penetration by the roots and slowing growth.

Vegetable transplants should be planted slightly deeper than they were originally grown. Tomatoes should be set with much of the stem below the soil surface (Figure 6-8) since they will form adventitious roots along the submerged stem. These stem roots create a larger root system and a more vigorous plant.

After planting, the transplants should be watered with starter fertilizer as discussed

Table 6-3 Ornamental Vegetable Varieties

Vegetable	Ornamental species or varieties	Remarks
Artichoke, globe	*Cardunculus* species (common name, cardoon)	Large gray leaves and thistlelike purple flowers; petioles and roots are edible
	Scolymus species (common artichoke)	Purple flower heads, gray foliage; smaller plant than cardoon
Artichoke, Jerusalem	Any	Mature plants are 6 ft (2m) or more tall with yellow sunflowers
Asparagus	Any	Feathery foliage and red berries in fall
Bean, bush	'Royalty' variety	Purple pods
Bean, pole	'Scarlet Runner' variety	Red flowers; shade- and cool-temperature-tolerant
Cauliflower	'Purple Head' and 'Royal Purple' varieties	Heads tinged lavender at maturity
	'Greenball' variety	Head green at maturity
Chard	'Rhubarb' variety	Red petioles
Corn	Indian corn, 'Rainbow' variety	Red, yellow, and white kernels
	Strawberry corn	Small, strawberry-shaped ear with red kernels
	'Gracillima' variety	Dwarf size
	'Japonica' variety	Leaves striped white and pink
Cabbage, ornamental	No varieties	Ruffled leaves are lavender, pink, and green; does not head
Kale, flowering	No varieties	Ruffled leaves, red- or green-tinged

Vegetable	Ornamental species or varieties	Remarks
Lettuce, leaf	'Oakleaf' variety	Dark green lobed leaves
	'Red Salad Bowl' variety	Red-tinged leaves
	'Ruby' variety	Red leaves
Pepper, bell	'Golden Calwonder' variety	Yellow fruits
	'Bell Boy Hybrid' variety	Red fruits
	'Midway' variety	Red fruits
Pepper, hot	'Fiesta' variety	Red, yellow, and green fruits
Eggplant	'Morden Midget' variety	Deep purple, miniature fruit
Sunflower	No varieties suggested	Enormous yellow flowers on 6- to 12-ft (1.8- to 3.6-m) stems; edible seeds excellent when toasted
Tomato	'Patio' variety	Dwarf plant to 1 ft (0.3 m) tall with golf-ball-sized tomatoes
	'Small Fry' variety	Vining plant producing large quantities of cherry tomatoes; good in a hanging basket
	'Tiny Tim' variety	6-in. (15-cm)-high plant with tiny tomatoes
	'Yellow Pear' variety	2-in. (5-cm) golden, pear-shaped tomatoes
	'Yellow Plum' variety	2-in. (5-cm) golden, plum-shaped tomatoes

Figure 6-7 Cutting through the wrapped roots of a pot-bound transplant.

in Chapter 5. The watering assures good soil contact with the roots and provides maximum soil moisture to prevent shock.

VEGETABLE GARDEN MAINTENANCE

Thinning

Thinning is the removal of excess seedlings which are spaced too closely for best growth. Thinning can be done either once or twice for each crop.

One-time thinning should be done as

Figure 6-8 Deep planting of a tomato previously transplanted encourages stem rooting.

soon as the leaves of neighboring plants touch and should remove as many seedlings as necessary to achieve the recommended final spacing for the vegetable (Figure 6-9).

Twice-over thinning is practical for vegetables grown for their leaves, such as chard, lettuce, and Chinese cabbage. The first thinning is done while the plants are seedlings, but it removes only enough plants to prevent severe overcrowding. The second thinning takes place 2 to 4 weeks later and thins the remaining plants to the final spacing. By the second thinning, plants are large enough to be harvested and eaten.

Weeding

Weed control is very important to a successful vegetable garden. The detrimental effect of weeds on plant growth is masked, showing up as a slowed plant growth rate due to the competition of the crop with the weeds for water, nutrients, and light. Many gardeners incorrectly attribute this sluggish growth to poor soil or adverse weather, not knowing the real cause.

Because seedlings have limited root systems, they are especially vulnerable to competition from vigorous weeds within the row. If unchecked, the weeds will eventually crowd out the desirable plants altogether.

If seedling plants are invaded by weeds, pulling is probably the only control method. Later, hoeing or mulching is effective. Because the roots of vegetable plants are shallow, hoeing for weeds should be a scraping rather than a digging operation. The scraping will cut weed seedlings at the soil line, and if they are annuals, they will not resprout.

Mulching

Mulching can considerably reduce the time spent weeding and will conserve soil moisture. In hot areas of the South and Southwest it

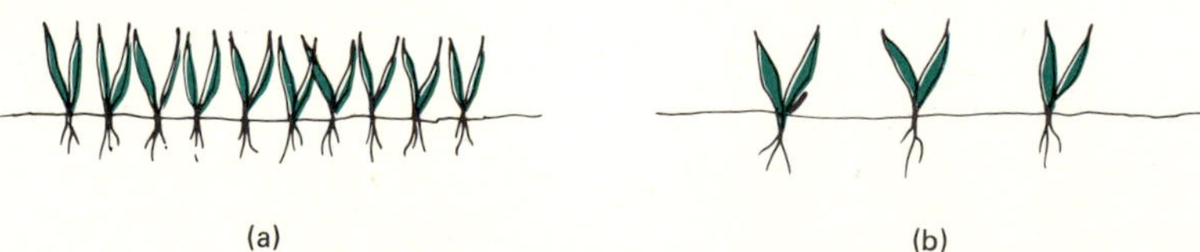

Figure 6-9 (a) Before thinning of seedlings. (b) After thinning.

helps keep soil from overheating and consequently is beneficial to root growth. Since vegetable gardens are tilled every year, many people find it easier to use organic mulches which will decay in the soil. Compost, leaves, and grass clippings from lawns which have not been recently treated with weed killer are satisfactory.

Mulch should be applied 2 to 3 inches (5–8 centimeters) thick as a general rule. Transplants should be mulched sparingly about 1 inch (3 centimeters) deep when young, gradually increasing the depth to 3 inches (8 centimeters) as the plants grow taller. For low-growing plants such as lettuce and radishes, mulch should be applied on either side of the row at a depth of 1.5 to 2 inches (4–5 centimeters). Vining crops such as melons should be mulched over the entire area where the vine will grow. Covering the area will help keep the fruits clean and reduce losses from rot diseases which occur when the vegetables lie on the soil.

Newspapers held down with soil are another suitable vegetable garden mulch. Although not attractive, they are readily available and inexpensive.

Plastic mulches are very useful in northern vegetable gardens because they absorb heat and radiate it into the soil underneath. They will speed up the growth of warm-season crops by heating the soil and can mature the crop a week or more ahead of normal. Either clear or black plastic mulch may be used. The former will raise the temperature more than the latter; however, weeds will grow beneath it.

For especially early harvests, a plastic mulch can be used with transplants of cucumbers and melons. The time gained by using transplants combined with the warmth provided by the plastic will encourage rapid growth and fruit set.

Plastic mulches can be applied in two ways. For row crops, strips of plastic are laid on either side of the row, with the plants between. For vegetables in hills, the area can be covered and a flap can be cut in the plastic so the seeds can emerge (Figure 6-10).

Hot Caps and Frost Protection Devices

The advantages of using plastic mulches to speed the growth of warm-season crops have already been discussed. The use of hot caps (Figure 6-11), paper or plastic domes set over plants in early spring, is another method of hastening growth. The caps admit light but insulate against heat loss, keeping the air surrounding the transplants several degrees warmer than the outside air. This slight rise in temperature is often enough to speed growth considerably during early spring. As an added feature, hot caps offer frost protection. They enable gardeners to set out warm-season transplants and seeds several weeks ahead of the last predicted frost date.

Frost protection of warm-season crops in fall is more difficult because of the size of the plants. If only light frost is expected, a layer of newspapers over the plants will often give adequate protection.

Figure 6-10 A black plastic mulch used in a row of tomatoes.

Watering

Watering the vegetable garden should begin as soon as seeds are planted and be done as often as necessary to prevent wilting. Do not assume, however, that every time a plant wilts, the garden needs watering. Many times large-leaved squash or cucumber plants wilt repeatedly during the hottest part of the day due to an inability to absorb enough moisture to compensate for the enormous water loss through the foliage. However, if they do not recover at night, water is definitely needed. If, after watering, they still do not recover, the plant may be infected by a wilt disease or nematodes.

Watering the vegetable garden should soak the soil to a depth of about 18 inches (45 centimeters). Perennial vegetables such as asparagus and artichokes will need to be irrigated more deeply.

Figure 6-11 Hot caps being used to cover transplants in the garden.

Fertilizing

If fertilizer is worked into the soil of the vegetable garden during preparation, it will often be sufficient to supply the needs of the crops to maturity. On sandy soils, on soils of low fertility, or if deficiency symptoms are noted, a midsummer sidedressing of balanced granular fertilizer may spur crop growth. Nitrogen fertilizers can stimulate leaf production at the expense of the fruit and should be used with care.

Training

Peas, runner beans, tomatoes, and cucumbers are vining plants which can be trained on stakes, strings, or wire. Training keeps the fruits from becoming dirty and can lessen the chances of fruit rot by preventing contact of the fruits with the soil. It also optimizes space usage in the garden and makes harvesting easier.

TEEPEE TRAINING

Teepee training (Figure 6-12) can be used for a group of several cucumber or bean plants. Ideally the teepee should be built first and then seeds planted at the base of each stake. The vines are tied to the stakes at 1-foot (0.3-meter) intervals.

STAKING

A single sturdy stake (Figure 6-13) will support one tomato plant or two to three runner beans. Beans will climb without help, but tomatoes should be tied loosely with pieces of cloth at intervals along the stem.

STRING TRELLISES

Runner beans and peas are trained easily on a trellis made of twine strung between poles (Figure 6-14). The poles should be pounded in deeply not more than 10 feet (3 meters) apart to provide a sturdy support for the trellis and vines.

Figure 6-12 Teepee training of runner beans.

TOMATO CAGES

Tomato cages using a wire-mesh cylinder either purchased or made at home (Figure 6-15) are an easy way to support tomatoes. Concrete reinforcing wire has an ideal strength and a good mesh size for tomato cages; many types of galvanized fencing (not chain link) are also suitable. It is only necessary that the mesh be sufficiently large to reach through to pick ripe fruit. Tying the vines is not required.

CROP ROTATION

Crop rotation is the planting of crops in different areas of the garden every year. It involves, for example, switching the locations of the beans and cabbage each year. Crop rotation deters the buildup of soil-carried disease organisms associated with particular crops. Although crop rotation is not essential, it is

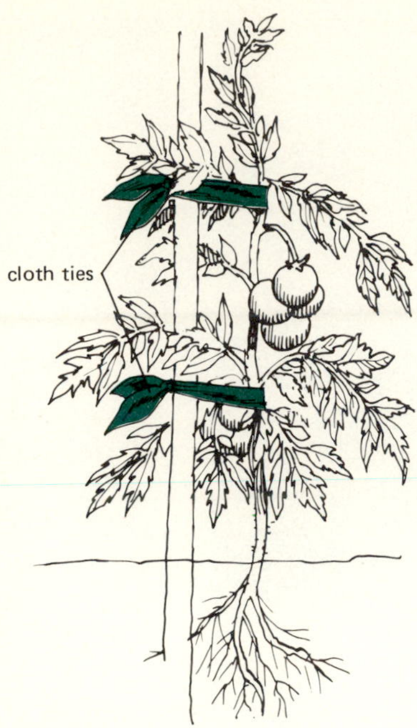

Figure 6-13 Staking is one of the most common ways to train tomatoes.

Figure 6-14 A string trellis used to support runner beans.

a sound practice that does not require much effort.

The easiest form of crop rotation is to simply reverse the vegetable garden plan each year, being careful to avoid shading problems due to placement of tall plants. The progression shown in Figure 6-16 is another rotation sequence.

COVER AND GREEN MANURE CROPS

Cover cropping (Figure 6-17) is used to maintain fertility in established vegetable gardens, but green manure crops are used to improve poor soil before a garden is planted. In most areas of the country, cover crops are planted in fall over the vegetable garden. They grow during winter and are turned under in spring, supplying organic matter and increasing the fertility of the soil as they decay. Green manure crops perform similarly but are planted at any time of year, grown halfway to maturity, and then turned under to decay.

Green manuring can be used in place of adding organic soil amendments and is much less expensive. The general practice is to sow the green manure crop, wait until it is 8 to 10 inches (20–25 centimeters) high, turn it under, wait 10 to 14 days, and then resow another green manure crop. The plowing and resowing are repeated as many times as necessary until the soil is of acceptable quality.

Figure 6-15 Tomato cages.

Members of the pea family such as alfalfa, clover, cowpeas, soybeans, and some varieties of vetch are among the best green manure and cover crops. They absorb nitrogen from the air and, when turned under to decay, will release that nitrogen into the soil. Winter-growing cover crops include wheat, rye, ryegrass, and buckwheat. Buckwheat is very tolerant of adverse soil conditions and is especially useful in very poor soils.

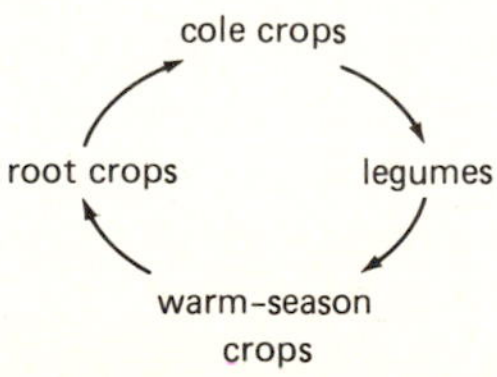

Figure 6-16 Suggested crop rotation for a home vegetable garden.

CULTURE OF COMMON GARDEN VEGETABLES

Artichokes

Two very different plants are called artichokes. The first is the globe artichoke (*Cynara scolymus*), a tender perennial generally grown only in areas where the ground does not freeze. The second is the Jerusalem artichoke (*Helianthus tuberosus*), a hardy perennial that can be grown throughout the United States and Canada.

Globe Artichokes

The globe artichoke is a large, thistlelike plant up to 6 feet (1.8 meters) in diameter. The large flower buds it produces in late spring and early summer are the edible portion of the plant (Figure 6-18).

The buds should be harvested before they begin to open by cutting the stem ½ to 1 inch (1–3 centimeters) below the base and then refrigerated until ready to eat. An artichoke will keep 2 to 3 weeks if it is kept in the coldest part of the refrigerator in a plastic bag.

CLIMATE REQUIREMENTS

The globe artichoke grows best in areas where winter temperatures are mild and summers are cool. Areas along the California coast, where summer fogs are common, are ideal. In hot-summer regions artichokes often produce buds which open rapidly and have a leathery texture. In zones 1–7, artichokes can be grown in containers in summer and moved indoors in the winter, or they can be dug after cold has killed the leaves and the crowns stored in damp sawdust until spring. A heavy mulch will generally provide sufficient protection in zones 8 and 9 to overwinter the crowns.

Figure 6-17 A cover crop in spring before being plowed under.

SOIL

Artichokes grow best in rich, well-drained soil with a neutral to slightly alkaline pH.

CULTURE

Artichokes are usually grown from divisions which consist of a woody stalk, a root, and leaves. This division should be planted vertically, with the base of the leaves slightly below the soil surface. In most areas late winter or early spring is the best time to plant. Allow 3 to 4 feet (1–1.2 meters) between plants. A starter fertilizer is beneficial at planting.

After the plants begin to grow, plenty of water and nitrogen fertilizer should be provided. After the buds are cut, the supporting stem will shrivel and should then be cut back to healthy growth. New shoots will sprout from the base to provide next year's harvest. In cold-weather regions nitrogen fertilizer should be withheld after midsummer to enhance cold tolerance.

Jerusalem Artichokes

The Jerusalem artichoke is a sunflower with coarse, hairy leaves which grows to a height of 6 feet (1.8 meters). The edible portions of the plant are the underground tubers which are produced abundantly in the fall. The tubers are harvested after frost or may be left in the ground all winter and dug as needed. After harvest the tubers can be stored from 2 to 5 months under high humidity and at temperatures as close as possible to freezing.

CLIMATE REQUIREMENTS

The Jerusalem artichoke can be grown in zones 2–9. However, the best quality and highest yields occur in the colder zones.

Figure 6-18 A field of globe artichokes.

SOIL

Jerusalem artichokes thrive in any well-drained soil. Highest yields are obtained in soils with a low nitrogen level, so Jerusalem artichokes may be grown in the poorest section of the garden.

CULTURE

Propagation is from tubers which may be purchased from nurseries or seed companies or can be found in the produce department of a grocery.

Tubers are planted 3 to 5 inches (8–13 centimeters) deep (depth is not critical) and 5 to 10 inches (13–25 centimeters) apart in spring. The plants are very vigorous and generally do not require any fertilization or care until harvest. When harvesting, remove all tubers except a few to start the next season's crop. If many tubers are left in the ground, the plant may become a weed problem.

Asparagus

Asparagus (*Asparagus officinalis*) is a perennial vegetable grown for its tender sprouts. Some gardeners grow asparagus for its tall, fernlike foliage and ornamental red berries. Sprouts appear early in the spring and should be harvested when they are 6–8 inches tall (15–20 centimeters)—before they begin to branch. The best way to harvest is to snap the tips by hand (Figure 6-19). The shoots will break at the point where they become fibrous and inedible. During warm weather, beds can often be harvested daily or even twice daily. If a frost threatens, all emerged sprouts should be harvested regardless of length since they are not frost-tolerant. Sprouts should not be allowed to form leaves until after the harvest season; new sprouts will stop developing as soon as leaf development occurs.

Asparagus should be refrigerated immediately after harvest because its quality diminishes rapidly in warm temperatures. Flavor will be retained best under high humidity and at a temperature just above freezing.

CLIMATE REQUIREMENTS

Asparagus will grow in all climate zones except the warmest sections of zone 10. Yield is often heaviest in areas with a cold-winter climate. In warm climates growth may continue all winter, depleting the carbohydrate reserves in the crown.

SOIL

Sandy soils high in organic matter are best for growing asparagus, but this vegetable will succeed in any soil. If grown in loam or clay soils, crowns should be planted 5–6 inches (13–15 centimeters) deep; in lighter soils, crowns should be planted 6–8 inches (15–20 centimeters) deep.

CULTURE

Asparagus can be grown from either seed or 1- to 2-year-old crowns. Seed will not produce a harvestable crop until the third year. Consequently, growing from crowns is the preferred method. They should be planted 8 to 12 inches (20–30 centimeters) apart in rows in the spring or fall. Harvesting should be delayed until the second year to allow the plants to become established. During this time plants should be watered regularly and fertilized with a balanced fertilizer. During the second year, spears should be harvested for a 2-week period and then allowed to leaf out. In the third year spears may be harvested until the plants show weakening by producing small-diameter spears or ones which branch close to ground level. Care should be taken not to harvest for too long a period since this will weaken the crowns and cause low yields the following year.

Beans

The common bean (*Phaseolus vulgaris*) includes many different garden beans known as snapbeans, stringless or string beans, wax beans, and green beans. Beans may be either climbing vines called *pole types* or low-growing, bushlike plants called *bush* or *dwarf types.* The edible pod is eaten while immature and should be picked before the seeds inside enlarge. Since quality declines rapidly after harvesting, beans should be refrigerated and kept in high humidity. For best quality they should be eaten within 24 hours.

CLIMATE REQUIREMENTS

Beans are warm-season annuals and will not tolerate cool temperatures or frost. Therefore beans should be planted only after the weather has warmed in the spring.

SOIL

Any well-drained soil is acceptable for growing beans. A pH of between 6.5 and 7.0 is best.

CULTURE

Beans should be planted 1 inch (2.5 centimeters) deep and 2–3 inches (5–8 centime-

Figure 6-19 Picking asparagus by the snap method.

ters) apart and do not require thinning. If bush beans are grown, staking is not necessary; however, if pole types are grown, stakes should be provided, preferably at planting time. Beans should be sidedressed with a balanced fertilizer prior to flowering if leaves are not a dark green color.

Summer irrigation is required in much of the South and Southwest to prevent blossom drop. This malady is aggravated by hot, dry winds.

Bush beans normally yield sooner than pole beans, but the harvest period will be shorter. Succession plantings should be made 2 to 3 weeks apart to prolong the harvest. Although pole beans take longer to begin producing, they will often bear until killed by frost if all pods are kept picked.

Beets

Beets (*Beta vulgaris* var. *crassa*) are one of the easiest vegetable crops to grow. The plants are only about 1 foot (30 centimeters) tall, with red-tinged leaves originating from a large red root. The root is the primary edible portion of the plant. The leaves are also edible; in fact they are more nutritious than the roots.

Beet roots can be harvested and eaten at any time. However, the quality is best before they reach 3 inches (8 centimeters) in diameter. Leaves can be harvested either when the plants are thinned or by picking individual leaves from the best plants. No more than two leaves should be removed from a single plant, or root formation will be retarded. Roots can be stored for 3 to 5 months in the coldest part of the refrigerator if tops have been removed.

CLIMATE REQUIREMENTS

Beets can be grown in all climate zones in the summer and in warmer zones as a winter crop.

SOIL

A light, well-drained soil with a pH above 6.5 is preferred; however, beets will grow in most garden soils.

CULTURE

Since beets tolerate light freezes, they should be planted as early as possible in the spring or year-round in mild-winter areas. For most varieties a "seed" will contain several embryos, resulting in several plants at each location. Thinning is thus necessary to space the plants 1 inch (2.5 centimeters) apart. If greens are desirable, thinning can be delayed until several 3- to 4-inch (8- to 10-centimeter) leaves have formed. The tenderest beets are produced when plants are growing quickly, so a sidedressing of a balanced fertilizer is desirable 4 to 6 weeks after planting. Irrigation should also be provided if rainfall is inadequate. When harvesting roots, pull the largest first so that more room is provided for those remaining.

Broccoli

See "Cole Crops."

Brussel Sprouts

See "Cole Crops."

Cabbage

See "Cole Crops."

Cantaloupe

See "Cucurbits."

Carrots

Carrots (*Daucus carata* var. *sativus*) are an easy-to-grow annual root crop for all climate zones. Roots are edible from the time they are pencil size until they are between 1 and 2 inches (3 and 5 centimeters) in diameter. After harvest carrots should be refrigerated in high humidity. If fully mature, they will keep for 4 to 5 months under these conditions. Younger carrots can generally be stored only 4 to 6 weeks.

CLIMATE REQUIREMENTS

Carrots are a cool-season crop that can be grown as a summer crop in all zones and as a winter crop in warm-winter regions.

SOIL

High-quality carrots are produced in deep loose soils with good drainage. In heavy soils roots will often be crooked or forked; short-rooted varieties are likely to produce the best results.

CULTURE

Seeds should be planted shallowly in soil that has been well prepared to eliminate clods. Where crusting of the soil surface is a problem, organic matter should be mixed with the soil to cover the seeds.

Carrot seeds are small and slow to germinate. To aid emergence and to mark where the carrots are planted, radish seed can be mixed with the carrot seed. The radish seed will germinate rapidly and break the surface crust as well as mark the location of the carrot row. The radishes can then be removed after the carrots germinate.

It is important to thin carrots to provide adequate room for the roots to develop. Thinning can be done several times to ultimately leave 1½ to 2 inches (3–5 centimeters) between carrots. The last thinning can usually be delayed until the thinnings are large enough to be eaten. If carrot leaves are a light green, a sidedressing with a balanced fertilizer will improve yield. As with most root crops, fast growth will produce a more tender and tasty root.

Cauliflower

See "Cole Crops."

Celery

Celery (*Apium graveolens*) is one of the most difficult vegetables to grow in the home garden and therefore is not recommended for the beginning gardener or one with limited space. It is harvested by cutting the plant at ground level or by removing individual stalks as needed. Celery will keep for several weeks if refrigerated in high humidity.

CLIMATE REQUIREMENTS

Celery requires a long growing season with cool temperatures (65–70°F, 18–21°C). If temperatures are too warm, it will become tough. If prolonged periods with temperatures below 50°F (10°C) occur, it will bolt before leaf and stalk development take place.

SOIL

Celery should be grown in rich soil high in organic matter.

CULTURE

Celery is generally grown from transplants 3 to 4 inches (8–10 centimeters) tall planted 10 inches (25 centimeters) apart. Seeds require 3 weeks to germinate and another month to reach transplanting size. Regular watering and frequent fertilizing with a high-nitrogen fertilizer will maximize growth. Plan on approximately a 5-month period from seeding to harvest.

Blanching excludes light from the stalks and decreases their chlorophyll content. It was a common practice in the past and can be accomplished by mounding soil around the lower stalks or otherwise shading them. Most gardeners today do not blanch celery since it results in a lower nutritional value and many varieties are "self-blanching."

Chard

Chard (*Beta vulgaris* var. *cicla*) (Figure 6-20) is probably the easiest to grow and the most useful leaf vegetable for the home garden. As the botanical name indicates, chard is actually a beet but is grown for its spinachlike leaves rather than for its roots. The culture is the same as for beets. Leaves are harvested individually by cutting the outer ones off at the base as they reach edible size. Chard will produce leaves year-round in mild areas and from early spring until late fall in colder areas. It is a vegetable that is practically pest-free and is highly recommended for the home garden.

Cole Crops

The term *cole crop* refers to several cool-season crops related to cabbage which have similar cultural requirements. Included are broccoli, Brussels sprouts, cabbage, cauliflower, collards, and kale as well as several less common vegetables.

CLIMATE REQUIREMENTS

All cole crops thrive under cool temperatures and will survive light frosts. In hot-summer areas they should be grown as a spring or fall crop; where winters are warm, they should be grown as a winter crop. Collards alone are tolerant of very hot summer weather.

SOIL

Any well-drained soil is acceptable. A pH between 5.5 and 6.5 is ideal, but a wide range is satisfactory in practice.

CULTURE

All cole crops germinate readily from seed planted ¼ inch (6 millimeters) deep or may be started from transplants. Abundant moisture and a sidedressing with a high-nitrogen fertilizer are beneficial. The most serious problem with growing cole crops is the control of caterpillars and aphids.

INDIVIDUAL REQUIREMENTS

Broccoli (*Brassica oleracea,* Italica Group). "Sprouting" varieties which form a large number of small heads rather than one large head should be selected for planting. The edible part is the flower head, which should be harvested before the buds begin to open. If all buds are kept picked, broccoli will continue to produce for many months.

Brussels Sprouts (*Brassica oleracea,* Gemmifera Group). The edible parts are the small axillary buds which form along the main stem. As the plant grows taller, the lower leaves can be removed until leaves remain only at the top of the plant. At this point the plant will somewhat resemble a palm tree with side growths looking like small cabbages.

Sprouts should be harvested when about 1 inch (2.5 centimeters) across but before they begin to open. If the weather is warm, they will often be soft and leafy rather than hard. Flavor and quality will improve as cool weather arrives in fall. If long storage of sprouts is desired, the entire plant should be pulled up in fall and replanted in damp sand in a cool cellar. Sprouts will store for several months in this manner.

Cabbage (*Brassica oleracea,* Capitata Group). Cabbage should be harvested when the heads are hard. If allowed to remain in the garden for too long, the heads will split, especially after a heavy rain. Since cabbage heads tend to mature at the same time, succession planting or the use of several different varieties with varying harvest dates is desirable. Cabbage will store for a month or more under cold temperatures and high humidity.

Cauliflower (*Brassica oleracea,* Botrytis Group). This is one of the more difficult cole crops to grow. For best results, the plants should be kept growing vigorously by watering and fertilizing; any slowdown in growth will cause the production of a small head of poor quality. Transplants which are not vigorous should be avoided because these will often produce a head only 1 inch (2.5 centimeters) or so in diameter immediately after transplanting.

Like broccoli, the edible parts of cauliflower are the immature flower buds. The head is harvested while the buds are still pressed closely together. Quality declines rapidly once the buds begin to separate. To produce white heads similar to those available in markets, the heads must be "blanched." This is accomplished by tying several of the upper leaves over the head as soon as it begins to form in order to shade it (Figure 6-21). There are also several "self-blanching" varieties available which will naturally cover the developing flower buds with leaves, but only if the weather is cool. Failure to blanch cauliflower will result in heads light green in color which, although more nutritious, have a stronger flavor. Cauliflower plants produce only one head and should be discarded after the head is harvested.

Collards (*Brassica oleracea,* Alcephala Group). This nonheading cabbage is tolerant of both very cold weather and the heat of summer; as a result it is one of the most reliable plants for the production of greens. Individual leaves can be harvested at any time, although during hot weather the flavor is likely to be strong.

Figure 6-20 Swiss chard.

Kale (Brassica oleracea, Alcephala Group*).* Kale is another easy-to-grow plant for greens, but it will not survive hot summers. It will, however, often survive the coldest of winters, producing tender greens as soon as the snow melts. To harvest, remove the outer leaves individually. Kale is less prone to insect problems than other cole crops.

Figure 6-21 Blanching cauliflower.

Collards

See "Cole Crops."

Corn

Sweet corn (*Zea mays* var. *saccharata*) is one of the most popular home-garden vegetables. Since hundreds of varieties of sweet corn are available, the gardener can pick from midgets less than 3 feet (1 meter) tall to giants 6 to 7 feet (1.8–2.0 meters) tall. However, regardless which variety is selected, popcorn or field corn should not be grown within 100 feet (30 meters) of sweet corn to prevent cross-pollination and lower quality.

Corn is harvested when the silk (female flowers) on the ears turns dark brown and the kernels are plump and filled with milky sap. Since with the exception of a couple of relatively new varieties, the sugar turns to starch after picking, sweet corn should be cooked immediately after harvest.

CLIMATE REQUIREMENTS

Varieties are available which make corn-growing possible in most areas. Since corn is a warm-season crop, it should be planted only after danger of frost is past.

SOIL

Corn will thrive in a wide variety of well-drained soils. A pH between 6.0 and 6.5 is best.

CULTURE

Corn should be planted 1 to 2 inches (2.5–5 centimeters) deep and 8 to 10 inches (20–25 centimeters) apart. Several short rows rather than one long row should be planned to ensure pollination. After corn has germinated, it is necessary to control weeds, provide water, and give one or two sidedressings of high-nitrogen fertilizer. Since corn tends to reach harvesting stage all at one time, several succession plantings should be made.

Cucumbers

See "Cucurbits."

Cucurbits

The group of vegetables called *cucurbits* includes many warm-season vine crops with similar growth requirements. Members of the group include cucumbers, squash, melons, and pumpkins. In all cases the edible portion of the plant is the fruit.

CLIMATE REQUIREMENTS

Cucurbits require a warm climate for optimum growth and fruit development. The best results are obtained when summers are long and hot. This does not preclude the growing of these plants in northern areas; however, plantings should be made only after all danger of frost is past and night temperatures are consistently above 45°F (7°C). Since melons require the highest temperatures and the longest growing season, melon varieties should be selected carefully for adaptability to the local climate. Winter growing of cucurbits is restricted to the warmest areas of California, Florida, and Texas.

SOIL

Any well-drained soil with a pH above 6.0 is acceptable.

CULTURE

Cucurbits can be started either from seed sown ½ inch (1.3 centimeters) deep in late spring or from transplants. Since they do not transplant easily, seedlings should be purchased only if they are grown in peat pots or other plantable containers.

Cucurbits can be planted in rows or hills. If only a few plants are to be grown, the hill method is best because of better space utilization. Vines can be either allowed to run along the ground or trellised. Trellising works very well for cucumbers but is generally not used for melons and squash because of the heavy fruits. Gardeners are often disappointed by the lack of fruit set early in the year on cucurbit plantings; however, with most varieties, this is natural. Cucurbits produce separate male and female flowers, and the normal pattern is for male flowers to be produced first, followed later by female flowers and corresponding fruit set.

INDIVIDUAL REQUIREMENTS

Cucumbers. Compared with melons, cucumbers *(Cucumis sativus)* require a shorter growing season and less heat and therefore are adapted to a wider range of climates. Two types of cucumbers are commonly grown in the home garden, depending on their proposed use. These are pickling cucumbers and slicers. Pickling cucumbers are bred for their qualities as pickles and are harvested while immature. Slicers are bred for eating fresh, are larger, and have thicker skins and a more attractive appearance. For the home gardener who desires cucumbers for both pickling and fresh uses, pickling varieties are best because they are of a high quality for fresh use as well as for pickles.

Bitterness in cucumbers is due to a number of factors including genetic composition, high temperatures, and lack of water. Though difficult to prevent, adequate water will decrease the chance of bitter cucumbers.

Melons. The most commonly grown melons are cantaloupes (more correctly, muskmelons, *Cucumis melo,* Reticulatus Group), watermelons (*Citrullus lanatus*), and honeydew melons (*Cucumis melo,* Inodorus Group). All require a long growing season, hot days, and plenty of moisture. The best areas for melon growing are warm regions of the country, although varieties are available which can be grown in most of the United States and Canada. Honeydews require the most heat, followed by watermelons and muskmelons. Since the climate requirements are rather stringent, selection of varieties adapted to local conditions is imperative. As a general rule, varieties which produce small or midget fruit perform best where summers are cool.

Often one of the most difficult tasks the home melon grower faces is determining when the fruit is ripe. Watermelons should be harvested when the stem of the melon has shriveled, whereas muskmelons should be picked when the stem slips easily from the fruit with slight pressure from the thumb. Honeydew melons are at their best when they develop a yellowish skin color and sweet smell.

Squash and Pumpkins. Squash and pumpkins (*Cucurbita maxima, C. pepo, C. mixta,* or *C. moschata*) are the easiest of the cucurbits to grow, and they produce an abundance of fruit. They are divided into two types, based on whether the immature or the mature fruit is eaten. Those squash eaten when the fruit is mature and possessing a hard skin are "winter squash." The name is derived from the fact that these types ripen in late fall and can be stored in a cool place through much of the winter. Pumpkins are essentially a type of winter squash. Summer squash, eaten while immature, include zucchini, yellow crookneck, and yellow straightneck squash. They are perishable and must be eaten soon after harvest.

Winter squash should be harvested when the rind becomes hard, evident by an inability to puncture it with a thumbnail. Usually a change in color will accompany the ripening process. Summer squash are picked before the skin begins to harden but not necessarily when they are very small. If summer squash become large, they are still edible by scooping the seeds from the center prior to cooking.

When selecting varieties of squash and pumpkins, consider selecting bush types instead of vine types since they require much less room and produce the same high-quality fruit.

Eggplant

Eggplant (*Solanum melongena* var. *esculentum*) is a warm-season vegetable related to tomatoes and peppers. The plant is bushy with heart-shaped leaves and purple star-shaped flowers. The edible portion is the egg-shaped fruit, which is generally large and purple in color and is harvested when it has reached the size appropriate for the variety but before the skin has lost its natural sheen. Storage of eggplant should be at 50–55°F (10–13°C).

SOIL

Any well-drained soil with a pH above 6.0 is suitable.

CLIMATE REQUIREMENTS

Eggplants require a long period of warm weather to set and mature fruit properly. When growing eggplant in northern climates, early varieties should be selected.

CULTURE

Eggplants are grown from transplants since the plants are very slow to develop from seed. A starter fertilizer should be used at transplanting, followed by a sidedressing of a balanced fertilizer after the first fruit has set. Since eggplants are subject to several soilborne diseases, they should be rotated in the garden to avoid being grown in the same area where related plants (tomatoes, peppers, potatoes) were planted.

Honeydew Melons

See "Cucurbits."

Kale

See "Cole Crops."

Lettuce

Lettuce (*Lactuca sativa*) is a cool-season vegetable. It is available in a number of types which form a firm head (called *head lettuce* or *iceberg*) (Figure 6-22), semiheading types such as Bibb (Figure 6-23), and types which produce no head at all, called *leaf lettuce* (Figure 6-24). Heading and semiheading lettuce is harvested by cutting the head from the roots when it is full grown and, in the case of head lettuce, firm. Leaf lettuce can be harvested by removing leaves individually or by cutting the plant at ground level. Head lettuce will keep 1 to 2 weeks in the coldest part of the refrigerator, but leaf lettuce should not be stored more than 1 or 2 days.

SOIL

Lettuce will grow in most soils, however, emergence of the seedlings in heavy soils may be hindered by soil crusting. To remedy this, mix sieved organic matter into the soil used for covering the seeds.

CLIMATE REQUIREMENTS

Lettuce requires cool temperatures and is therefore best grown as either a spring or a fall crop. It can also be used as a winter crop in mild-winter areas, provided no hard freezes occur. During hot weather lettuce will become bitter and tend to flower. As a preventive

Figure 6-22 Head lettuce.

measure, plant only during cool weather and choose heat-resistant varieties.

CULTURE

Leaf and semiheading varieties are generally seeded shallowly and later thinned to a 10-inch (25-centimeter) spacing. Head lettuce may be either seeded or transplanted. Transplants should be spaced at least 10 inches (25 centimeters) apart; seedlings should be thinned to the same spacing as soon as they begin to crowd. If not thinned, the lettuce will be prone to rot and may fail to head.

Difficulty in germinating may be due to overly warm soil. If this is suspected, place the seed in water in the refrigerator overnight prior to planting. Lettuce should be kept well watered and should be sidedressed with high-nitrogen fertilizer once during its growing period.

Figure 6-23 Semiheading Bibb lettuce.

Figure 6-24 'Salad Bowl' leaf lettuce.

Melons

See "Cucurbits."

Muskmelons

See "Cucurbits."

Onions

Onions (*Allium cepa*) are a popular, easy-to-grow vegetable crop. They can be harvested at almost any stage of maturity, depending on the proposed use. Young onions, regardless of variety, can be harvested before bulbing for use as green or "bunching" onions. In this stage they are perishable and should be refrigerated. If mature onions with bulbs are desired, a variety with good storage characteristics should be selected, and the onions should be allowed to remain in the garden until the tops fall over. This signals that the bulb is mature. After digging out the bulb, cut the leaves off about 1 inch (2.5 centimeters) above the bulbs. The bulbs should be dried outside if the weather is warm and dry or in slatted crates or mesh bags in a warm indoor location. After several weeks of drying, the onions can be stored in a cool, dark location.

SOIL

Onions grow best in a sandy soil high in humus. When onions are growing in heavy soils, large amounts of organic matter should be added to the soil.

CLIMATE REQUIREMENTS

Onions are classified as a cool-season crop, but they are tolerant of a wide range of temperatures including frost. The best-quality bulbs will be produced when temperatures are cool during the early part of the growing season and moisture is abundant.

The most important factor affecting onion growing is photoperiod. Onions form bulbs only when the night is shorter than the critical photoperiod for that variety. If the critical photoperiod is reached too soon after seeding, the onions will bulb before the plants are large enough, and the resulting bulbs will be very small. If the critical photoperiod is never reached, the onions will never form bulbs.

To avoid bulbing problems, select only varieties recommended for the area. In areas where both winter and summer crops can be raised, a different variety must be selected for each, due to the difference in day length between the seasons.

CULTURE

Onions can be grown from seed, transplants, or small bulbs called *sets*. Green onions can be easily grown from seed, but bulb types will be ready for harvest sooner if sets or transplants are used. They may be planted 3 to 4 inches (8–10 centimeters) apart or closer together with every other plant pulled for use as a green onion.

Regular watering and one or two side-dressings of balanced fertilizer during the growing season are recommended.

Peas

Peas (*Pisum sativum*) are a cool-season garden crop that shows a spectacular improvement in quality over purchased produce when home grown and picked immediately prior to eating. The peas should be harvested when they begin to fill out the pod but before becoming tough. Storage of peas is not recommended, but if unavoidable, they should be kept in the coolest part of the refrigerator.

SOIL

Peas will thrive in any well-drained soil with a pH above 6.0.

CLIMATE REQUIREMENTS

Peas require cool temperatures and abundant moisture to grow well. They will tolerate moderate freezes but not high temperatures. They are best grown as an early spring or late fall crop or, in mild-winter areas, as a winter crop. Where warm weather makes growing peas difficult, a heat-resistant variety such as 'Wando' should be chosen.

CULTURE

Pea seeds are sown about 1 inch (2.5 centimeters) apart and are not thinned. A common practice is to sow peas in double rows about 8 inches (20 centimeters) apart and train two rows on a single string trellis.

It is important to supply water regularly to peas and to pick the pods as soon as they are mature to lengthen the harvest. A side-dressing with balanced fertilizer is recommended when the plants are 4 to 6 inches (10–15 centimeters) tall.

Peppers

Peppers (*Capsicum annuum*) are bushy plants with long shiny green leaves and white star-shaped flowers. Many are well adapted to pot growing and very attractive when bearing fruit.

Peppers are known as either "sweet" or "hot." The sweet peppers include bell and sweet banana types. The hot peppers usually have fruits which taper to a point and can be green, red, or yellow. Sweet peppers may be picked any time after the fruit has attained full size and the color characteristic of the particular variety. Bell peppers which are not picked soon after they reach maturity will eventually turn red; they are still edible and not hot. Hot peppers can be picked when the full size and proper color have been reached or can be allowed to dry on the plant for winter use. However, picking the fruits as they ripen will increase the total fruit production.

A common problem encountered by home gardeners raising both hot and sweet peppers is cross-pollination. Varieties of peppers cross freely, and although the fruits will each retain their proper taste, the seeds will bear the characteristics of both parents. The seeds of sweet peppers grown near hot peppers should be removed before eating.

Another problem, poor fruit set, is usually weather-related. Both excessively high temperatures and cloudy, wet weather will prevent fruit set.

Fresh peppers should be stored in the refrigerator and will retain good quality there for a week or more.

SOIL, CLIMATE REQUIREMENTS, AND CULTURE

Same as for eggplant.

Pumpkins

See "Cucurbits."

Radishes

Radishes (*Raphanus sativus*) are the fastest-growing garden vegetable and will produce an edible root in 21 to 30 days. These quick-maturing radishes are the red globe-shaped type. The second type, called *Chinese* or *icicle* radishes, takes longer to mature, but produces larger roots, which are white. Globe radishes are suitable for planting in the spring and should be picked as soon as they reach edible size, or they become tough and pithy. Icicle varieties are preferred as a fall crop where the weather is warm.

SOIL

Sandy soils are ideal, but radishes are tolerant of most garden soils.

CLIMATE REQUIREMENTS

Radishes are cool-season crops and will grow best in temperatures below 70°F (21°C). Under high-temperature conditions they tend to be hot and to flower.

CULTURE

Seed should be sown in early spring to avoid the warm temperatures which slow growth and cause bolting. Plants should be thinned to 1 inch (2.5 centimeters) apart soon after germination. Several sowings made 1 week apart are recommended to assure a steady supply. Radishes can also be seeded along with carrots or leeks or between transplants of other crops since they mature quickly and can be harvested before crowding occurs.

Squash, Summer

Includes yellow straightneck, yellow crookneck, and zucchini. See "Cucurbits."

Squash, Winter

Includes acorn, butternut, spaghetti, and other thick-rind types. See "Cucurbits."

Swiss Chard

See "Chard."

Tomatoes

Tomatoes (*Lycopersicon lycopersicum*) are the most popular of all garden vegetables. Although they are actually herbaceous perennials, they are grown as annuals in the vegetable garden and develop into branched bushes or vines with compound leaves and yellow flowers.

Botanically, tomatoes are related to eggplants and peppers but are easier to grow than either. The fruit will be most flavorful when allowed to ripen fully on the plant. Green tomatoes can be eaten pickled or fried.

SOIL

Tomatoes will succeed in a wide range of soil types provided the drainage is adequate. However, they should not be planted in the same spot year after year due to the increased danger of wilt diseases and nematodes. A pH between 6.0 and 6.8 is best.

CLIMATE REQUIREMENTS

Like peppers and eggplants, tomatoes are a warm-season crop. For fruit to set, temperatures must be warm since pollen is usually not shed when night temperatures are below 59°F (15°C) or above 80°F (27°C). In areas with cool summers or short frost-free growing seasons, early varieties should be selected since they not only mature more quickly but grow better at cool temperatures. Varieties which produce small fruits such as cherry tomatoes tolerate a wide range of temperatures.

CULTURE

Although tomatoes can be raised easily from seed planted directly in the garden, they are usually raised from transplants for an earlier harvest. Spacing of the plants will depend on whether the variety grown is "determinate" or "indeterminate." Determinate tomato varieties grow into bushy plants which can be either staked, caged, or allowed to sprawl along the ground. Indeterminate types grow long and vinelike and definitely require support.

When planting tomatoes from transplants, a starter fertilizer is beneficial. Avoid additional fertilization until after the first fruit has reached ½ inch (1.3 centimeters) in diameter, after which a sidedressing with balanced fertilizer can be made.

Although pruning is sometimes advocated for home-grown tomatoes, most experts now agree that pruning only slows fruit production and increases sun scald of the fruit as well.

Turnips

Turnips (*Brassica rapa*) are a cool-season root crop related to cole crops. Though the root is the primary edible portion, the leaves are also edible; in fact some varieties are grown mainly for their leaves. Turnip leaves do not store well and should be picked immediately prior to eating. Roots, on the other hand, store for long periods if the tops are removed and if kept in a cool, humid environment. They should be harvested when they are 2 to 3 inches (5–8 centimeters) in diameter.

SOIL

Turnips grow best in a light- to medium-weight soil high in organic matter but will produce large crops in almost any soil as long as the drainage is adequate.

CLIMATE REQUIREMENTS

Turnips are planted as a spring or fall crop in most areas or as a winter crop in mild-winter areas. In locales with hot summers, fall plantings are most successful.

CULTURE

Turnips are grown from seed that germinates rapidly in the garden. The young plants grow vigorously and will need to be thinned so that they are about 3 inches (8 centimeters) apart when they are 3 inches (8 centimeters) tall. Thinning will not be required if the turnips are grown for greens only.

During the growth period the plants should be kept well watered and sidedressed once or twice with a light application of balanced fertilizer. This will help the plants to grow rapidly and will assure tender roots.

Watermelon

See "Cucurbits."

Zucchini

See "Cucurbits."

MINOR VEGETABLES

Chinese Cabbage

Chinese cabbage (*Brassica rapa,* Pekinensis Group), sometimes called *celery cabbage,* is a cool-season annual grown for its leaves. Seeds should be sown early in spring or as fall approaches to avoid hot weather and flowering. Seedlings should be thinned to 8 inches

(20 centimeters) apart when they begin to crowd; the heads are later harvested individually. This is a very fast-growing leaf vegetable prized for Chinese dishes.

Endive

Endive (*Cichorium endivia*) is grown for its slightly bitter leaves, which are used fresh as a salad green. Seed should be planted in early spring, and leaves can be harvested whenever they are large enough to eat. For better flavor the mature outer leaves can be tied over the inner leaves to blanch them.

Kohlrabi

Kohlrabi (*Brassica oleracea,* Gongylodes Group) is a relative of cabbage grown for its enlarged stem. Plants are harvested when the enlarged portion is 2 to 4 inches (5–10 centimeters) in diameter. See "Cole Crops" for cultural information.

Leeks

Leeks (*Allium ampeloprasum*) (Figure 6-25) are a nonbulbing relative of the onion with a mild flavor. They are generally grown from seed and require a long, cool growing season. The white underground stem is eaten at any size. Soil can be mounded around the plants to blanch more of the stem, thus producing more edible area per plant.

Mustard

Mustard (*Brassica juncea*) is a fast-growing, cool-season, vegetable grown for its strong-flavored greens. It is seeded directly in the garden in early spring, and individual leaves are picked when they reach 3 to 4 inches (8–10 centimeters). Planting early is important since mustard will flower and die as soon as the weather becomes warm.

Okra

Okra (*Abelmoschus esculentus*) is a large, fast-growing plant that produces edible seed pods. To grow well, okra requires hot weather, well-drained soil, and a relatively long growing season. The pods should be picked before they reach 3 inches (7.6 centimeters) in length.

Parsnips

Parsnips (*Pastinaca sativa*) produce white, carrot-shaped roots with a sweet, nutlike flavor. Cool-season vegetables, they should be seeded in deeply worked soil very early in the spring and grown like carrots. The roots can be harvested at any size and can be allowed to remain in the ground for winter and spring eating.

Parsnip seeds are among the shortest-lived of all vegetable seeds, so storage of old seed is not recommended.

Potatoes

Potatoes (*Solanum tuberosum*) are grown for their underground tubers and will grow in most climate zones. However, quality will be best in a climate with cool nights.

Potatoes are grown from "seed potatoes," small pieces of a potato which contain at least one "eye" or bud per section. Only certified seed potatoes which are guaranteed to be disease-free should be used.

"New potatoes" are thin-skinned, immature potatoes harvested when the plants begin to flower and the potatoes are still small. They should be eaten immediately after picking in summer. To have potatoes for storing for winter use, delay harvest until the tops of the plants have died in fall.

Unless a large garden space is available, potatoes are not considered worth raising in the home garden. The quality will not differ substantially from that of supermarket pota-

Figure 6-25 Leeks.

toes, and since supermarket potatoes are inexpensive, monetary savings are usually minimal. In addition, home-grown potatoes will require careful attention to control insects and diseases.

Rhubarb

Rhubarb (*Rheum rhabarbarum*) is a perennial grown for its leaf stalks, which have an acid, fruitlike flavor. It will grow in all zones except 10. One- or two-year-old crowns should be planted in spring and light harvesting started the following spring. The leaves should not be eaten, as they contain oxalic acid and are poisonous.

Flower stalks should be removed, as they appear to divert the strength of the plant into vegetative growth. Fertilizing once per year in early spring is sufficient, and clumps should be divided about once every 4 years.

Rutabagas

Rutabagas (*Brassica napus*, Napobrassica Group) are grown for their large turniplike roots, which have a flavor similar to that of a strong turnip. Their culture is the same as for turnips.

Snow Peas, Chinese Pea Pods, Edible Podded Peas (*Pisum sativum* var. *macrocarpon*)

Culture the same as for peas.

Spinach

Spinach (*Spinacia oleracea*) is a nutritious cool-season vegetable grown for its strongly flavored leaves. It should be seeded in early spring directly in the garden and harvested before the weather becomes hot because the plants flower in warm weather.

Sweet Potatoes

Sweet potatoes (*Impomoea batatus*) can be grown in warm climates with a frost-free growing period of 120 days or longer. They are grown from transplants spaced 24 to 30 inches (0.6–0.7 meter) apart and can be harvested as soon as they reach edible size. If plans include storing the sweet potatoes for winter use, harvest should be delayed until frost kills the tops. The roots are then dug, cured at 85°F (30°C) for 3 weeks, and stored at 55°F (13°C). Care should be taken that the roots are not bruised and that curing is performed properly; otherwise sweet potatoes will not store well.

Selected References for Additional Reading

Abraham, G. *The Green Thumb Book of Fruit and Vegetable Gardening.* Englewood Cliffs, N J: Prentice-Hall, 1970.

Bush-Brown, J., and Bush-Brown, L. *America's Garden Book.* New York: Charles Scribner's Sons, 1966.

Crockett, J. U. *Vegetables and Fruits.* New York: Time-Life Books, 1972.

Knott, J. E. *Handbook for Vegetable Growers.* rev. ed. New York: Wiley, 1962.

Ortho Books. *All About Vegetables.* San Francisco: Regensteiner Press (Chevron Chemical Company), 1973.

Tiedjens, V. A. *The Vegetable Encyclopedia.* New York: Crown, 1943.

Chapter 7

Growing Tree Fruits and Nuts

Tree crops are raised for their edible fruits or seeds. Most fall into the designations of fruits or nuts, but a few, such as the avocado and olive, grow on trees but fit neither category.

Many homeowners would like to grow fruit or nut trees for edible produce. The species and the number they should select will depend on many factors, among them the following.

1. *Climate.* Minimum winter temperature as well as summer temperature and length will determine which species and varieties can be grown. A few tree crops such as avocado and citrus are semitropical and will tolerate only temperatures to the mid-20°F (2°C) range. Others are adapted to temperate climates with widely varying winter and summer temperatures. They will not survive without a cool winter period and will fail to produce fruit and even die in a warm-winter area.

Unexpected spring frosts in temperate areas can also cause problems with some fruits which flower early. Apricots are notably vulnerable because of their early bloom period.

Summer temperature and length can also affect ability to successfully grow some fruit trees. During some years the plant will not receive temperatures warm enough for proper fruit ripening and sugar formation, or the growing season will not last long enough for the fruit to develop completely.

2. *Property size.* The size of the property will also determine how many fruit and nut trees can be raised. Owners of small properties frequently incorporate fruit and nut trees into the public and private areas in lieu of ornamental trees. Others use dwarf varieties which grow only about 8 feet (2.4 meters) high. Although this is a solution with most fruit trees, no nut trees are available in dwarf form.

3. *Expected maintenance time.* Although some species have little problem with pests and diseases, others, such as the apple, will require spraying every 2 weeks in addition to yearly pruning and fertilization.

SITE SELECTION

A property may have several possible sites for planting fruit and nut trees. If so, the following criteria (in order from most to least important) should be taken into account when the spot is selected.

Soil Drainage

Few plants will grow in an area with poor drainage, and trees are no exception. Water must drain away rapidly to allow air to penetrate to the roots. If it does not and the roots are in soggy soil for an appreciable period, the tree will die. "Stone fruits" (those with a single, hard pit such as cherry, plum, peach, and nectarine) are the most susceptible to poor drainage.

Amount of Sunlight

Fruit trees should have as much sun as possible to bear heavily and ripen their crops. Six hours of direct sun per day is considered minimum for average fruit production. Although a sunny spot is also desirable for nut trees and will speed their growth considerably, it is not as essential as for fruit trees. The nut trees, being ultimately larger, will compete for sun, whereas dwarf fruit trees will not.

Land Slope and Exposure

In all but subtropical and tropical climates, situating fruit and nut trees on the upper part of a slope is ideal because it partially protects against frost injury. As discussed in Chapter 1, cool air flows downhill, leaving the higher area of the slope several degrees warmer. The few degrees difference may prevent a frost that kills the season's crop. Unless it is unavoidable, a fruit area should not be planted in a low spot or at the bottom of a slope. Flat land is preferable to those areas.

The direction in which the slope should ideally face will depend on the weather conditions in the area. South-facing slopes warm earlier in the spring, and trees will start growth sooner. However, this also means that they may be in a tender state and susceptible to late spring frosts. Consequently, a north-facing slope is suggested in climate areas with spring frost problems, but southern slopes are best in the areas of the country which are completely or nearly frost-free. South slopes are also advisable where summer temperatures are likely to be too cool for complete fruit maturation, since a south-facing slope receives direct sun the entire day and will be several degrees warmer.

INCORPORATING TREE CROPS IN THE LANDSCAPE

On small properties, a homeowner may have to work fruit and nut trees into the landscape design in lieu of setting aside a separate orchard area. How the trees can be incorporated will depend on the unique features of the property, but the following suggestions can aid in planning.

Substituting Nut Trees for Deciduous Shade Trees

Since nut trees are large, they work well as substitutes for shade trees. Many have a pleasing fall color, although the flowers are inconspicuous. Large fruit trees also make handsome shade trees. However, if spraying is a necessary part of the maintenance program, the job will be harder with a large tree.

Use of Dwarf Trees as a Screen

Three or more dwarf trees planted close together (about 6 to 8 feet or 1.8 to 2.4 meters apart) will eventually grow to form a screen. Staggering the plants in two rows (Figure 7-1) will create the screen effect faster without crowding the trees.

Figure 7-1 Fruit trees used as a screen in a landscape.

Use of Dwarf Fruit Trees as Patio Trees

Fruit trees are very ornamental with lovely spring blossoms and heavy sets of ripe fruit. Dwarf types can be conversation pieces, but semidwarfs should be planted if shade is desired.

Espaliered Fruit Trees

The flat training method called *espalier* (Figure 7-2) can fit several dwarf fruit trees into the smallest yard. The tree will still bear a worthwhile amount of fruit, although probably not as much as it would growing in a normal form.

TREE SELECTION

In selecting an ornamental plant, the variety may be picked for its outstanding flowers or vivid fall color, but neither is crucial to its success in the landscape. But choosing a proper variety is very important in fruit and nut growing. The variety can lead to failure, so it must be chosen carefully. Read not only the nursery catalog description but also state Agricultural Extension Service publications for information on the best varieties for the area. Look for the following information in making a variety decision.

Winter Hardiness

Some fruit varieties are bred for growing in the South, whereas others of the same species

Figure 7-2 Espaliered apple tree.

are bred for the North. A particular southern peach variety may not survive a northern winter, much less produce the expected size and quality of fruit.

Required Maturity Period

Varieties of fruits bred for cold climates ripen more quickly than those bred for warm climates where the summer is longer. A variety should be chosen that will mature and ripen within the limitations of the frost-free growing season (Chapter 6). Moreover, fruits advertised as maturing extremely early in the season will often be inferior in quality to later varieties.

Disease Resistance

Selecting disease-resistant varieties can eliminate much of the labor of spraying against fungus- and bacterium-caused diseases. Apple varieties, for example, are available which are resistant to scab, fire blight, and cedar rust, three common disease problems. Chinese chestnuts are immune to chestnut blight, whereas the American chestnut is so susceptible that only a few trees are alive in the United States today.

Fruit Use

Some fruit varieties are developed primarily for fresh eating, whereas others (originally bred for commercial orchard use) are best for drying or canning. Prune versus fresh-eating plum varieties are examples. Any fruit bred for processing can, of course, be eaten fresh, but its quality may be poorer than that of a fresh-eating variety.

Yield

The quantity of produce that can be expected from a fruit or nut tree is associated with both the variety and the care the tree receives. All commercial varieties have average or better yields, and some are exceptional producers.

Fruit Appearance

Size, coloring, and shape are included under fruit appearance. Many gardeners have a preference for yellow over red apples, for example, or a variety that produces mammoth fruits. But appearance should not be allowed to outweigh the more important factors of hardiness, disease resistance, maturity period, and yield.

Pollination Requirements

Many fruits and nuts require cross-pollination with another variety or, with nuts, a tree of the opposite sex to bear a crop. Many homeowners have grown healthy trees for years without a crop, due to failure to recognize the pollination requirement of the variety.

SOURCES OF FRUIT AND NUT TREES

Unless you live near a nursery specializing in fruit and nut trees, the best selection is available from mail-order nurseries. Several firms specialize exclusively in fruits and nuts and offer many varieties of each species. Some are listed under "Selected References" at the end of this chapter.

PLANTING FRUIT AND NUT TREES

Planting fruit and nut trees is the same as planting an ornamental tree. Deciduous trees are normally sold leafless and bare-rooted for planting in the fall or spring. Evergreens such as olive, citrus, and avocado are sold in plastic sleeves or containers and can be planted at any time of the year.

The hole should be dug large and deep enough to accommodate the root system without crowding. After settling the plant in the hole, it should be filled around with a mixture of the removed topsoil and an equal amount of organic matter such as peat moss or compost. Watering thoroughly afterward is essential.

Most fruit trees and many nut trees are grafted, and the placement of the graft union aboveground at planting is important (Figure 7-3). This placement is necessary to retain the special qualities of the rootstock. For example, it may have disease resistance or cause the top to be dwarfed instead of full size. If the scion (top portion) comes in contact with soil, it will send out roots of its own, and the desirable rootstock characteristic will be lost.

Staking for support may be necessary for fruit and nut trees for the first year or two until the root system grows large enough to anchor the plant. Many dwarf varieties are very shallow and fibrous-rooted, so permanent staking or trellising to prevent blowing over may be advisable. A short stake attached to the trunk is usually sufficient. Trees should be tied to the stakes using a wire covered with a section of rubber hose or pieces of rag. Bare wire or string may injure the bark by rubbing or girdling the trunk.

Figure 7-3 Graft union of a fruit tree.

Normally fruit and nut trees are pruned before shipping, but if not, they will have to be pruned after planting. This is discussed under "Pruning and Training."

DWARF FRUIT TREES

Nut trees are available only in "standard" (normal) sizes. However, most fruit trees come in three sizes: standard, semidwarf, and dwarf. Standard trees grow to the normal size for the species and are large in most cases. A standard apple tree can grow 30 feet (9.1 meters) across and to an equal height, although 20 to 25 feet (6–7.6 meters) is more typical.

Semidwarf trees range from 10 to 15 feet (3–4.6 meters) and dwarfs from 5 to 12 feet (1.5–3.7 meters). Most fruit trees sold to home gardeners are semidwarf or dwarf for various reasons. First, because they are shorter, picking, pruning, and spraying are easier, and a ladder is not usually required. Dwarfed trees also bear sooner. An average of 5 to 6 years from planting is not unusual before a standard tree bears its first crop, but a dwarf will bear the second year after planting and may even produce a few fruits the first year. Dwarf and semidwarf trees occupy less space, and a homeowner with limited land can grow more species and varieties of fruit in an area than with standard trees.

Dwarf trees are produced in one of two ways. The most common is by grafting onto a "dwarfing rootstock" which prevents the variety from growing to full size. The grafting combines the dwarf characteristic of one variety with the exceptional fruit of another to produce a plant with the desirable characteristics of each.

Sometimes three varieties are involved, a rootstock, a scion, and an *interstock* (Figure 7-4). An interstock would be used, for example, when the stock and scion are "incompatible" (will not graft together), but each is compatible with the interstock. Grafting with an interstock might also be used to combine superior fruiting (in the scion), dwarfing (in the interstock), and a vigorous root system (in the rootstock). Many nurseries use this technique commonly to avoid the poor root systems inherent with many dwarfing rootstocks.

A second type of tree dwarfing is called *genetic dwarfing,* whereby the fruiting portion of the tree remains small due to its genetic makeup rather than because of a dwarfing rootstock. However, genetically dwarf trees are also frequently grafted. The genetically dwarf top is grafted onto a particularly vigorous or disease-resistant root system for the qualities the rootstock can contribute to the tree.

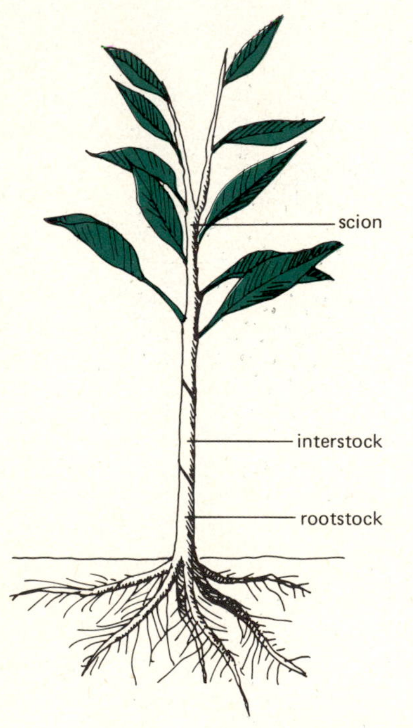

Figure 7-4 An interstock grafted between the rootstock and scion on a fruit tree.

POLLINATING FRUIT AND NUT TREES

In most fruit and all nut trees, flowers must be pollinated and fertilized to set a crop. Exceptions to this rule are rare but include fruits such as navel oranges, persimmons, and some figs and pears. These fruits are able to set "parthenocarpically," that is, without fertilization.

The remaining tree crops require pollination, either self or cross. Self-pollination is the rule on most nut trees (see the nuts listed at the end of the chapter for exceptions), the majority of which bears separate male and female flowers on the same tree, which are pollinated by wind. Self-pollination is also found in a few fruit trees, but more often fruit trees are incapable of pollinating themselves and are called *self-sterile.* Pollination with another compatible variety flowering at the same time is necessary if a crop is to be produced.

The causes of self-sterility are primarily the following.

1. *Incompatibility of the pollen with the female flower parts.* This can be compared to an allergic reaction in which the pollen grains do not function correctly when in contact with the female flower parts of the same plant. Pollen may fail to germinate or grow too slowly to the egg, even though it will perform normally with other varieties.
2. *Nonviable pollen.* The pollen is not vigorous and does not germinate.
3. *Failure of the pollen and female flower parts to attain maturity at the same time.* Pollen is shed after the female flower parts have

passed their receptive stage or before they have attained it.

4. *Dioecious plants.* In the case of a few tree crops such as the date, male and female flowers are produced on separate plants, and there is simply no pollen produced with which the tree can self-pollinate.

In summary, cross-pollination with a "compatible" variety flowering at the same time is a prerequisite for successful growing of many tree crops and should be carefully taken into account when purchasing trees.

MAINTENANCE OF TREE CROPS

Watering

Watering of a fruit or nut tree varies with the species. Some, such as the olive, are very drought-resistant and adapted to arid climates, whereas most others will produce their best crops only with regular rainfall or irrigation.

As a general rule, a newly planted tree should be watered deeply every 1 to 2 weeks its first growing season. Decrease the frequency in subsequent years to about every 3 weeks, depending on the soil water-holding capacity and natural rainfall.

Watering with a drip irrigation system (see Chapter 11) conserves both water and labor. Many commercial orchards use drip irrigation systems, and they can be highly recommended for home use.

Fertilization

Tree crops are normally fertilized yearly on a precautionary basis rather than because the tree has shown any distinct nutrient deficiency. Without seasonal soil tests it is impossible to apply the exact amount of fertilizer required, and yearly fertilizing is used to prevent possible deficiencies.

Methods for applying fertilizer include the drill hole method explained in Chapter 5 and the surface topdressing method. The drill hole method is used when the area under the tree is covered with grass or groundcover plants which would be injured by the fertilizer application.

General recommendations specify applying granular-type fertilizers in spring before or just as new growth starts. About ½ cup (120 milliliters) of high-nitrogen fertilizer per inch (2.5 centimeters) of trunk diameter will supply an adequate amount of nutrition. Little other fertilization is needed except in sandy soils which do not hold nutrients well. In these cases the applications should be repeated in midsummer.

Micronutrient fertilizers are frequently used to prevent iron deficiency on citrus growing in alkaline soil. Either foliar or soil applications should be made repeatedly until the foliage regains its healthy green color.

Pruning

The three main pruning styles used on tree crops are the central leader, modified leader, and vase shape. Espalier training styles are less commonly used. The pruning style selected depends mostly on the species of tree; some adapt more easily to one style than another. In all cases the object of the training is to produce a tree with well-placed, strong branches capable of supporting a heavy weight of fruit without breaking.

CENTRAL LEADER FORM

Central leader form is used on most nut trees, on sweet cherries, and occasionally for apples and pears. The pruning starts at planting, if necessary, with the selection of the central leader and several scaffold branches spaced widely around and down the trunk. Figure 7-5 shows a young tree pruned to a central leader style. Pruning after the first several years should consist of removing rootstock suckers

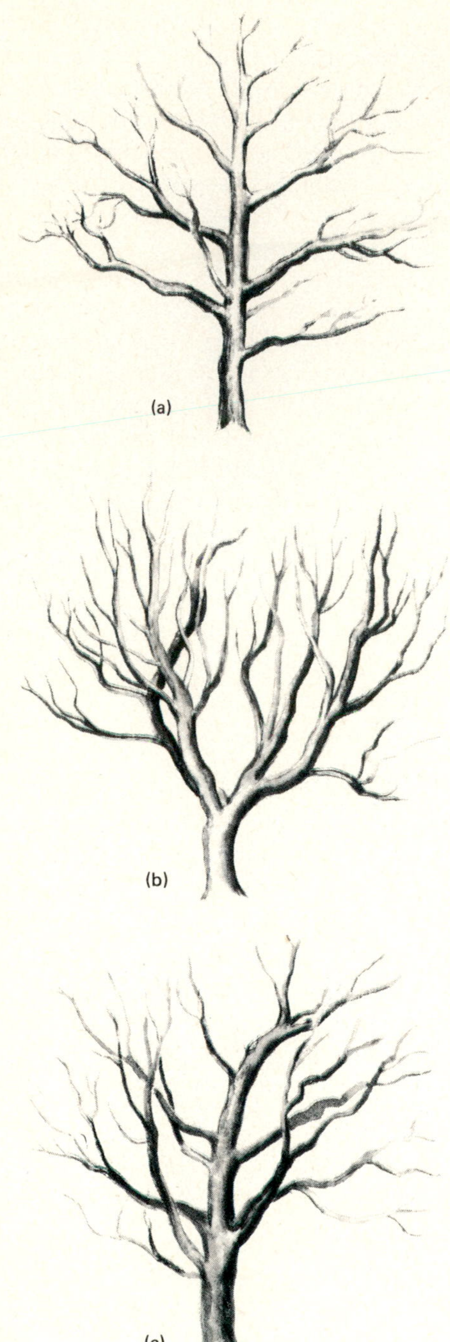

Figure 7-5 Fruit trees pruned to central leader (a), vase (b), and modified central leader forms (c).

(Figure 7-6) and competing branches arising from the trunk.

MODIFIED CENTRAL LEADER FORM

In a modified central leader (Figure 7-5) form, the tree is first trained to a central leader form for 2 to 3 years. The next year the leader is cut back to a strong side branch, leaving three to five scaffold branches on the tree. With the leader gone, the side branches grow stronger and longer, and the height of the tree is kept in check. Modified central leader form is suggested for apples, pears, plums, and sour cherries.

VASE FORM

Vase form, called *open-center form,* allows two to five scaffold branches to grow from a short trunk (about 1 to 2 feet, or 0.3 to 0.6 meter). The branches originate close together on the trunk, and the central leader is removed to a side branch.

A vase-shaped tree (Figure 7-5) is pruned

Figure 7-6 Pruning suckers from the base of a tree.

to a final form directly after planting by selecting the lowest, strongest branches on the tree. Vase pruning opens the center of the tree to admit light and encourage fruiting and keeps the bearing branches of the tree low. Vase pruning is common on peach, apricot, and nectarine.

ESPALIER PRUNING

Espaliers can take many forms (Figure 7-7). Training must start as soon as the tree is planted; it consists of pruning, bending, and tying the branches as they mature to achieve a symmetrical shape.

Although espaliered fruit trees do not produce as large a crop as conventionally pruned trees, the amount is acceptable considering the small space they use. Apple and pear trees are among the best for espalier training. Peach and plum trees are more difficult but possible. Cherries and nut trees should not be espaliered because their production becomes limited as a result.

Training

Whereas pruning is the primary method for training a tree, several other supplementary techniques help create the desired form. Many fruit trees have branches which grow upward, instead of outward, with narrow branch crotches. As the branches enlarge, they exert pressure at the crotch, causing it to become weak with a tendency toward splitting under a snow or fruit load. However, when young, the tree can be trained to develop wide crotches by two methods.

Tying (Figure 7-8) involves bending flexible young branches down and securing them to the ground with a stake. The branch should be tied for one or two seasons and frequently checked to make sure it is not being injured. After the rope is removed, the branch will grow outward instead of upward.

Spreading (Figure 7-9) uses forked or nail-tipped pieces of wood wedged tightly between young branches of a tree to widen the crotch angle. As with tying, the spreader is left in place for one or two seasons to encourage the crotch to remain open.

Fruit Thinning

Fruit thinning is the removal of a portion of the fruits on a tree while they are still small. Many deciduous fruit trees (nut trees are excluded) produce more fruit than can fully mature. If unthinned, the fruit will be un-

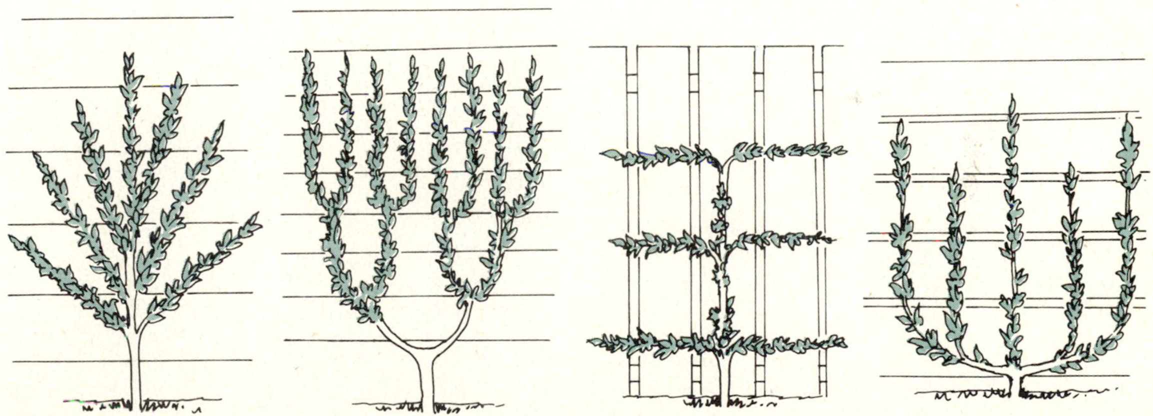

Figure 7-7 Sample espalier forms for fruit trees.

Figure 7-8 Tying a fruit tree branch to the ground to encourage a wide branch crotch to form.

dersized and frequently poorly colored. In addition, the excessive fruit load may weaken the tree. Bearing in alternate years may result, with the fruitless year devoted to regaining the vigor of the tree. Since most gardeners want fruit every year, the advantages of thinning are obvious.

Fruits should be thinned when they are ½ to ¾ inch (13–20 millimeters) in diameter, and after the natural drop of young fruit has occurred. Twisting each fruit will remove it quickly and easily.

The number of fruits that should be left on a tree depends on the species. Table 7-1 lists approximate distances which should be left between fruits along a branch for moderate fruit load. In practice, fruits are not spaced evenly but are close together on strong branches and far apart on weak ones. The spacing guidelines should be taken only as a guide and not an absolute. The largest, healthiest fruits should always be left on the tree, regardless of their relative spacing on a branch.

If a gardener has a large orchard of, for example, apples, hand thinning may be impractical. In that case the trees may be thinned chemically with a hormone application shortly after bloom. The gardener should

Table 7-1 Suggested Fruit Load after Thinning for Common Fruit Species

Species	Distance between fruits
Apples	6–8 in. (15–20 cm)
Apricots	1½–2 in. (40–50 cm)
Avocados	Thinning not necessary
Cherries	Thinning not necessary
Citrus, figs, mulberries	Thinning not necessary
Nectarines, olives	4–5 in. (10–13 cm)
Peaches	6–8 in. (15–20 cm) for early varieties, 4–5 in. (10–13 cm) for later ones
Pears	Thinning not necessary but can be thinned lightly to increase size
Persimmons	Thinning not necessary
Plums, Japanese	2–4 in. (5–10 cm)
Plums, European	Thinning not necessary
Quinces	Thinning not necessary

Figure 7-9 A branch spreader in place in a tree.

consult his county Cooperative Extension Agent for specific recommendations.

RENOVATING NEGLECTED FRUIT TREES

When buying an older home, it is common to inherit at the same time one or more neglected fruit trees. The tree frequently will be bearing sparsely, and the fruit produced will be small and of poor quality. It may have dead limbs or be diseased.

The question of whether or not to save the tree is a personal choice. It will probably never bear as high-quality fruit as a newer variety, but fruit quality can be improved to an acceptable grade if the tree is given some attention. The tree may be valuable for shade and flowers and thus be worth renovating from this standpoint.

Renovation pruning should be spaced over 3 years to avoid stimulating excess sucker growth. The first pruning should remove dead wood, suckers, and some of the crossing branches. Refuse and leaves drop in the fall and serve as a source of disease reinfection. They should be removed, and a disease and pest control program should be started.

During the next two seasons excess branches can be gradually removed to open up the head, admitting light and improving fruit production. Fertilizer can be withheld until after pruning is complete but should be applied immediately if deficiency symptoms are present.

FAILURE TO BEAR

The failure of a tree to bear may be caused by one or a combination of the following factors.

Age

The age at which a tree can be expected to produce its first crop varies with the species. Most dwarf and semidwarf types bear by the third or fourth year after planting.

Lack of Pollination

Either lack of a nearby pollinating variety or wet, cold weather during the pollination period can be a cause.

Insufficient Winter Chilling

In climates with mild winters the winter temperatures may not be sufficiently low to provide the necessary chilling period. Deciduous fruit and nut trees are most susceptible.

Insufficient Sunlight

A tree receiving insufficient sunlight will produce lush vegetative growth but few flowers and fruits.

INDIVIDUAL CROPS AND THEIR GROWING REQUIREMENTS

Almonds (*Prunus* var. *dulcis*)

Almond growing is restricted mainly to USDA zones 7 to 9 and to the western states. The primary problem of growing in colder zones is not minimum winter temperature but an early bloom period which hinders pollination and sometimes frosts the blossoms. The spring rains and high humidity of the East are also a hindrance to pollination and nut production in that climate. Lack of winter chilling excludes almonds from zone 10 (tropical).

Almond trees are available in only one size, and they grow 20 to 30 feet (6–9.1 meters) high. They should be pruned to a modified leader form and can be expected to bear the third or fourth year after planting. Cross-pollination with another variety is required in almost all cases.

The almond is produced inside a shell, which is in turn encased in a peachlike hull. In fall the hulls split, dropping the ripe nuts to the ground.

Apples (*Malus pumila* Descendants)

Apples are grown in zones 4 through 9. Special "tropical" varieties must be selected for zone 9 since apples have a chilling requirement. Most apples are self-fertile. They are available in dwarf, semidwarf, and standard sizes.

For greatest strength, apples should be pruned to a modified leader system. Dwarfs can be expected to bear the first or second year after planting.

Apples are highly susceptible to disease and insect problems, and regular spraying will usually be required for good fruit. However, disease problems can be minimized by selection of resistant varieties.

Apricots (*Prunus armeniaca*)

Like apples, apricots require winter chilling. They can be grown in zones 5 through 8 and some parts of zone 9. Apricots do not have a cross-pollination requirement and are available in standard and semidwarf sizes.

Apricots bear 1 to 2 years after planting

and are among the most ornamental of fruit trees. However, their early bloom period exposes them to frost damage. Training can be done in either a modified leader or a vase shape, and pruning should be done yearly to restrict height and increase spread.

Avocados (*Persea americana*)

Avocados are strictly subtropical (zones 9–10) and are hardy only to 20–24°F (−7 to −5°C). Their deep green, evergreen foliage and round forms make them valued as ornamentals.

Avocados divide into three races: the Guatemalan, with large fruit; the West Indian, which has large fruit but is more tropical; and the Mexican, which is hardier but has smaller fruit. The trees are available only in standard size. When mature they range from 30 to 60 feet (9.1–18.3 meters) tall (depending on the variety) and may be of equal width. Avocados tend to biennial bearing, and although cross-pollination is not essential, it will result in a better fruit set.

Avocados require little pruning except for limbing up obstructing lower branches and removing crossing or dead limbs. Flowers appear from winter through spring, depending on the variety, and the fruit will mature 7 to 12 months later. About 3 years is average from planting until the first set of fruit.

Cherries (*Prunus avium* and *P. cerasus*)

The sweet cherry (*Prunus avium*) is known for fresh eating, whereas the tart cherry (*Prunus cerasus*) is most useful for pies and other cooking uses. Both have a chilling requirement. The sweet cherry can be grown in zones 5 to 7, yet the tart cherry is more cold-hardy and will survive to zone 4.

Sweet and tart cherries are available in dwarf, semidwarf, and standard sizes. While tart cherries do not require cross-pollination to produce fruit, sweet cherries have very particular pollination requirements. Careful attention must be paid to selecting compatible varieties from among the many sweet cherry varieties available.

Pruning of cherries should be modified leader for sweet types and modified leader or vase for sour varieties.

Chinese Chestnuts (*Castanea mollissima*)

The chestnut that was originally grown for eating in the United States was the American sweet chestnut, a majestic tree of poetic fame. But in the early 1900s a blight disease killed nearly all the native chestnuts, and now the Chinese chestnut is the primary source of nuts. This blight-resistant tree will grow in zones 4 through 8 and averages 40 to 50 feet (12.2–15.2 meters) but develops at a more rapid pace after it becomes established.

The chestnut will set nuts without cross-pollination, but a heavier set will result if another tree is situated nearby for pollination. Bearing will begin after 5 to 8 years if the plant is from seed or after 2 years after planting a grafted nursery plant. The nuts are perishable and should be stored in the refrigerator in a plastic bag to prevent drying.

Citrus (*Citrus* spp., *Fortunella* spp.)

All citrus are evergreen shrubs or trees adaptable to growing in zones 9 and 10. A few such as the kumquat will survive in zone 8. The major citrus grown in this country are: grapefruit (*Citrus* × *paradisi*), kumquat (*Fortunella margarita, F. japonica*), orange (*Citrus sinensis*), lemon (*Citrus limon*), lime (*Citrus aurantiifolia*), and tangerine or mandarin orange (*Citrus reticulata*).

Most are available in standard, semidwarf, and dwarf sizes and do not require cross-pollination. Citrus require little pruning to maintain a full, round-headed appearance and fruit production. Suckers and crossing limbs may have to be removed occasionally.

Because of their year-round growth cycle, citrus should be fertilized twice per year instead of once. They bear when very young and retain their fruits in excellent condition on the plant for many months.

Figs (*Ficus carica*)

The fig is a deciduous tree or large shrub grown in zones 7 through 10. It can even be grown as far north as zone 5 with winter protection. Although the top will die back to the ground, it will resprout each spring from the roots and produce a crop.

Depending on the variety, the fig can grow from 30 to 80 feet (9.1 to 24.4 meters) at maturity. No dwarf forms are available. The fruit sets parthenocarpically (without pollination) when the plant is about 4 years old. In zones 9 and 10 two crops per year (one in June and one in August–November) can be expected. In colder zones only one will set and is matured in late summer.

Since fig trees can grow very large, pruning is necessary to keep them low enough to harvest the fruit. Young trees should be trained to a low, flat, vase shape, but after achieving this form, no seasonal pruning is recommended.

European Filberts or Hazelnuts (*Corylus avellana*)

The filbert (Figure 7-10) is a small tree or shrub which grows 10 to 15 feet (3–4.6 meters) in height and equal that in width. It is hardy in zones 4 through 9.

Filberts bear separate male and female flowers on the same plant and are pollinated by wind. Although they will set some nuts without cross-pollination, a better set can be expected with it.

Figure 7-10 Filberts are excellent nuts for cultivation on small properties.

Little pruning is required for filberts except for removal of suckers and thinning to permit light to enter the center of the plant. Filberts bear when 2 to 3 years old, and a harvest of 5 to 10 pounds (2.3–4.5 kilograms) of nuts is average for a mature 10- to 15-year-old plant with cross-pollination. The nuts are harvested after drop.

Hickory Nuts (*Carya ovata* and *C. lacinosa*)

Hard-shell hickories are related to pecans; in fact there are many hybrids between them called *hicans*. They are found in zones 4 through 8. Hickory trees are rugged looking, with taproots and a strong central leader form. They bear at from 3 to 4 years of age if grafted but may take 10 to 15 years from seed.

A mature hickory is quite tall (up to 70 feet; 21 meters), and pruning is practical only when the plant is young. Many varieties are self-sterile and require cross-pollination. The nuts are harvested from the ground.

Mulberries (*Morus alba, M. rubra, M. nigra*)

Mulberry trees of one of the above species grow in any zone of the United States. Many are used as ornamentals; fruitless types are sold for this purpose.

The species planted will largely determine its ultimate height. The white mulberry (*Morus alba*) will grow to 80 feet (24.4 meters) and produce white, pinkish, or purple fruits. *Morus nigra* is the shortest at 30 feet (9.1 meters), and the southern red mulberry (*Morus rubra*) is intermediate at 60 feet (18.3 meters).

Mulberries are monoecious, with flowers of separate sexes. Pruning should aim to maintain the naturally spreading branches low so that the fruits will be within reach when ripe. Dwarf forms are occasionally available.

Nectarines (*Prunus persica* var. *nucipersica*)

Nectarines are a mutation of peaches and not a hybrid between a peach and plum, as is popularly believed. They have been cultivated since the days of the Roman empire.

Culture of nectarines is the same as for peaches, except that they are more susceptible to fruit rots and may require more spraying. They do not require cross-pollination and should be trained to a vase shape. Both dwarf and standard sizes are available. Standard trees are moderately small at maturity, 12 to 15 feet (3.7–4.6 meters). USDA zones 5 through 8 are suitable for growing nectarines.

Olives (*Olea europaea*)

Olives will grow in zones 9 and 10 or wherever winter temperatures do not fall below 12°F (−11°C). They require hot summer temperatures to mature their fruits, but they are grown for their ornamental value alone in coastal climates where the high humidity decreases the fruit produced.

Olive trees are small, 25 to 30 feet (7.6 to 9 meters), single- or multitrunked trees with silver-gray foliage. They transplant easily (even at a very old age) and are moderate to fast growing. Pruning does not create any particular form, but it should develop strong scaffold branches and remove suckers which would eventually cause a bushlike form. Dead wood and twigs are common and should be removed yearly.

Cross-pollination is not a prerequisite for olive production. Biannual bearing is common but can be minimized by fruit thinning to leave two to three olives per foot of branch.

Peaches (*Prunus persica*)

Peaches grow in zones 5 through 9, although care must be exercised in selecting a variety with a low winter chilling requirement for zones 8 and 9. They are generally trained to a vase shape atop a short 18-inch (45-centimeter) trunk. Pruning should be done every year to thin the head and keep the top low. Bearing begins at about the third year and will peak during the eighth to twelfth years, with no cross-pollination required. Compared to other fruit trees, peaches are short-lived; trees tend to decline rapidly after 10 to 12 years as a result of winter injury and wood rot infection.

Pears (*Pyrus communis*)

By choosing varieties carefully, pears can be raised in zones 4 through 9, although zones 5 through 8 provide the best climate. Most pear varieties are vulnerable to the bacterial disease "fire blight," and highly susceptible varieties such as 'Bartlett' should not be planted east of the Rocky Mountains.

The trees are long-lived, and although some are self-fertile or set fruit parthenocarpically, most need cross-pollination. Pears are available in dwarf, semidwarf, and standard sizes and should be trained to a modified central leader. They naturally grow very upright, and spreading of the branches may be required.

Pecans (*Carya illinoinensis*)

The pecan, like many other nut trees, has a taproot and a strong central leader growth habit. It is available in only one size and will easily grow to 60 or more feet (18 meters) tall when mature.

Zones 6–8 are best for pecan growing. The tree is hardy even further north, but the growing season is usually not long enough to mature the nuts.

Pecans begin bearing at 4 to 7 years. Trees are monoecious, but more and larger nuts will be borne if cross-pollination is provided. A biannual bearing tendency may develop in older trees. Compared to other nut trees, pecans require relatively high maintenance.

Persimmons (*Diospyros kaki*)

The persimmon is a deciduous tree useful as both an ornamental and a fruit tree in zones 6 through 10 and occasionally in zone 5. The fruits which are apple size and brilliant orange, mature after the leaves drop. Some people question the value of persimmon fruits because of their unpleasant astringency when they look and feel ripe. Astringency is partly varietal, but it can be avoided by ripening the fruits until they are extremely soft before eating.

Persimmons are available only in standard or semidwarf size, with standards growing into roundheaded trees 20 to 60 feet (12.2–18.3 meters) high at maturity. The species has brittle wood and should be pruned to modified leader form for strength. Annual pruning may be used to maintain a small size. Most bear fruit within 2 to 4 years after planting, and most do not require cross-pollination.

Plums (*Prunus domestica* and *P. salicina*)

Plums are of two types, the European (*Prunus domestica*) and the Japanese (*P. salicina*). The European group includes the blue and prune plums including the well-known Damson variety. Japanese plums take in the larger red and yellow plums popular for fresh eating. They are very early bloomers and should be rejected in favor of Europeans in climates where frosts frequently kill fruit tree flowers.

Plum trees are available in both semidwarf and standard sizes. Their training method and pollination requirements vary by type. Europeans are most often trained to a modified central leader and will usually set fruit without a pollinating variety. Japanese usually require crossing with another Japanese variety and are normally trained to be vase-shaped.

Zones 5 through 9 are suitable for plum growing. The trees are long-lived, produce outstanding spring flowers, and can be expected to bear the first crop 3 to 4 years after planting.

Quinces (*Cydonia oblonga*)

The quince was a much more popular fruit a few years ago than it is now. It is used today primarily for preserves and as a flavor enhancer to apples in baking. The fruit is greenish-yellow when mature and about the size and shape of an apple. It is covered with a peach-like fuzz, which is wiped off before the fruit is used. The fruit is picked while hard, preferably after a light frost, and should be treated gently to avoid bruising.

Quinces grow in zones 5 to 9 on spreading trees which grow to 15 feet (4.6 meters) in height. They are slow growing, and little pruning is required beyond the initial vase shape training. The fruit is produced on the current season's growth.

Bearing age for quinces is 3 to 4 years, and about one bushel of fruit per tree can be expected at peak bearing. Cross-pollination is not required, but spraying for diseases and insects is usually necessary because fire blight is a common problem.

Walnuts (*Juglans regia* and *J. nigra*)

Walnuts are of two types: English (Carpathian) (*Juglans regia*) (Figure 7-11) and black (*Juglans nigra*) (Figure 7-12). The English type is the smooth-shelled walnut raised commercially and sold in nut baskets. The black walnut is a rough-shelled, native American tree more commonly grown for shade than nuts. Both are hardy in zones 5 through 9 and reach an average height of 50 feet (15.2 meters) or more when full grown. However, the summer in the colder zones may be insufficient to develop kernels of English walnuts, resulting in empty nuts.

Normally, walnut trees are not pruned except to remove crossing and dead branches; they naturally grow in a vase or modified leader shape. Cross-pollination is not necessary, and the nuts are harvested after they drop. Small crops can be expected starting 2 years after planting.

Care should be taken to plant walnut trees away from ornamental plantings and vegetable gardens. This is necessary because the roots secrete a chemical that is toxic to many other plants.

Figure 7-11 English walnuts.

Figure 7-12 Black walnuts.

Selected References for Additional Reading

Bountiful Ridge Nurseries. *Retail Catalog.* Princess Anne, MD 21853.

California Nursery Company. *Retail Catalog.* Niles District, Fremont, CA 94536.

Carlson, R. F., et al. *North American Apples: Varieties, Rootstocks, Outlook.* East Lansing, MI: Michigan State University Press, 1970.

Childers, N. F. *Modern Fruit Science: Orchard and Small Fruit Culture.* 6th ed. New Brunswick, N J: Horticultural Publications, Rutgers University, 1975.

Cumberland Valley Nurseries. *Retail Catalog.* McMinnville, TN 37110.

Hansen's New Plants. *Retail Catalog.* Fremont, NB 68025.

Hilltop Orchard and Nurseries. *Retail Catalog.* Rt. 2, Hartford, MI 49057.

Hume, H. H. *Citrus Fruits.* New York: Macmillan, 1957.

Jaynes, R. A., ed. *Handbook of North American Nut Trees.* Knoxville, TN: Northern Nut Growers Association, 1975.

Kraft, K., and P. Kraft. *Fruits for the Home Garden.* New York: Morrow, 1975.

Lundy's Nursery. *Retail Catalog.* Rt. 3, Box 35, Live Oak, FL 32060.

Riotte, L. *Nuts for the Food Gardener: Growing Quick Crops Anywhere.* Charlotte, VT: Garden Way, 1975.

Scheer, A. H., and E. M. Juergenson. *Approved Practices in Fruit and Vine Production.* 2d ed. Danville, IL: Interstate, 1976.

Stark Brothers Nurseries. *Retail Catalog.* Louisiana, MO 63353.

Chapter 8

Bush and Other Small Fruits

The "small fruits" are small perennial plants which produce fruit annually. Included in this group are strawberries, grapes, raspberries, blackberries (called "brambles"), and blueberries. Currants, gooseberries, and cranberries are less well-known members. Melons are annual fruits, but they are included in Chapter 6 under "Vegetables."

Because of their size, small fruits can be grown on almost any property. On a large property, enough fruit can be grown for canning and freezing as well as fresh eating. The small fruits are easier to grow than many tree crops. Cross-pollination is seldom required, and spraying for disease and insects is uncommon. Only one or two species are regionally limited, and most bear within 1 or 2 years after planting. By carefully choosing varieties to mature serially throughout the growing season, a harvest of small fruits can be assured from early summer until frost.

PLANNING THE SMALL FRUIT GARDEN

Site Selection

Rapid soil drainage and full sunlight are the two site prerequisites for successful small fruit growing. An area where water stands after a rain is unsatisfactory and will kill plantings quickly. Without full or nearly full sun, the plants will grow only vegetatively and will fail to bloom or ripen fruit. Beyond these considerations, virtually any site is acceptable. Areas with good garden soil are ideal, but even poor-soil areas can be amended to a satisfactory condition with compost or other organic soil additives.

Many gardeners prefer to plant all their fruits in one area. If this is the case, care should be taken to space and arrange the plants to prevent the taller species from shading the smaller species (strawberries). Since the sun is always slightly in the southern sky, this requires placing the taller plants north of the intermediate and short ones. Figure 8-1 gives two sample layouts for fruit garden areas.

Small fruits can also be incorporated in a landscape. Currants, gooseberries, and brambles can be used as hedge plantings which will grow to 4 to 5 feet (1.2–1.5 meters) tall when mature. Highbush and rabbiteye blueberries grow even taller, reaching 10 feet (3 meters). Low-growing strawberries produce lovely white blossoms in spring and make an excellent border for a perennial flower bed. They can also be used as groundcover plants in small sunny areas or in planters around a patio or deck area. Finally, vining grape plants can be trained to an overhead arbor or espaliered against a wall or fence.

Plant Selection

Of the small fruits discussed in this chapter, all but the cranberry are highly climatically adaptable. The choice is therefore a matter of personal preference and, most important, selecting a variety adapted to your climate. The county or state Cooperative Extension Service will furnish literature on the small fruit varieties most successful in your state.

The following sections discuss each of the eight small fruits and their growing requirements.

BLACKBERRIES (*RUBUS* SPP.)

Blackberries are one of the bramble fruits. They look essentially like raspberries except the fruits are larger, and when picked, the center core remains in the berry (with a raspberry it pulls out at picking). They withstand more heat and drought but less cold than raspberries.

Blackberries fall into two categories based on growth, habit, height, and climate adaptation. "Erect" blackberries grow 3 to 5 feet (0.9–1.5 meters) high and grow without supports. The canes of "trailing" blackberries reach 6 to 8 feet (1.8–2.4 meters), and the plants must be grown on a trellis or staked for support. Erect blackberry types are most often grown in the East, Southwest, and Midwest and are the more cold-hardy of the two types, surviving winter temperatures of down to −20°F (−29°C). Trailing types are the traditional blackberries of the Pacific Coast states and southern states, although they are occasionally grown as far north as Michigan.

There is often name confusion with blackberries regarding the names boysenberries, dewberries, and loganberries. Boysenberries are a type of trailing blackberry, *Rubus ursinus* 'Boysen.' They are raised mostly in the West and South, and the fruits are very large. Dewberries are a synonym for the trailing blackberry, whether wild or cultivated. Loganberries (*Rubus ursinus* 'Logan') are thought to be red-fruited mutations of a wild blackberry and are raised exclusively on the West Coast. Whereas most blackberries and other brambles are thorny, a few thornless varieties are available.

BLUEBERRIES (*VACCINIUM* SPP.)

Blueberries will grow from northern Minnesota to Florida and to the West Coast; however, they have a soil pH requirement of between 4.2 and 5.5. Since soils which are naturally this acidic are uncommon, pH modification is essential for blueberry growing in most cases. Chapter 5 discusses the use of ammonium sulfate and other compounds for lowering soil pH. Blueberries also grow best in soils high in organic matter. The addition of organic matter such as compost or peat moss will satisfy this need and help to keep the soil pH low.

The species of blueberry that should be raised depends on the regional climate. Table 8–1 lists the three main cultivated blueberry

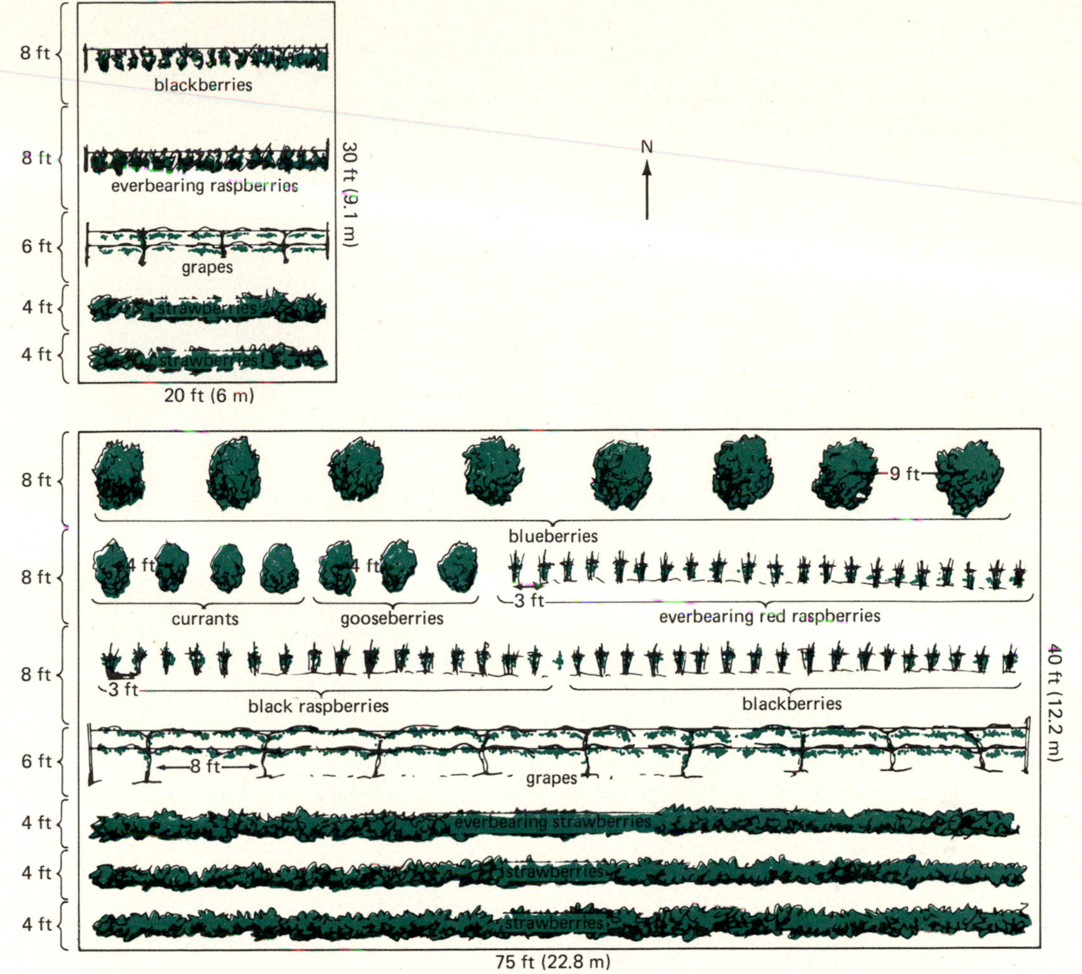

Figure 8-1 Two sample layouts for small fruit gardens.

species and their identifying characteristics and geographic range. Two varieties of blueberries should be planted for cross-pollination, as many are nearly self-sterile.

Besides the cultivated species, additional species such as the dryland blueberry (*V. pallidum*), mountain blueberry (*V. membranaceum*), and lowbush blueberry (*V. angustifolium*) grow in the United States. However, they are normally wild, and varieties selected for home growing should be chosen from the three cultivated species.

CRANBERRIES (*VACCINIUM MACROCARPON*)

Cranberries have exacting growth requirements: moist, peat-bog conditions and a pH of 4.2 to 5.0. Since these conditions are hard to create, cranberries cannot be recommended as a home fruit.

CURRANTS (*RIBES* SPP.)

Currants (Figure 8–2) are very resistant to cold temperature and will survive in even the

Table 8-1 Cultivated Blueberry Species for Home Gardens

Species	Botanical name	Characteristics	Range
Rabbiteye blueberry	*Vaccinium ashei*	6–8′ (1.8–2.4 m) with large fruits; moderately heat- and drought-resistant	Southern states and other areas having a frost-free growing period of at least 160 days
Highbush blueberry	*Vaccinium corymbosum*	10–15′ (3–4.6 m) tall (unpruned); not drought-resistant; sand and peat soil best; hardy to −20°F (−29°C); requires winter chilling and should only be grown north of zone 8	Zones 5–7; East Coast, Pacific Northwest, Midwest
Evergreen blueberry or evergreen huckleberry	*Vaccinium ovatum*	Evergreen with ornamental foliage used by florists; grows to 15′ (4.6 m) tall; fruits are small, shiny, black, and strong-flavored	Pacific Coast from central California to British Columbia

coldest areas. Their growing area spans across the northern United States wherever moist, cool conditions exist. In the South they grow less satisfactorily due to prolonged summer heat.

The plants are small attractive shrubs ranging in height from 3 to 5 feet (0.9–1.5 meters). The plants should be set 4 to 5 feet (1.2–1.5 meters) apart when grown in the garden or about 3 feet (0.9 meter) apart when used as an ornamental hedge.

The most popular currants are the red varieties (*Ribes sativum*), which produce large clusters of fruits and are easy to grow. White (*Ribes sativum*) and black (*Ribes nigrum*) currants are relatively rare, but the black varieties are popular in Europe, where the fruits are made into juice.

GOOSEBERRIES (*RIBES HIRTELLUM* AND *R. UVA-CRISPA* VARIETIES)

The growing requirements of gooseberries are similar to those of currants: moist soil and average-to-cool summers such as are found in the Midwest, North, and northern Pacific states. In warm areas planting in partial shade will reduce summer heat and allow the crop to be raised successfully.

Gooseberries are relatively unknown to most home gardeners. They are shrubs about 3 feet (0.9 meter) high and densely covered with thorns (with the exception of one or two varieties). The fruits are the size of large grapes and are green to red when mature, depending on the variety . They can be eaten fresh, as well as made into pies, jam, or jelly.

Figure 8-2 White currants.

GRAPES (*VITIS* SPP. AND HYBRIDS)

Grapes can be grown for fresh eating, winemaking, or making into jellies and juice. When a variety is selected, the choice should be based on climate and the main use for which the fruit will be raised. Native American grape varieties (*Vitis labrusca*) are the easiest to raise and very productive. The familiar Concord grape is one American type used commonly for jellies and juice, but others are available which are of fresh-eating or winemaking quality.

The French-American hybrid grapes combine the vigor of American varieties with the winemaking quality of European grapes (*Vitus vinifera*). The resulting varieties are excellent for wine use and are raised throughout the country for that purpose.

Finally, muscadine grapes (*Vitis rotundifolia*) are the least common, being confined to the southern states where temperatures do not fall below 10°F (−12°C). The fruits are large and are borne in small clusters.

RASPBERRIES (*RUBUS* SPP.)

Raspberries are the second bramble fruit, growing wild in some areas of the country. Red (*Rubus indaeus*) and black (*Rubus occidentalis*) raspberry varieties are the most common, with the red having a slightly tarter taste than the black. Novelty varieties in yellow and purple are also available.

Besides being available in numerous colors, raspberry varieties (along with strawberries) can be spring-bearing or "everbearing." Everbearing plants produce two harvests per year: one at the normal time in early summer and a second in fall.

STRAWBERRIES (*FRAGARIA CHILOENSIS × ANANASSA* VARIETIES)

Strawberries are one of the most popular small fruits. Although they are perennial, the productive life of a strawberry planting is limited to about 2 to 4 years. The first year following planting, a full crop will be harvested. The second year the plants will bear only two-thirds as much, and the third year only one-third as much. The decline is due to invasion of the plants by viruses which lessen their productivity, hybrid vigor decline, and competition from weeds. The bed should be dug out and replanted.

Like raspberries, strawberries are available in spring-bearing and also everbearing varieties. The everbearers produce lightly throughout the summer until fall, but the quality of the berries is inferior to that of spring-bearing plants.

Amount to Plant

Unlike tree crops, in which one or two plants assure a plentiful harvest, small fruit plants must be ordered in larger quantities. Table 8-2 lists the small fruits, the estimated yield from each plant at maturity, and the suggested number of plants for a family of five.

Preparing and Planting the Area

SOIL IMPROVEMENT

The amount of soil improvement in the fruit-growing area will depend on the present condition of the soil. If the area selected has been recently cultivated as a flower or vegetable garden, little additional soil preparation may be necessary.

However, if an unworked area is chosen, considerable effort may be needed to create a favorable soil environment for the fruits. If practical, a soil test should be taken, and the correct amount of fertilizer and pH-altering materials added to the area based on the results of the soil analysis. Organic matter should be incorporated deeply to loosen a clay soil or to improve the water-holding capacity of a sandy one. If a sod area is selected, the sod should be turned under the fall prior to planting so that it will decay and can be

Table 8-2 Small Fruit Cultivation

Fruit	Minimum planting distance between: Plants	Rows	Life of plants (years)	Estimated yield per mature plant	Suggested number of plants for family of five
Blackberry, erect	4–5′ (1.2–1.5 m)	6–8′ (1.8–2.4 m)	10–12	1 qt (0.9 liter)	15–20
Blackberry, trailing or semitrailing	6–10′ (1.8–3 m)	6–8 (1.8–3 m)	8–10	4–10 qt (3.8–9.5 l)	8–10
Blueberry	6–8′ (1.8–2.4 m)	8–10′ (2.4–3 m)	20 plus	3–4 qt (2.8–3.8 l)	8–10
Currant	4′ (1.2 m)	6–8′ (1.8–2.4 m)	12–15	3–4 qt (2.8–3.8 l)	4–6
Gooseberry	4′ (1.2 m)	6–8′ (1.8–2.4 m)	12–15	4–5 qt (3.8–4.7 l)	4–6
Grape	8–10′ (2.4–3 m)	8–10′ (2.4–3 m)	20 plus	¼–½ bushel (8.8–17.6 l)	5–10
Raspberry, spring-bearing	3–4′ (1–1.2 m)	6–8′ (1.8–2.4 m)	8–10	1–1½ qt (0.9–1.4 l)	20–25
Raspberry, everbearing	3′ (1 m)	8′ (2.4 m)	8–10	1 qt (0.9 l) spring ½ qt (0.5 l) fall	15–20
Strawberry, spring-bearing	1½′ (0.46 m)	4′ (1.2 m)	3–4	½–1 qt/ft of row (1.5–2.7 l/m of row)	100
Strawberry, everbearing	1–1½′ (0.3–0.46 m)	1–1½′ (0.3–0.46 m)	2–3	½ qt (0.47 l)	100

worked into the soil in spring. The success or failure of the crops will depend highly on the soil conditioning prior to planting.

PLANTING

Most small fruits are planted in spring and begin producing fruit the following season. The ground is cultivated in early spring, and the plants are transplanted while dormant, using the spacings listed in Table 8-2, and then watered.

Small fruit varieties are commonly ordered from mail-order nurseries and will arrive without any soil around the roots to decrease shipping cost. Planting is best done immediately, but if it must be delayed, the shipping box should be opened and the plants moistened if they are dry. The plants should then be heeled in, or the roots covered with plastic, and set in a cool (30–60°F; −1–15.6°C) area.

PRUNING AND TRAINING SMALL FRUITS

All the small fruits will produce a better crop if they are pruned and trained. On shrub fruits, pruning may be needed only to remove weak and unproductive wood, whereas on the grape, pruning will be needed to control the productivity and shape of the vine.

Brambles

The pruning of brambles (except everbearers) revolves around the knowledge that the fruit is borne only on 2-year-old canes. These canes grow from the roots the first year, form flower buds that fall, bloom and fruit the following summer, and then die. The fruit is borne on side shoots which grow from buds on the parent cane (Figure 8-3).

On everbearing raspberries, two crops are produced per cane (Figure 8-4). The first crop is borne on the tips of first-year canes in fall, followed the next summer by fruiting further down the cane.

TRAINING

Most brambles (with the exception of trailing blackberries) can be planted as a hedge and allowed to grow without support. But training will help in harvesting by keeping the plants from growing into an impenetrable thicket.

Hill system training (Figure 8-5) is useful for all brambles. With this method a single

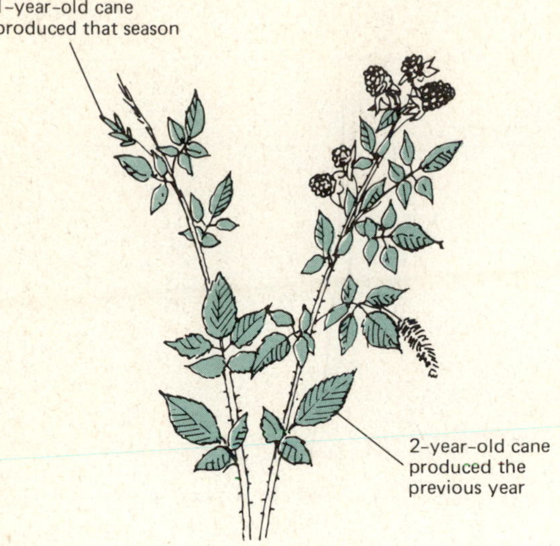

Figure 8-3 **Two-year-old fruiting and one-year-old nonfruiting cans on a typical bramble (red raspberry).**

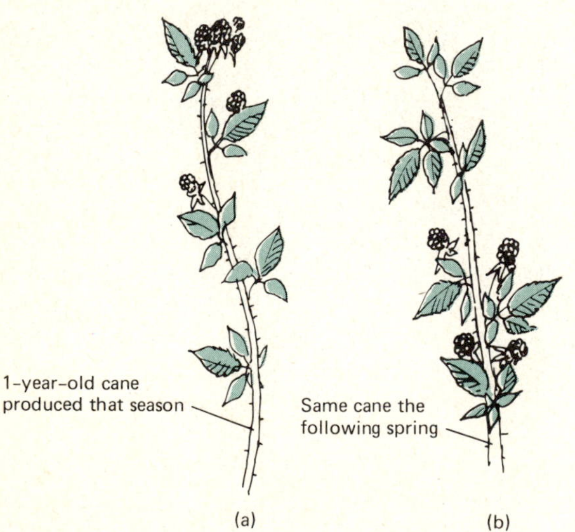

Figure 8-4 **Fruiting habit of an everbearing raspberry. (a) In the fall. (b) In the summer of the following year. Note that fruit is produced on both 1-and 2-year-old canes.**

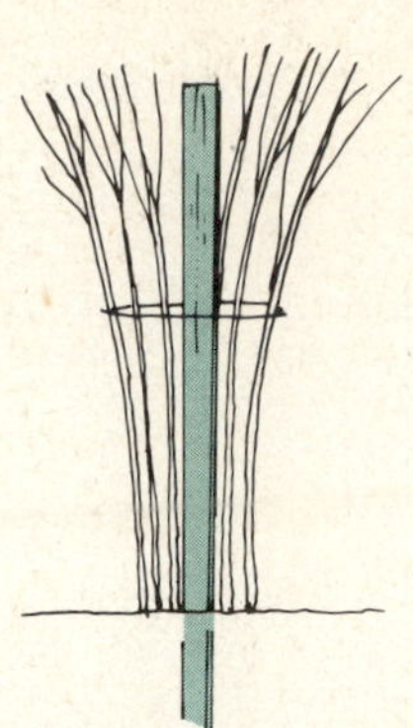

Figure 8-5 **The hill system for training brambles.**

stake driven into the ground supports the plant. Each spring after dormant pruning, the fruiting canes are tied to the stake in one or two places. The plant then maintains an upright form and is easily managed.

Trellis training is more complicated but is more practical for large plantings. A two-wire trellis similar to that used for grapes (Figure 8-6) is constructed, and the fruiting canes are tied to the wire in early spring.

PRUNING

Brambles should be pruned two to three times per year, depending on the species.

Red and Yellow Raspberries and Trailing Blackberries (Figure 8-7). These brambles species need two prunings annually: one in early spring before the break dormancy and the second in summer after harvest is complete.

The spring pruning removes weak, short canes to divert the energy of the plant to fruiting on the strong ones. These remaining strong canes are then cut back to a uniform height of 4 to 5 feet (1.2–1.5 meters) and tied to the stake if desired.

The second pruning (in late summer) consists of cutting the canes which fruited that year back to ground level. Since they die nat-

Figure 8-6 A trellis system for training brambles.

urally soon after fruiting, this pruning does not remove any valuable canes but only encourages the growth of new fruiting canes from the roots.

Black and Purple Raspberries and Erect Blackberries (Figure 8-8). The first pruning for these brambles is in early spring, the same time as for reds, yellows, and trailers. However, the species will look different. Red and yellow raspberries and trailing blackberries produce only long, unbranched canes. But blacks, purples, and erect raspberries will have developed side shoots on their fruiting canes during the past season (Figure 8-8). Spring thinning of weak canes will be needed, along with shortening of these side branches (laterals) on the fruiting canes. For black raspberries, side shoots should be cut back to contain from 8 to 12 buds or to 8 inches (20 centimeters). The side shoots of purples should be left longer (10 to 18 inches; 25–45 centimeters), and any spindly canes removed entirely. Erect blackberries should also be left about 18 inches (45 centimeters) long.

The second pruning for these brambles takes place in early summer before fruiting and involves only the canes produced that spring. These canes are pinched (Figure 8-8) once they reach a height of about 3 feet (1 meter).

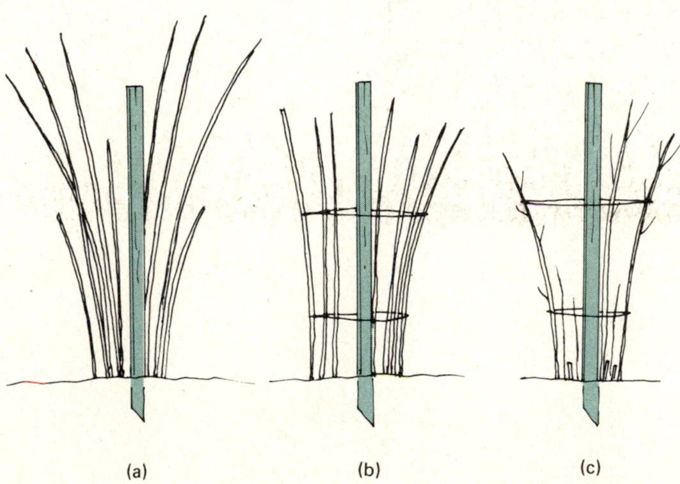

Figure 8-7 Pruning technique for red and yellow raspberries and trailing blackberries. (a) Before pruning (the plant is dormant). (b) Early spring dormant pruning removes weak canes and shortens all remaining canes to 4 to 5 ft (1.2–1.5 m). Canes are tied to the support. (c) Postharvest pruning removes fruited canes to make room for new season's canes.

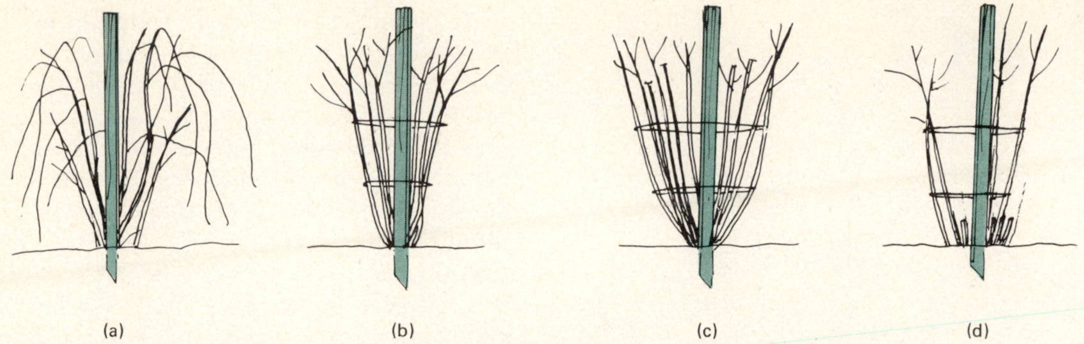

Figure 8-8 Pruning technique for black and purple raspberries and erect blackberries. (a) Before pruning (the plant is dormant). (b) Early spring dormant pruning removes weak canes and shortens side shoots. Canes are tied to the support. (c) Spring pruning after new canes have grown to a height of 3 ft (1 m) pinches tips to encourage branching. (d) Postharvest pruning removes fruited canes.

Pinching stops the canes from becoming excessively tall and encourages growth of side shoots. Pinching should be used only on black and purple raspberries and erect blackberries. Red and yellow raspberries and trailing blackberries form weak laterals when pinched which are easily winter-killed.

The third pruning on vigorous brambles duplicates that used for reds and consists of removing the old canes after they have finished fruiting.

Everbearing Raspberries (Reds, Yellows, and Purples). Everbearing raspberries fruit first on the tips of the current year's canes in fall and further down those same canes the next spring. They should *not* be pinched and should not be pruned to the ground after fall fruiting.

Instead, only dormant spring pruning to eliminate thin weak canes and tip back the canes which fruited in fall should be done. After the spring fruiting of these canes, they can be removed (Figure 8-9).

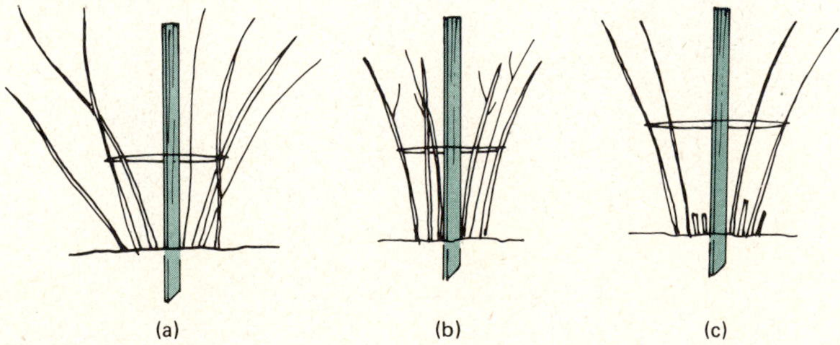

Figure 8-9 Pruning technique for everbearing raspberries. (a) Before pruning (the plant is dormant). (b) Early spring dormant pruning removes weak canes and tips back canes which fruited during the previous fall. (c) After summer fruiting, those canes which fruited are removed.

Blueberries

PRUNING

The blueberry bears fruit on the growth produced the previous season, and the amount of pruning will vary with the species.

Highbush. Highbush blueberries overbear and, if not pruned anually, will yield small berries. In addition, the energy expended in producing the overabundant crop will inhibit vegetative growth and drastically reduce the next year's yield. Accordingly, highbush blueberries should be pruned annually in early spring before new growth starts.

Upright-growing varieties will benefit from thinning out the older center branches (Figure 8-10). This will reduce the berry load, admit more light, and stimulate growth. Drooping lower branches on spreading varieties should be cut back to avoid breakage during fruiting.

Rabbiteye. Rabbiteye blueberries do not require pruning. The bushes are strong enough to mature large crops of fruit without damaging the plant. An exception to the no-pruning rule should be made for older bushes, which benefit when older stems are thinned out lightly.

TRAINING

No trellising or other training is used on blueberries, which grow naturally into large, attractive shrubs bearing white, bell-shaped flowers in spring.

(a)

(b)

Figure 8-10 (a) Before dormant pruning of a highbush blueberry. (b) After pruning.

Currants and Gooseberries

Currants and gooseberries, like brambles, grow into shrubs from canes produced by the roots. The fruit is borne on 1-, 2-, and 3-year-old canes, with older canes becoming progressively less productive.

PRUNING

Pruning both currants and gooseberries should be done once per year before spring growth begins. Pruning should remove wood more than 3 years old. Excess weak wood should also be removed and the bush shaped into a compact shrub 3 to 5 feet (1–1.5 meters) high.

TRAINING

No training (other than pruning to a shrub form) is practiced on currants and gooseberries.

Grapes

Grapes require pruning once per year in late winter. If left unpruned, they overproduce and weaken.

TRAINING AND TRAINING SYSTEMS

There are many methods of training grape vines. The most commonly used are listed here.

Four-Arm Kniffen System. This system uses a two-wire trellis in which the wires are strung between posts in the grape row. It is popular throughout the United States for vigorous native American grapes such as 'Concord'. To train a vine, the strongest new shoot of the vine should be tied to the bottom wire and the rest of the shoots pruned off the first winter after planting (Figure 8-11). The next summer the shoot should reach the top wire, where it is again tied and cut (Figure 8-11) to encourage side shoots. The side shoots which will then develop are trained and tied out on the wire, with a single cane per wire.

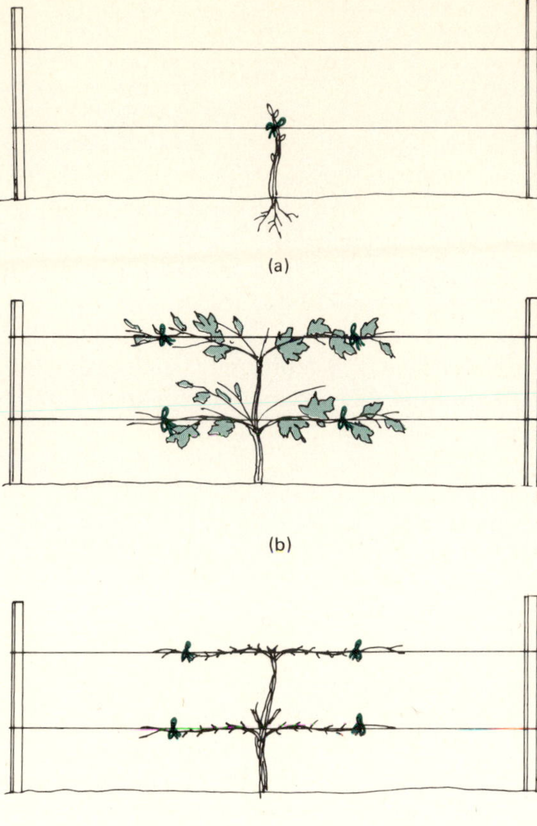

Figure 8-11 Training a grapevine using the four-arm Kniffen system. (a) First winter after planting. (b) The following summer. (c) The following winter after pruning.

After this is done, annual pruning as described in the following section will be sufficient.

Arbor Training (Figure 8-12). Arbor training produces both fruit and shade in the landscape. Since an arbor-trained vine will need to grow to a large size, only a vigorous grape variety should be used.

Arbor training a vine should start at planting. The vine should be trained as described under "Four-Arm Kniffen System." During each growing season that follows the

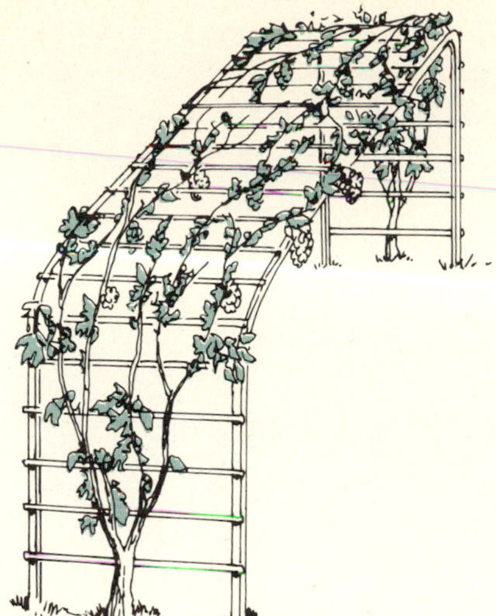

Figure 8-12 A grapevine trained to an overhead arbor.

main trunk shoot is continually trained upward. Gradually it will produce side shoots (fruiting canes) to cover the arbor. These fruiting canes should be spaced by pruning and tied to the support and replaced annually as described in the following section.

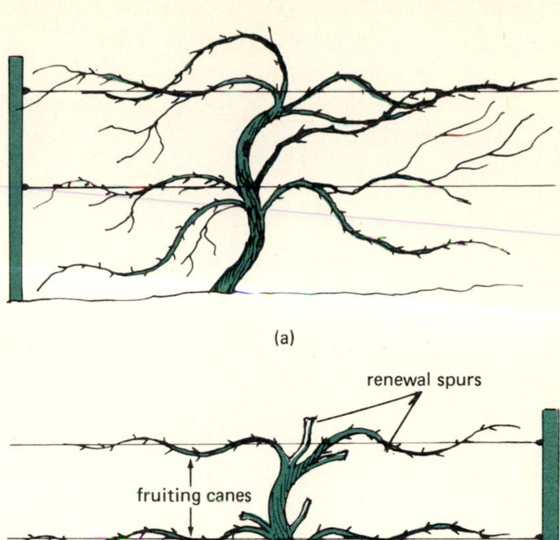

Figure 8-13 Before annual pruning of a grapevine trained by a four-arm Kniffen system. (b) After pruning.

ANNUAL PRUNING

Grape vines bear their fruit only on 2-year-old canes, that is, canes which grew the previous summer. Heavy annual pruning should remove most of the wood growing from the trunk and leave only four shortened canes for fruit bearing, one to grow in each direction along the trellis wires (Figure 8-13).

The canes chosen to remain for fruit bearing should be slightly larger than pencil thickness for best fruit production; they should not necessarily be the largest canes on the vine. Each should be cut to a length that allows for 15 buds. In addition to these four canes, four canes shortened to two to four buds should be left near the two wires. These short canes (called *renewal spurs*) will produce the fruiting canes for the following season.

RENOVATION PRUNING

Grape vines which have been neglected for many years require severe pruning to produce quality fruit again. The plants should be cut back severely before growth starts, removing as much of the accumulated older wood as possible and following the specifications for annual pruning. Any old wood that cannot be removed at this pruning may be taken out with the regular pruning the following year.

Strawberries

TRAINING

Three systems of training are used for strawberries: the matted row, spaced runner, and hill systems. Regardless of which system is used, flowers which form on plants during the year they are planted should be removed instead of being allowed to develop fruit. This will increase the harvest the following year.

Matted Row (Figure 8-14). This system involves transplanting the young strawberries 2 feet (0.6 meter) apart in a row, with the rows 3 to 4 feet (0.9–1.2 meters) apart. The

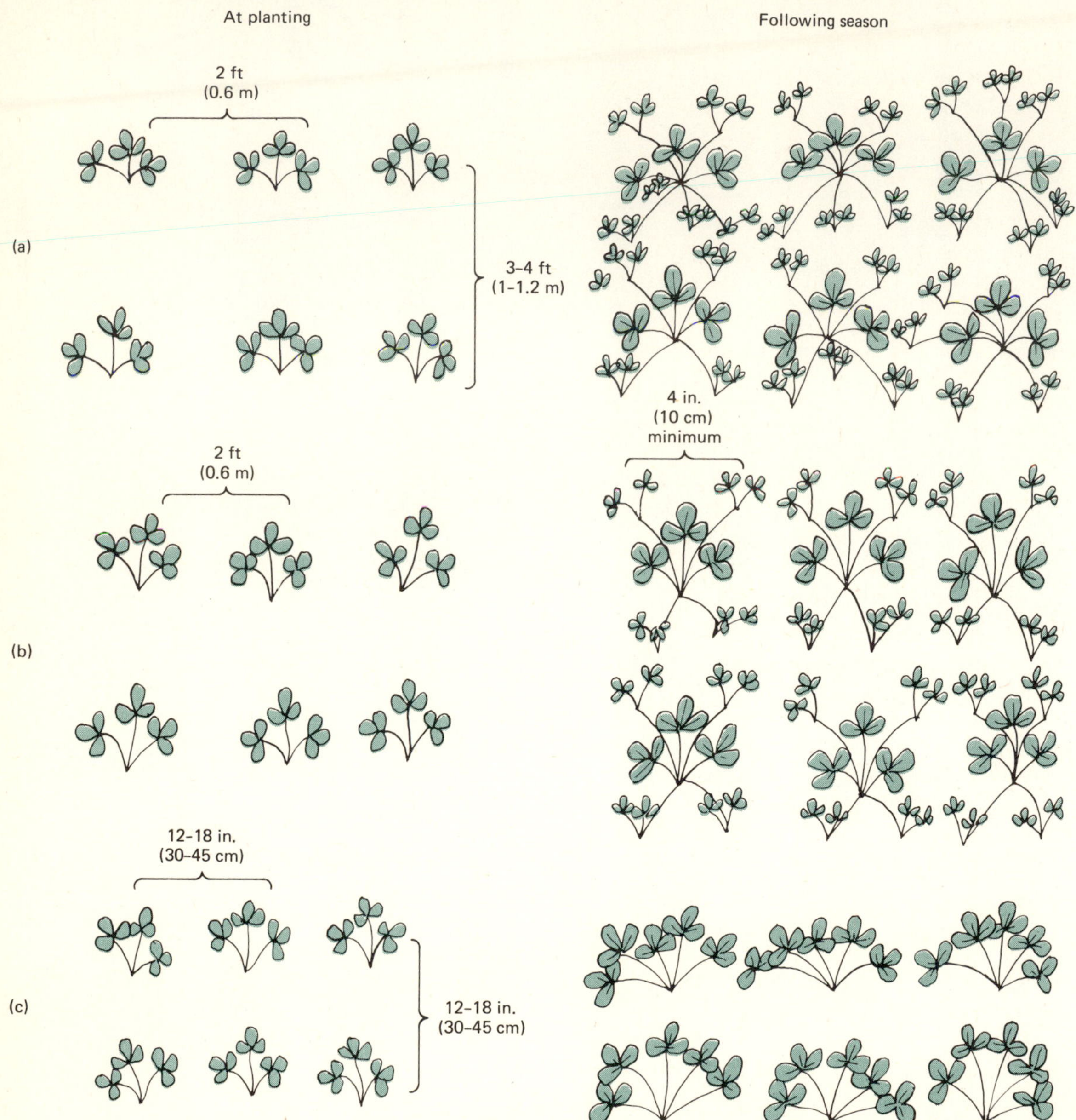

Figure 8-14 Strawberry training systems. (a) Matted row system. (b) Spaced runner system. (c) Hill system.

runners produced in late spring and summer are allowed to root at random, forming a dense mat about 2 feet (0.6 meter) wide. This training system is the least productive but the easiest.

Spaced Runner (Figure 8-14). In the spaced runner system, mother plants are spaced at the same distance as in the matted row system, but the runners are trained to root not closer than 4 inches (10 centimeters) apart. After a row 2 feet (0.6 meter) wide forms, all new runners are removed. This system leaves more room for each individual plant and results in larger and better-quality berries than the matted row. Although there is more work involved, the yield is usually greater.

Hill System (Figure 8-14). The hill system is the best training method for everbearing strawberries and is also useful for spring bearers. The plants are set 1 to 1½ feet (30–45 centimeters) apart in rows 1 to 1½ feet (30–45 centimeters) apart. All flower blossoms and runners which appear the planting year are pinched off, resulting in a strong mother plant that will bear profusely the following season. Yields from strawberries trained to a hill system can sometimes double that from a matted row.

RENEWING THE STRAWBERRY PATCH

Properly renewed, a strawberry planting can last up to 4 years, although the yield will not approach that picked the first season.

Renewal of the patch should begin as soon as harvesting is complete. The old foliage is mowed off to 1 inch (2.5 centimeters) above the crown, and the debris is raked off. A balanced garden fertilizer should be broadcast over the area, all weeds removed, and the rows narrowed to 1 foot (0.3 meter) wide by hoe or rototilling. The patch should be watered thoroughly. New runners for the next year's crop should appear within 1 month.

GENERAL MAINTENANCE OF PERENNIAL SMALL FRUITS

Fertilizing

Most small fruits are shallow-rooted and will benefit from an annual spring application of balanced fertilizer. Sidedressing with granular fertilizer no closer than 6 inches (15 centimeters) from the crown of the plant is advisable for brambles, blueberries, grapes, currants, and gooseberries. Strawberries can be broadcast fertilized and should be watered thoroughly afterward to keep fertilizer granules from burning the leaves. Blueberries should be fertilized annually with an acid-forming fertilizer (such as ammonium sulfate) to keep the pH of the soil in an acceptably low range.

Actual fertilizer rates will vary with the natural soil fertility. General recommendations for plants fertilized with a balanced formula of not more than 15 percent nitrogen are listed in Table 8-3.

Table 8-3 Fertilizer Recommendations for Small Fruits[a]

Fruit	Amount of fertilizer annually
Brambles	½ cup (120 ml) per plant
Blueberries	½–1 cup (120–240 ml) per plant
Currants	¼ cup (60 ml) per plant
Gooseberries	¼ cup (60 ml) per plant
Grapes	1 cup (240 ml) per plant
Strawberries	1 cup (240 ml) per 20 ft (6 m) of 2-ft (0.6-m)-wide row

[a]Recommendations are made for mature plants and are based on a balanced (1-1-1 ratio) fertilizer containing less than 12 percent nitrogen.

The amount of supplemental watering which small fruits require will depend on the natural rainfall. Provisions for supplemental watering should be made in any climate though, since the shallow-rootedness of these fruits makes them particularly susceptible to drought damage.

Figure 8-15 Brambles mulched in for winter in a cold climate.

Small fruits should be mulched in summer with a plastic or organic mulch for weed control and soil moisture retention. In cold climates mulching in late fall for protection against winter injury is also advisable. During a cold winter, mulching can mean the difference between losing an entire fruit planting and retaining a living crown and roots to restore the plant in spring.

Fall mulch should be applied late in the year after temperatures have dropped into the 40°F (5°C) range or lower. If the plants are mulched sooner, the mulch may retain sufficient heat near the roots to continue growth, and resulting tender shoots will be killed.

The shrub-type small fruits (brambles, blueberries, currants, and gooseberries) should be mulched deeply around and covering the crowns and lower stems of the plants (Figure 18-15). Deep mulching will prevent frequent freezing and thawing of both the roots and the lower stem areas.

Strawberries are mulched by completely covering them with straw or a similar lightweight material such as wood shavings. The mulch should be raked from the patch promptly when new growth begins in the spring to admit light for photosynthesis.

Grapes should be mulched thickly in a circle extending out about 4 feet (1.2 meters) from the trunk. This will assure that the entire root zone is protected.

Selected References for Additional Reading

Columbia Basin Nursery. *Retail Catalog.* P.O. Box 458, Quincy, WA 98848.

Farmer's Seed and Nursery Company. *Retail Catalog.* Faribault, MN 55021.

Hill, L. *Fruits and Berries for the Home Garden.* New York: Alfred Knopf, 1977.

Kelley Brothers Nurseries. *Retail Catalog.* 940 Maple St., Dannville, NY 14437.

New York State Fruit Testing Association. *Retail Catalog.* Geneva, NY 14456.

Riotte, L. *The Complete Guide to Growing Berries and Grapes.* Charlotte, VT: Garden Way, 1974.

Savage Farm Nursery. *Retail Catalog.* P.O. Box 125, McMinnville, TN 37110.

Schneider, G. W., and C. C. Scarborough. *Fruit Growing.* Englewood Cliffs, N J: Prentice-Hall, 1960.

R. H. Shumway Company. *Retail Catalog.* Rockford, IL 61101.

Striblings Nursery, Inc. *Retail Catalog.* 1620 W. 16th St., Merced, CA 95340.

Waynesboro Nurseries. *Retail Catalog.* Box 987, Waynesboro, VA 22980.

See also Cooperative Extension Service publications from individual states.

Chapter 9

Flower and Herb Gardening

Flower gardening is a popular type of home gardening, and many gardeners specialize strictly in growing annual and perennial ornamentals. Closely allied to flower gardening are herb, shade, rock, and container gardening. Each is discussed within this chapter, along with roses and flowers for cutting and drying.

TYPES OF GARDEN FLOWERS

Annuals

Every spring there is a brisk sale of annual seedlings for transplanting into home gardens. Although they live for only one season, during that brief time they more than justify their expense by flowering continuously and profusely with minimal care (Table 9-1, Figures 9-1, 9-2, and 9-3).

Commercial breeding of annuals for brighter colors, larger flowers, and different sizes is a continual process. There is an annual to fit almost every garden situation: sun or shade, poor or good soil, 3 inches (8 centimeters) or 3 feet (1 meter) tall. New varieties are evaluated and introduced yearly. Those judged most valuable are given the All-America Selection (AAS) award for their beauty and adaptability to diverse United States climates.

Most annuals are purchased as greenhouse-grown transplants in packages of four, six, or eight plants. By buying seedling plants, the gardener saves several weeks of growing time and can transplant nearly mature plants into the garden for immediate bloom. A few annuals which either do not transplant well or mature very rapidly are grown from seed. Among these are nasturtium, California poppy, and balsam.

Most annuals are not frost-tolerant.

Figure 9-1 'Dolly' marigold. Marigolds are among the most popular and dependable annuals.

Table 9-1 Selected Flowering Annuals

Common name	Botanical name	Cool season/ warm season	Remarks
Ageratum	*Ageratum houstonianum*	Warm or cool	Pink, white, or most commonly blue flowers like small powderpuffs; not frost-tolerant
Snapdragon	*Antirrhinum majus*	Warm or cool	Frost-tolerant; does not transplant well
Calendula	*Calendula officinalis*	Warm or cool	Yellow and orange shades of flowers of a daisylike shape
Bachelor button	*Centaurea cyanus*	Warm or cool	Blue, white, pink colors; frost-tolerant
Marguerite daisy	*Chrysanthemum frutescens*	Warm	White or yellow daisies; very fast growing to 3 ft (1 m) and floriferous
Cosmos	*Cosmos bipinnatus*	Warm	Red through white and orange; many tall and some short varieties
Cape marigold	*Dimorotheca sinuata*	Warm	Frost-tolerant; sow in place; daisylike flowers in orange and yellow
California poppy	*Eschscholzia californica*	Warm or cool	Frost-tolerant; does not transplant well
Gaillardia	*Gaillardia pulchella*	Warm	Orange, yellow, red combinations; good in poor soils
Balsam	*Impatiens balsamina*	Warm	Spike flowers in many colors; fast growing from seed
Sweet pea	*Lathyrus odoratus*	Cool	Pink, white, red, or lavender flowers; vining or bush types
Sweet alyssum	*Lobularia maritima*	Warm or cool	Matlike habit; white or purple fragrant flowers; frost-tolerant
Stock	*Matthiola incana*	Cool	Pink, white, red, lavender, or yellow spike flowers; very fragrant
Petunia	*Petunia × hybrida*	Warm	All colors, bushy or vining varieties
Moss rose	*Portulaca grandiflora*	Warm	Succulent and low-water-requiring; many colors of delicate flowers; creeping habit
Salvia	*Salvia splendens*	Warm	Red and, occasionally, pink or purple shades; many heights available
Marigold	*Tagetes erecta and patula*	Warm	White, orange, yellow, red, and combinations; many heights available
Black-eyed Susan vine	*Thumbergia alata*	Warm or cool	Reliable vine; flowers orange-yellow with black centers; perennial in zones 9 and 10

However, a few, called *cool-season annuals,* are tolerant of light frost and grow better in temperatures in the 50–60°F (10–16°C) range. These annuals (Table 9–1) can be planted in spring before the last frost date. In mild-winter areas they are planted in fall for winter and spring bloom.

Perennials

Many garden flowers are herbaceous perennials with foliage that dies yearly but with a crown and roots which renew the plant the next season. Only two common perennials, roses and tree peonies, are woody.

Figure 9-2 A planting of petunias will bloom continuously.

Unlike annuals, which flower continuously, most perennials have a 2- to 3-week bloom period and flower less profusely. However, they increase in size every year and, in established clumps, create a massive display of flowers and foilage. Perennials are more expensive than annuals. They are sold individually in containers or can be ordered from mail-order nursery catalogs and shipped while dormant. Annuals must be replanted yearly, but perennials are permanent and will continue to flower every year with minimal care. For this reason, they are the mainstay of a flower garden (Table 9-2, Figure 9-4).

Figure 9-3 Impatiens, a favorite annual for shady locations.

Figure 9-4 Poppies, a garden perennial.

Biennials

Only a few popular flowers are biennials. Foxglove and sweet William are two of the better-known examples. Biennials are sold as 1-year-old plants ready for blooming or can be grown from seed.

Cold-hardy Bulbs

Spring-flowering tulips and summer-flowering lilies are two examples of cold-hardy bulbs. This group also includes plants like crocus which are not bulbs in a botanical sense but which function similarly.

Hardy bulbs differ from other perennials in several ways. Unlike perennials which can be planted from spring until fall, bulbs should be moved only in late summer and fall when dormant. They then generate roots in fall and winter and flower the following year. Also, unlike other perennials which retain their foliage throughout the growing season, bulb foliage dies in midsummer, leaving only the underground bulb through the latter part of summer.

The usual source of garden bulbs is a local nursery or mail-order catalog. Although the "major bulbs" like tulip, hyacinth, and daffodil are best known, there are many "minor bulbs" which are easily grown and are a worthwhile addition to any garden. Table 9.3 lists a number of these lesser-known bulbs.

Table 9-2 Selected Flowering Perennials

Common name	Botanical name	USDA zones	Height	Bloom period	Remarks
Astilbe, false spirea	*Astilbe* spp.	3–9	18–36″ (0.5–1 m)	Mid- through late summer	White, pink, lilac, or red feathery plumes
Aster, michelmas daisy	*Aster* spp.	1–10	6″–4′ (15 cm–1.2 m)	Late summer to frost	Many colors
Bellflower	*Campanula* spp.	2–10	1–3′ (0.3–1 m)	Late spring to summer	White or blue bell-shaped flowers
Chrysanthemum	*Chrysanthemum × morifolium*	3–10	12–48″ (0.3–1.2 m)	Late summer to fall	All colors except blue; pinch off flower buds to delay bloom until fall
Feverfew	*Chrysanthemum parthenium*	4–10	2–3′ (0.6–1 m)	Early summer to fall	Small, white flowers with yellow centers; produced profusely
Shasta daisy	*Chrysanthemum × superbum*	2–10	24–36″ (0.6–1 m)	Midsummer to fall	White daisy with yellow center; profuse flowers
Delphinium	*Delphinium elatum*	1–10	24–48″ (0.6–1.2 m)	Mid- to late summer	Blue, white, purple spikes
Pinks	*Dianthus* spp.	2–10	8–18″ (0.2–0.5 m)	Early through late summer	Pink through white and red carnationlike flowers
Leopard's bane	*Doronicum cordatum*	4–10	24″ (0.6 m)	Early spring	Yellow daisy-like flowers; foliage dies in summer
Daylily	*Hemerocallis* spp.	2–10	15″–4′ (0.4–1.2 m)	Early to late summer, depending on variety	Yellow through wine-red flowers; not suitable for cutting, as flowers close at night
Bearded iris	*Iris* spp.	5–10	12–36″ (0.3–1 m)	Early summer	Many colors

Common name	Botanical name	USDA zones	Height	Bloom period	Remarks
Siberian iris	*Iris* spp.	6–10	24″ (0.6 m)	Spring to early summer	Flowers differently shaped than bearded iris
Lythrum	*Lythrum salicaria* and *virgatum*	2–10	24–48″ (0.6–1.2 m)	All summer	Pink through red–purple flowers; tolerant of poor growing conditions including wet soil
Bee balm	*Monarda didyma*	3–7	36″ (1 m)	Midsummer	Pink, lavender, red, or white flowers; leaves have mint smell
Peony	*Paeonia* spp.	4–10	36″ (1 m)	Late spring to early summer	Large globular flowers in white through red
Oriental poppy	*Papaver orientale*	2–8	24–36″ (0.6–1 m)	Early summer	Pink through white and red carnationlike flowers
Hardy phlox	*Phlox paniculata*	4–10	24–40″ (0.6–1 m)	Midsummer	Large flower heads in many colors
False dragonhead	*Phystostegia virginiana*	6–10	36–48″ (1–1.2 m)	Midsummer to fall	Pink or lavender spikes; very vigorous

Summer or Nonhardy Bulbs

Although perennial in mild climates, nonhardy bulbs cannot survive winter temperatures in most parts of the country. Consequently, they are planted in spring for summer flowering and stored in fall for replanting the following year. Tuberous begonias, dahlias, cannas, and gladiolas are included in this group (Figures 9-5 and 9-6). They should be dug from the garden immediately after the first killing frost, dried in the shade, and stored in dry peat moss in a cool 50–60°F (10–16°C) area. Table 9-3 lists selected summer bulbs.

PLANNING THE FLOWER GARDEN

Site Selection

Good soil drainage is an important factor in selecting a site for a flower garden. Few plants

Table 9-3 Selected Bulbs and Related Plants

Common name	Botanical name	Cold hardiness (USDA zones)	Time of bloom	Remarks
Ornamental onion, allium	*Allium* spp.	4–10	Late spring	Many species in white, blue, and pink; globe-shaped flower heads
Belladona lily	*Amaryllis belladonna*	5–10	Summer	Pink lilylike flowers
Anemone	*Anemone* × *hybrida*	6–9	Midspring	3–15″ (7–38 cm) tall with pastel flowers
Tuberous begonia	*Begonia* × *tuberhybrida*	9–10 perennial, 2–10 summer bulb	All summer	Single and double flower varieties and many colors; good in hanging baskets
Caladium	*Caladium* × *hortulanum*	9–10 perennial, all others summer bulb	Foliage only all summer	Grown for foliage only, which is multicolored
Canna	*Canna* hybrids	7–10 perennial, all others summer bulb	All summer	Tall (36–48″) (1–1.2 m) background plants in red, yellow
Glory of the snow	*Chinodoxa luciliae*	3–10	Early spring	Blue, pink, or white flowers; less than 1′ (30 cm) tall
Autumn crocus	*Colchicum autumnale*	4–10	Fall	Planted in August for fall flowers; leaves appear the following spring
Crocus	*Crocus* spp.	3–10	Early spring	Plants 2–4″ (5–10 cm) tall
Dahlia	*Dahlia* hybrids	9–10 perennial, summer bulb all other zones	Summer	Easy to grow; many colors and sizes
Winter aconite	*Eranthis* spp.	4–9	Early spring	Yellow buttercuplike flowers; plants 2–4″ (5–10 cm) tall
Fritillaria	*Fritillaria imperialis*	3–10	Midspring	Unusual and attractive, 2½–4′ (0.7–1.2 m) tall
Gladiola	*Gladiolus* × *hortulanus*	8–10 perennial, summer bulb other zones	Summer	Tall, to 48″ (1.2 m), with spike flowers
Amaryllis	*Hippeastrum* hybrids	9–10 perennial, houseplants all others	Winter	Lilylike flowers on tall stem
Hyacinth	*Hyacinthus orientalis*	4–10	Midspring	Fragrant blooms in many colors, 6–12″ (15–30 cm) tall
Bulbous iris	*Iris* spp.	5–10	Spring	Purple, blue, yellow, white
Garden lily	*Lilium* spp.	3–10	Early summer	Many species and varieties, 12–36″ (0.3–1m) tall

Common name	Botanical name	Cold hardiness (USDA zones)	Time of bloom	Remarks
Grape hyacinth	*Muscari* spp.	2–10	Early spring	Blue, white, or pink flowers on 3–4″ (7–10 cm) plant
Daffodil and narcissus	*Narcissus* spp.	4–10	Spring	Many species available; 12–18″ (30–45 cm) tall
Oxalis	*Oxalis* spp.	6–10, depending on species	Winter in 9–10, spring all others	White, pink, yellow flowers on clover-shaped leaves; plants 2–5″ (12 cm) tall
Ranunculus	*Ranunculus*	8–10 perennial, all others summer bulb	Midspring	Delicate flowers in pastel colors on 12–18″ (30–45 cm) stems
Scilla, squill	*Scilla* spp.	4–10	Spring (depending on species)	Blue and other colors; plants about 6″ (15 cm) tall
Tulip	*Tulipa* spp.	3–7	Early to late spring, depending on variety	Multitude of varieties, 3–18″ (7–45 cm) tall
Calla lily	*Zantedeschia aethiopica*	8–10 perennial, summer bulb all other zones	9–10 winter, other zones summer	Outstanding foliage and white to yellow flowers: 36–48″ (1–1.2m) tall

Figure 9-5 Dahlia. A summer bulb in much of the country, but a perennial in USDA zones 9 and 10 (see Figure 1-5).

Figure 9-6 Tuberous begonia, a well-known summer bulb for shade.

can be grown in wet soil, and it is particularly unsuitable for bulbs. When drainage is questionable, a raised bed is a solution that will not only aid drainage but elevate the flowers for better viewing.

The amount of sunlight an area receives will determine which plants can be successfully grown. A full-sun area is preferable for growing the greatest variety of flowers. Many flowers can be grown in partially or fully shaded locations, but they bloom less profusely and produce more foliage. Others weaken from the lack of light and die. Accordingly, if a shady location is chosen for a flower garden, plant selection must be made carefully and limited to species adapted to shade (Table 9-4).

Finally, the site chosen for a flower bed should be readily viewable. The private living area (see Chapter 10) around a patio or deck is a good choice.

Flower Bed Styles

THE FLOWER BORDER

A border of flowers with shrubs or a fence used as a background is a popular traditional bed

Table 9-4 Selected Shade Garden Plants

Common name	Botanical name	Classification	Remarks
Ferns	*Adiantum, Athryrium, Polystichum,* and the like	Perennial	Many cold-hardy species of varying appearance
Ajuga	*Ajuga reptans*	Perennial	Low growing with blue to pink flowers in spring, depending on variety
Columbine	*Aquilegia* × *hybrida*	Perennial	Unusual-shaped flowers in blues, yellows, pinks, and so on; plants to 30″ (0.7 m) tall
Astilbe	*Astilbe* × *hybrida*	Perennial	Pink or white plumes atop fernlike foliage; plants up to 3′ (1 m) tall
Tuberous begonia	*Begonia* × *tuberhybrida*	Summer bulb	Large weeping plants with outstanding flowers in many colors
English daisy	*Bellis perennis*	Perennial	Small plant which blooms in spring to early summer with white to red flowers

Common name	Botanical name	Classification	Remarks
Bergenia	*Bergenia × schmidtii*	Perennial	Pale pink flowers on an evergreen groundcover plant
Caladium	*Caladium × hortulanum*	Summer bulb	Grown for its foliage, which can be a mixture of red, pink, white, and green
Periwinkle	*Catharanthus roseus*	Annual in all but the mildest climates	Continual blooms in pink to white; grows 8–18″ (20–45 cm) tall, depending on variety
Kaffir lily	*Clivia minata*	Summer bulb	Red flowers atop deep green straplike leaves
Coleus	*Coleus × hybridus*	Annual	Grown for colorful foliage in red, green, yellow, and all shades between
Lily of the valley	*Convallaria majalis*	Perennial	Fragrant bell-shaped white flowers; spreads to form a dense carpet
Cyclamen	*Cyclamen persicum* and other species	Cold-tender perennial	Clumping plant producing nodding flowers in pink, red, lavender, or white
Bleeding heart	*Dicentra spectabilis*	Perennial	Distinctive red and white flowers on plants 1–3′ (0.3–1 m) tall
Fuchsia	*Fuchsia × hybrida*	Annual in all but the mildest climates	A reliable shade performer with flowers in shades of white through red; cascading and upright forms are sold
Coral bells	*Heuchera sanguinea*	Perennial	Continual blooms in red to white; rise to 2′ (0.6 m)
Plantain lily	*Hosta fortunei*	Perennial	Solid green or variegated leaves with fragrant lavender flowers
Impatiens	*Impatiens walleriana*	Annual in most climates	Very reliable bloomer producing red, pink, white, or orange flowers
Forget-me-not	*Myosotis sylvatica*	Annual	Bright blue flowers on a plant up to 2′ (0.6 m) tall
Primrose	*Primula* spp.	Perennial	Spring flowers in many colors in a tight head
Violet	*Viola odorata*	Perennial	Tiny white, yellow, or blue flowers on plants to 6″ (15 cm) tall
Pansy	*Viola × wittrockiana*	Perennial	Low-growing favorite; many varieties available
Calla lily	*Zantedeschia aethiopica*	Summer bulb in all but the mildest climates	Large glossy leaves and white funnel-shaped flowers

Figure 9-7 A flower border.

style. The border can be as much as 10 feet (3.2 meters) wide and is designed to be viewed from only one side. Accordingly, tall plants are used in the rear, moderate-sized ones in the center, and short ones along the front (Figure 9-7).

THE FREESTANDING BED

A freestanding or island flower bed (Figure 9-8) is well adapted to modern landscape designs and is meant to be viewed from all directions. Tall plants should be placed in the center of the bed and gradually shorter plants stepped down to the edge. If a curving bed style is to be used, a garden hose can be moved in different ways to create a pleasing outline. The edge of the bed can then be marked by spading along the hose.

THE FORMAL BED

The formal bed (Figure 9-9) originated in Europe and was used for flowers and herbs around

Figure 9-8 A freestanding flower bed.

Figure 9-9 A formal flower bed.

palaces and large manor houses. It is a tightly restricted, symmetrical style and not commonly used today. Many gardeners prefer formal beds for herb gardening; compact annuals can be easily included also.

Plant Selection

Most gardeners prefer a mixed garden of herbaceous and woody perennials, bulbs, annuals, and sometimes herbs. A mixture combines the virtues of each: the dependability of perennials, the floriferousness of annuals, and the earliness of the hardy bulbs.

HEIGHT

Knowing the ultimate size of the flowers will enable a gardener to select plants of a variety of heights and to locate them in the correct place in the bed. Many annuals are available in several heights, and varieties of the same species can be used as either a foreground or a background plant.

LIGHT LEVEL

A few versatile flowers grow in any light levels, but most are restricted to either shade or sun. Choosing plants which are proven performers under the light available will maximize the beauty of the garden.

CONTINUAL BLOOM

Planning for continual bloom involves choosing a variety of plants so that flowers will appear in succession all season. Bulbs are the usual choice for early spring and chrysanthemums the choice for fall blossoming. The other perennials have 2- to 3-week blooming periods from late spring through summer, and their flowering sequence relative to each other can be found in reference books on flower gardening. Since annuals flower continuously, they tie together the bloom periods of perennials.

PREPARING AND PLANTING THE BED

The preparation needed for a flower garden will depend on the present condition of the soil. Recently gardened soils may need little preparation other than the incorporation of fertilizer. Areas which have been uncultivated or in turf will require more work. The area should be spaded to a depth of about 18 inches (59 centimeters) (the average root depth for perennials) incorporating a large volume of amendment such as compost or peat moss. About ½ cup (100 milliliters) of balanced fertilizer per 10 square feet (1 square meter) should be spread over the area and mixed in the spaded layer.

If the bed adjoins a lawn area, invasion by grass will be a constant problem. Edging material outlining the perimeter will reduce this problem and give a neater appearance. A row of bricks set flush with the ground level or slightly higher is one attractive edging (Figure 9-10). Thin edging boards (bender boards) or plastic edging material is also satisfactory.

Figure 9-10 Bricks used as an edging for a flower bed.

Flowers should be transplanted into the soil at the same level at which they were originally growing, and watered deeply. Bulbs should be planted according to Figure 9-11 or by following the directions on the package. Mark bulb areas with small stakes to avoid disturbing them accidentally when working in the bed.

Annuals grown from seed should be sown according to package directions and thinned to final spacing before they become

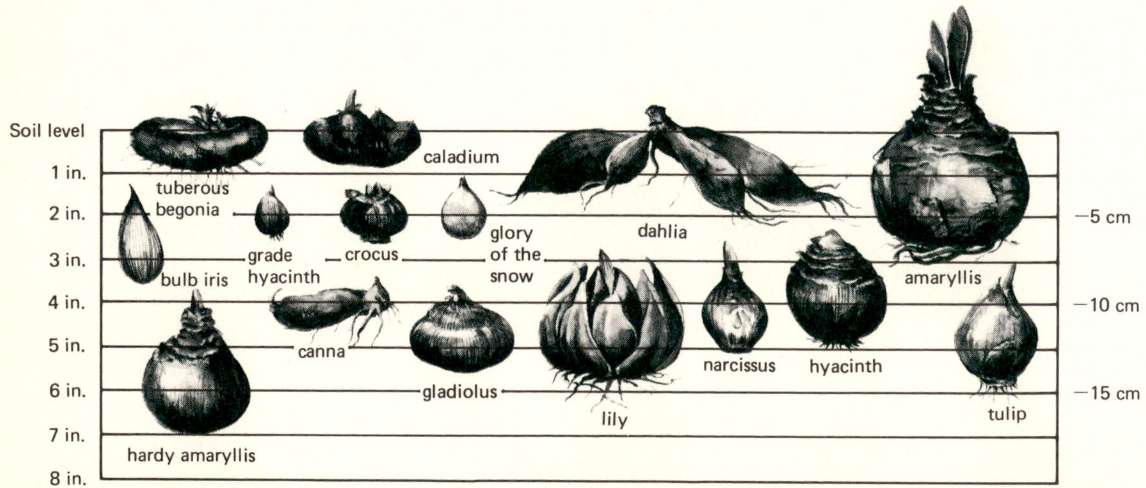

Figure 9-11 Suggested planting depths for common bulbs.

crowded. Only the largest, most vigorous plants should be kept. The others should be pulled out or transplanted.

FLOWER GARDEN MAINTENANCE

Watering

When natural rainfall is not sufficient, watering will be a frequent part of flower garden maintenance. Watering should soak the root zones of the plant as often as necessary to prevent wilting. Because annuals are relatively shallow-rooted, annual beds will require more frequent but shallower irrigation than perennial or bulb beds. In mixed beds, watering should be deep enough to soak the deepest-rooted plants.

Mulching

Mulching with compost, grass clippings, or other materials will conserve water, reduce weed problems, and give a neat appearance to the bed. Mulch should be applied in spring as soon as annuals have been transplanted and new growth of perennials appears. Over the summer the mulch can be renewed as necessary.

In cold-winter areas an additional layer of mulch can be laid down over the bed after the cold weather begins. This layer will reduce freezing and thawing of the bed, a common cause of winterkill. The mulch layer should be pulled back from the crowns of perennial plants in spring as new growth starts.

Fertilizing

Frequency of fertilization of perennials will depend on climate and soil type. An annual fertilization in spring will be sufficient for perennial beds in cold-winter climates grown in loam to clay soils. On sandy soils which lose nutrients rapidly, and in warm climates where growth continues all year, fertilization should be more frequent. Fertilizing every 2 months may be necessary on perennial beds, topdressing with a granular fertilizer.

Fast-growing annuals will benefit from frequent fertilizer applications. Topdressing every 4 to 6 weeks starting at transplanting will assure that sufficient nutrients are available for rapid growth.

Bulb fertilizing can be done at planting or when the bulb is actively growing. Some gardeners mix about a teaspoon (15 milliliters) of high-phosphorus granular fertilizer or bone meal into the soil beneath each bulb as they plant it. Others wait until foliage appears and fertilize by topdressing. Either method is satisfactory.

Pruning

Pruning of annuals, perennials, bulbs, and herbs encourages maximum bloom and keeps the garden neat. Pinching or clipping off dead flower heads should be routine, particularly with annuals. This prevents energy from being diverted into seed making, enhancing bloom production.

Dead leaves should be removed from all plants as necessary to improve the appearance of the garden and to eliminate a source of disease. The leaves of hardy bulbs will yellow and die naturally in midsummer, and they can be cut off at ground level at that time. Leaves should never be cut back until they are almost fully yellow, since to do so would remove the source of energy for the next year's blooms.

Division

The crowns of herbaceous perennials increase in size every year and, after several years, may become overcrowded. Bloom size and number may decrease due to this overcrowding, and the clump should be divided into several smaller pieces and replanted. Chapter 4 gives instructions for division of perennials and bulbs.

Staking and Supporting

Tall flowers and those with profuse or heavy flowers require staking to keep them from breaking and to improve their appearance. Staking will be especially necessary where frequent rains weight the blooms or where wind is common.

One or several stakes may be required to support a plant, depending on its size. Figure 9-12 shows ways of supporting plants by staking. Regardless of which method is used, the stakes should be placed behind or close to the plant so they are hidden by the foliage. The plant should be secured with twine, strips of cloth, or covered-wire twist tabs.

Figure 9-12 Two methods of staking to support a flower.

SPECIALTY GARDENS

Rose Gardens

Roses are the most popular garden flower, so gardens devoted exclusively to rose growing are very common. Many types of roses are grown; the most well-known types are discussed here.

HYBRID TEAS

Hybrid tea roses produce large single blossoms on moderate-sized, bushy plants and are favorites for cut-flower use. They are the most prone to disease problems and are not always reliably cold-hardy (Figure 9-13).

FLORIBUNDAS

Floribundas bear their blooms in clusters through summer and have blooms generally smaller than those of teas (Figure 9-14).

GRANDIFLORA

Grandiflora is the name given to the roses which are crosses between floribundas and

Figure 9-13 A tea rose.

hybrid treas. They produce fewer blooms per cluster than floribundas, but each blossom is nearly as large as a hybrid tea bloom.

CLIMBING AND RAMBLING ROSES

These have long canes which grow from the base of the plant. They grow to a very large size and are used for covering fences and garden arches (Figure 9-15).

TREE ROSES

Tree roses can be hybrid tea, grandiflora, or floribunda roses. They are grafted onto long-cane hardy species to achieve a tree form. Tree roses are frequently grown in pots for patio decoration in summer. In freezing areas they are then buried, still potted, in a sheltered place for overwintering.

MINIATURE ROSES

Despite their frail appearance, miniature roses are very cold-hardy. They produce thumbnail-sized blooms singly or in clusters and can be raised indoors in winter on a sunny windowsill.

ROSE CULTURE

Roses are often high-maintenance plants, requiring regular spraying for diseases and insects and annual pruning. Their cultivation is the subject of many books which deal in depth with the subject. The following gives only a brief overview of rose culture.

Soil. Quick- and thorough-draining soil is essential for roses. Large quantities of amendment should be added if the soil contains a large percentage of clay. Soil should be worked and amended to a depth of 20 to 24 inches (50–60 centimeters), and pH should be slightly acid, in the range of 6.5 to 7.0.

Fertilizing. Granular balanced fertilizer should be applied in spring and after each bloom flush.

Watering. Soil should be kept moist to

Figure 9-14 A floribunda rose.

a 2-foot (60-centimeter) depth. Rose foliage should not be wet since this encourages the spread of fungus diseases such as powdery mildew and blackspot.

Overwintering. Being woody, rose canes live over the winter and renew the foliage in spring. The canes of some types are not reliably cold-hardy though. Bush roses should be mounted with soil to 1 foot (0.3 meter) deep in areas where winter temperatures drop below 10°F (−12°C). Rambler and climber canes should be pinned to the ground and covered with soil.

Pruning. Pruning roses should remove only dead wood, weak canes, and crossing branches. Death of the top portions of canes during winter is a very common problem and will necessitate shortening the canes in spring to living wood. Unless the rose is very vigorous, pruning should be minimal. The removal of large quanities of wood will slow growth and reduce the number of flowers produced.

Rock Gardens

Rock gardening combines rocks and plants to create a natural-appearing area for the cultivation of species native to rocky or alpine regions. In some cases a gardener can make use of naturally rocky land with some reorganization of the available rocks, but more often large rocks must be brought to the site and arranged artistically to give the appear-

Figure 9-15 A climbing rose.

ance of a natural formation. Herbaceous perennials, succulents, bulbs, annuals, and even shrubs and trees (see Table 9-5) are then planted in the small crevices and soil pockets formed between the rocks. Most rock garden plants are small or even minute compared to other garden plants (Figure 9-16). When shrubs or trees are used, they are generally dwarf varieties.

Planting of a newly created site should be postponed until after the soil has settled among the imported rocks. Little soil amending beyond the addition of small amounts of organic amendment will be needed, since most rock garden plants are native to soils of low natural fertility. However, if acid-loving plants are to be grown, pockets between rocks can be amended with peat moss to create the proper soil environment.

Care of a rock garden consists of weeding, dividing large plants, and watering as needed when the soil becomes dry. In general, rock garden plants will need little moisture. Fall care in freezing areas consists of covering the area (with the exception of evergreens) with a mulch layer of hay or other material. The mulch minimizes heaving of the ground and uprooting of the plants.

Container Gardens

Container gardens (Figure 9-17) are a popular way of growing flowers and vegetables without land. Almost any plant can be grown in a pot, including trees, vines, vegetables, flowers, and shrubs.

Because of the limited volume of soil available for root growth in containers, the soil should be of the best possible type. Artificial media (discussed in Chapter 15) are the best choices, or a mixture of natural soil and large amounts of soil amendment. Pure garden soil is seldom satisfactory because of its tendency to pack and drain poorly when confined to a pot.

Supplying water to a container garden is the most time-consuming chore. Water is used up very quickly from the soil and must be replenished often. Daily watering will be necessary during hot weather or when clay or

Table 9-5 Selected Rock Garden Plants

Common name	Botanical name	Type of plant	Remarks
Bugleweed	*Ajuga genevensis* and *repans*	Perennial groundcover	Grows 2–6″ (5–15 cm) high, depending on variety, and thrives in sun or shade; produces blue flowers
Sea pink or thrift	*Armeria maritima*	Evergreen perennial	Small grasslike clumps of deep green leaves produce pink ball-like flower heads
Basket-of-gold	*Aurinia saxatilis*	Perennial groundcover	Low growing and covered with bright yellow flowers
Crocus	*Crocus* spp.	Hardy bulb	Bell-shaped flowers on tiny plants in early spring
Broom	*Cytisus* spp.	Small shrubs, deciduous or evergreen	White to yellow pea-shaped flowers on bright green plants
Garden pink	*Dianthus* spp.	Herbaceous perennial	Flowers pink, white, red, purple on short plants, to 12″ (30 cm) tall
Heaths and heathers	*Erica* spp., *Calluna* spp.	Evergreens	Bell-shaped flowers on plants 1–2′ (30–60 cm) tall; need acid soil
Candytuft	*Iberis sempervirens*	Evergreen groundcover	Deep green leaves and white flower heads
Iris	*Iris reticulata* and other small species	Hardy bulb	Yellow, white, or purple flowers in very early spring on plants 4″ (10 cm) tall
Juniper	*Juniperus* spp.	Needle evergreen	Many types available with blue-green to deep green foliage; choose low or groundcover varieties
Alyssum	*Lobularia maritima*	Annual in all except the mildest climates	White, pink, or purple flowers cover the creeping matlike plant; reseeds naturally
Grape hyacinth	*Muscari* spp.	Hardy bulb	Purple, blue, or white flowers on tiny plants, up to 4″ (10 cm) tall

Common name	Botanical name	Type of plant	Remarks
Creeping phlox or moss pink	*Phlox subulata*	Herbaceous perennial	White or pink flowers produced profusely over a matlike plant
Dwarf alberta spruce	*Picea abies* (dwarf varieties)	Dwarf evergreen tree	Very compact, slow growing, with deep green foliage
Mugo pine	*Pinus mugo* var. *mugo*	Needle evergreen shrub	Shrublike pine growing to 18″ (45 cm) high but spreading widely
Lavender cotton	*Santolina chamaecyparissus*	Herbaceous perennial	Finely cut silver-gray foliage is its outstanding trait; yellow to white flowers in summer
Stonecrop	*Sedum* spp.	Perennial succulent	Species vary from 1″ (2 cm) to 2′ (60 cm) high and have bright flowers
Hen-and-chickens	*Sempervivum* spp.	Perennial succulent	Small rosette plants in green or with tinges of purple or red
Thyme	*Thymus* spp.	Herbaceous perennial or annual	Dark green to gray leaves make up a creeping plant
Species tulip	*Tulipa kaufmanniana*	Hardy bulb	Small plants, about 6″ (15 cm) high, with the traditional tulip flowers

small pots are used.

Frequent fertilization is essential for container-grown plants. A soluble fertilizer (see Chapter 15) applied with the irrigation water is one solution. Topdressing with slow-release or granular fertilizer is also satisfactory.

Overwintering perennials, trees, and shrubs in the North in containers will necessitate winter protection. Being aboveground, the roots will be subjected to lower temperatures than roots growing in the earth, and death from cold is likely. If possible, the pots should be planted in the ground in fall and mulched heavily. If they are not movable, they can be overwintered in a garage or placed in a sheltered location next to a building.

Figure 9-16 A rock garden.

Cutting and Dried Flower Gardens

Although many gardeners harvest flowers directly from their display beds, others prefer cultivating a separate area for flowers to be cut and brought indoors. A cut-flower garden is usually located in the service area of the landscape out of view and includes varied annuals and perennials which lend themselves to arrangement and which will last when cut.

Flowers in the same area can be grown for drying and using in permanent arrangements. These should be harvested at their peak of bloom, bunched loosely with the stems together, and hung upside down in a dry area with good air circulation. Hanging out of direct sunlight will help preserve the natural color. Table 9-6 is a list of flowers suitable for drying and cutting.

Herb Gardens

Herb gardens are among the oldest types of gardens and have recently come back in favor. They are both ornamental and utilitarian, supplying fragrant leaves and seeds for decorating, cooking, teas, and perfuming.

The growing requirements of herbs are less stringent than those of other garden flowers. Full sun is required for best growth, but herbs will perform well in relatively poor, rocky soils and in drought. A few, including the mints, will even grow in wet soils. Table 9-7 lists the more common herbs, their hardiness, size, and uses.

Table 9-6 Selected Garden Flowers for Cutting and Drying

Common name	Botanical name	Annual/perennial	Remarks
Yarrow	*Achillea filipendulina* and *millefolium*	Perennial	Yellow, white, or pink umbrella-shaped flowers; use fresh or dry
Snapdragon	*Antirrhinum majus*	Annual	Spike flowers in many colors; select tall varieties for fresh use
Aster	*Callistephus chinensis*	Annual	For fresh use, with large heads in pink through red and purple; pick disease-resistant varieties
Cockscomb	*Celosia cristata* varieties	Annual	Red or yellow blooms in plume or cockscomb shape; use fresh or dried
Bachelor button	*Centaurea cyanus*	Annual	Blue, pink, or white button-sized flowers for fresh use
Marguerite daisy	*Chrysanthemum frutescens*	Annual in all but the mildest climates	Very fast growing and produces enormous amounts of daisy flowers in white and yellow; for fresh use
Chrysanthemum	*Chrysanthemum* × *morifolium*	Annual or perennial, depending on type	Many colors and forms for fresh use
Larkspur	*Colsolida orientalis*	Annual	Spike-edged white, pink, or purple flowers for fresh use
Cosmos	*Cosmos bipinnatus*	Annual	Tall-growing plant which may need staking; pink through red colors for fresh use

Common name	Botanical name	Annual/perennial	Remarks
Dahlia	*Dahlia* hybrid	Summer bulb	Both dwarfs and full-size varieties are useful; long lasting for fresh use and available in most colors
Gladiola	*Gladiolus* hybrid	Summer bulb	Tall spike flowers in many pastel shades; fresh use only
Globe amaranth	*Gomphrena globosa*	Annual	Cut fresh or for dried use; large cloverlike blossoms in white, pink, red, or purple
Baby's breath	*Gypsophila elegans* and *paniculata*	Annual or perennial	Fresh or dried use, in white or pink; grows into a shrub 3′ (1 m) tall
Strawflower	*Helichrysum bracteatum*	Annual	Use fresh or dried; 2″ (5 cm) blooms in yellow, red, white, purple, and shades between
Helipterum	*Helipterum manglesii*	Annual	Daisylike flowers in white or pink are used for drying
Statice	*Limonum sinuatum*	Annual except in very mild climates	Papery flower clusters in many colors; useful fresh or for drying
Bells of Ireland	*Molucella laevis*	Annual	Green bell-shaped flowers on a 3′ (1 m) plant; use fresh or dry
Narcissus	*Narcissus* spp.	Hardy bulb	Many types available; useful for fresh cutting in early spring
False dragonhead	*Phystosegia virginiana*	Perennial	White, pink, or purple flowers in spikes; a rampant grower for fresh-cut use
Blue salvia	*Salvia farinacea* and *Salvia patens*	Annual in all but mild climates, but reseeds readily	Deep blue spike flowers for use fresh or dried
Pincushion flower	*Scabiosa fischeri*	Annual	White, pink, or blue flower heads on slender stems; for fresh use
African marigold	*Tagetes erecta*	Annual	Yellow, orange, white flowers for fresh use; choose tall varieties; easy to grow
Immortelle	*Xeranthemum annuum*	Annual	Fluffy, papery flowers in pink through red; use fresh or dried
Zinnia	*Zinnia elegans*	Annual	Flower heads 1–4″ in diameter (2–10 cm), depending on variety; available in any color; for fresh use

Figure 9-17 A patio container garden featuring a variety of plants.

Table 9-7 Herb Garden Plants

Common name	Botanical name	Annual/perennial hardiness	Height	Uses
Garlic	*Allium sativum*	Tender perennial zones 9–10 (treat as annual)	To 2′ (0.6 m)	Bulbs used for seasoning
Chives	*Allivum schoenoprasum*	Perennial zones 2–10	To 2′ (0.6 m)	Leaves used for seasoning, very ornamental
Dill	*Anethum graveolens*	Annual all zones	To 4′ (1.2 m)	Leaves used fresh for seasoning and seeds dried for pickling
Tarragon	*Arthemisia dracunculus*	Perennial	To 2′ (0.6 m)	Leaves used for seasoning
Caraway	*Carum carvi*	Biennial zones 3–10	To 2′ (0.6 m)	Seeds used for seasoning
Coriander	*Coriandrum sativum*	Annual	To 3′ (1 m)	Leaves used fresh for seasoning, and seeds also

Common name	Botanical name	Annual/perennial hardiness	Height	Uses
Cumin	*Cuminum cyminum*	Annual	To 6′ (15 cm)	Seeds used as seasoning
Bay	*Laurus nobilis*	Woody shrub zones 7–10	To 2′ (0.6 m)	Leaves used for seasoning
Lavender	*Lavandula angustifolia*	Perennial zones 5–10	To 3′ (1 m)	Leaves used for sachets
Mint	*Mentha* spp.	Perennial zones 3–10	To 2′ (0.6 m)	Leaves used for seasoning and tea
Sweet basil	*Ocimum basilicum*	Annual all zones	To 2′ (0.6 m)	Leaves used for seasoning
Sweet marjoram	*Origana marjorana*	Tender perennial zones 9–10 (treat as annual)	To 2′ (0.6 m)	Leaves used for seasoning
Oregano	*Origana vulgare*	Perennial zones 3–10	To 2½′ (0.8 m)	Leaves used for seasoning
Parsley	*Petroselinum crispum*	Biennial treated as annual	To 15″ (0.4 cm)	Leaves used for seasoning and as garnish
Anise	*Pimpinella anisum*	Annual	To 2′ (0.6 m)	Seeds used for seasoning
Rosemary	*Rosemarinus officinalis*	Tender perennial zones 6–10	To 5′ (1.5 m) (prostrate forms common)	Leaves used for seasoning
Sage	*Salvia officinalis*	Perennial zones 3–10	To 3′ (1 m)	Leaves used for seasoning
Thyme	*Thymus vulgaris*	Perennial zones 3–10	To 1′ (0.3 m)	Leaves used for seasoning
Ginger root	*Zingiber officinale*	Tender perennial zone 10 (treat as annual or houseplant)	To 3′ (1 m)	Root used for seasoning and Chinese cooking

Selected References for Additional Reading

American Rose Society. *The American Rose Annual* (issued yearly by the American Rose Society, P.O. Box 30,000, Shreveport, LA). Kingsport, TN: Kingsport Press, 1977.

Bohm, C. *Rock Garden Flowers, A Concise Guide in Colour.* London: Hamlyn, 1970.

Browne, R. A. *The Common-Sense Guide to Flower Gardening.* New York: Funk and Wagnalls, 1968.

Crockett, J. U. *Annuals.* Time-Life Encyclopedia of Gardening. New York: Time-Life Books, 1971.

Crockett, J. U., and O. Tanner. *Herbs.* Time-Life Encyclopedia of Gardening. Alexandria, VA: Time-Life Books, 1977.

Doerflinger, F. *The Complete Book of Bulb Gardening.* Harrisburg, PA: Stackpole Books, 1973.

Fogg, H. G. W. *Dictionary of Annual Plants.* New York: Drake, 1972.

Hay, R., and P. M. Synge. *The Color Dictionary of Flowers and Plants for Home and Garden.* New York: Crown, 1972.

Miles, B. *Bulbs for the Home Gardener.* New York: Grosset and Dunlap, 1976.

Parcher, E. S. *Shady Gardens: How to Plan and Grow Them.* Boston: Branden Press, 1972.

Parks Seed Company. *Annual Retail Catalog.* Greenwood, SC.

Pierot, S. W. *What Can I Grow in the Shade?* New York: Liveright, 1977.

Schenk, G. *How to Plan, Establish and Maintain Rock Gardens.* Menlo Park, CA: Lane, 1964.

Sunset Books Editors. *Annuals.* 2d ed. Menlo Park, CA: Lane, 1974.

Truex, P. *The City Gardener.* New York: Alfred A. Knopf, 1964.

Wayside Gardens. *Retail Catalog.* Greenwood, SC.

Wilson, H. V. P. *Successful Gardening in the Shade.* New York: Doubleday, 1975.

Chapter 10

Home Landscape Planning and Installation

The importance of planning your landscape cannot be overemphasized. As land becomes more expensive and the lots for houses smaller, haphazard planting of shrubs and trees not only creates a cluttered effect but becomes impractical. In the end it is nearly as expensive and labor-consuming as a landscape based on a plan would have been.

A thoughtfully planned landscape offers many rewards. Most obviously it enhances the design and beauty of the home, thus increasing its value. But a landscape has other functions which are not as immediately apparent. For example, it can extend the living area of the home by a patio or deck screened for privacy and protected from wind and sun. An outdoor play area can provide an outlet for overflowing childhood energy and the noise that normally accompanies it.

From an ecological standpoint, well-placed trees absorb and filter summer sun, reducing the need for artificial air conditioning. Groundcovers prevent erosion of topsoil. In winter, evergreens can shield a house from heat-robbing winds and lessen heating bills.

Where smog is a problem, plantings of shrubs and trees absorb pollutants. They provide relief from noise pollution by reflecting and absorbing sound.

Finally, trees and shrubs shelter squirrels, birds, and other wildlife, and their berries and nuts serve as food. By adding a source of water such as a pool or fountain, all the requirements of small wildlife can be met and many species will inhabit the yard.

THE PROFESSIONAL LANDSCAPE

Landscaping a home is often expensive, because unless one is blessed with a home in a naturally beautiful setting with enough land to assure privacy, extensive planting and construction will frequently be necessary.

If cost is not a strong consideration, a professionally designed and executed landscape is easiest. The plan will be artistic, the shrubs and trees transplanted correctly, and the entire package will come with a guarantee.

But for many homeowners the cost of a professional landscape is prohibitive. For these people, two alternatives should be considered:

(1) paying for a professional plan but installing the landscape personally and (2) planning and installing the landscape yourself. Each has advantages and drawbacks.

A professional design will still cost several hundred dollars, but it is usually money well spent since most people lack the experience and boldness to design a professional-looking landscape. It is, however, possible for a homeowner to design an attractive landscape, provided considerable time and effort are devoted to the project. The following section gives directions and suggestions for the do-it-yourself landscape designer. References listed at the end of the chapter should be used to supplement this section.

THE HOMEOWNER-DESIGNED LANDSCAPE

Needs Analysis

The first step is to compile a needs analysis, that is, an inventory of what should be accomplished by means of the landscape and what the landscape will include, based on the references and life style of the owner's family.

To facilitate the needs analysis, the property should be divided into three areas (Figure 10–1). The "public area" includes all property viewable from the street: the drive, walk, front door, parking area, and sometimes the sides of the house. The "private living area" includes the patio or deck, play areas for children, barbeque and outdoor eating areas, and any game areas. The "service" or "utility area" is the area reserved for necessary but unattractive features such as recreational vehicle storage, dog kennel, and clothesline. A needs analysis for the public area should answer such questions as: "Is privacy needed from the street?" and "Does the entryway need lighting for night entry?" To design the private living area, the life style of the owners should be taken into consideration, asking such questions as: "Does the family entertain frequently outdoors?" and "Is a play area needed for children?"

Figure 10-2 is a checklist covering most of the options one would consider including in a typical landscape. Desired options should be checked on the list, and the completed form compared to the final design to make sure it meets all requirements.

Site Analysis

Site analysis, the second stage of planning, is an inventory of the property in terms of house architecture, views, soil, land slope, lot size, existing landscape, and the like. It lists both good and bad points of the property, so that they will be dealt with effectively in the design. For example, are there established trees which should be incorporated into the design or an air conditioning unit which should be screened from view? Are there poorly draining areas where nothing grows or a steep slope where erosion is a problem? Figure 10-3 serves as a checklist much like the needs analysis sheet.

The Preliminary Drawing

The preliminary drawing shows the basic layout and permanent features of the property and includes the information from the site analysis. A scale drawing of the lot, house, and valued plants should be made on a large sheet of graph paper. This drawing is the skeleton of the design. Sheets of tracing paper can then be laid on top for experimenting with different designs.

The drawing need not look professional to be serviceable; it is more important that it be accurate and include all the necessary

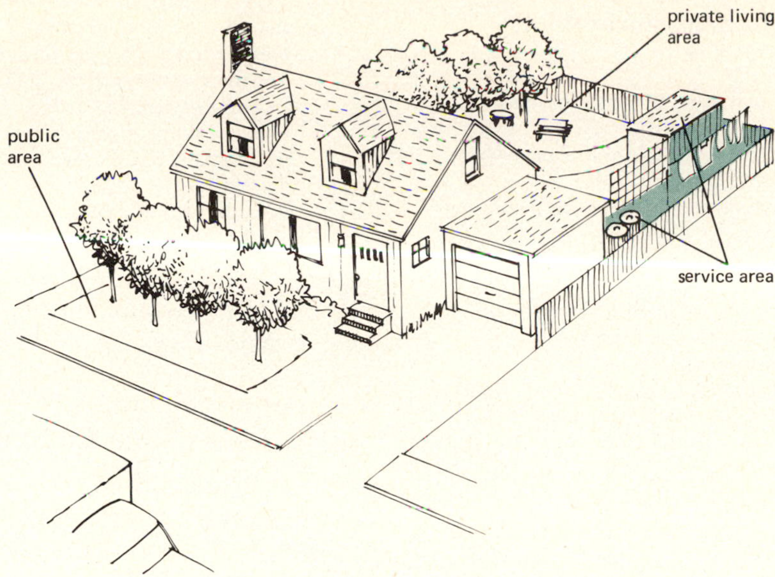

Figure 10-1 The public, private living, and service areas of a typical landscape.

information. Unless there are trees or particularly valued shrubs which will definitely remain, it is best not to include plants in the preliminary drawing. They limit creativity and may influence the designer to see the property only in terms of the present design. However, permanent features such as retainer walls, power lines, good and bad views, and the windows and doors of the house should

Figure 10-2 Needs analysis checklist.

Public area
- night lighting
- privacy from street
- extra parking
- improved walkway/entryway
- entry court
- private living area
- irrigation system
- shade
- lawn
- terracing for changes in level

Private living area
- patio/deck (specify size)
- roof for patio/deck
- shade
- screening from neighbors
- irrigation system
- night lighting
- plants attractive to birds
- settings for ornaments (statuary, birdbath, and the like)
- ornamental or swimming pool
- barbecue grill/fire pit
- children's play area (specify included equipment)
- game area (specify games and space needed)
- view emphasis
- wind control
- terracing for changes in level
- auxiliary patio/deck (specify location)
- lawn

Service/utility area
- recreational vehicle storage
- vegetable garden
- fruit/nut trees
- pet area
- tool storage shed
- clothesline
- trash can storage
- irrigation system
- firewood storage
- cut-flower garden
- greenhouse

Figure 10-3 Site analysis checklist.

Overall lot
1. Is the overall soil condition good, average, or poor?
2. Are there any areas with poor drainage where water stands after a rain?
3. Are there steep or gradual slopes to incorporate into the design?
4. Are there good or objectional views, and from where can they be seen?
5. Is the property higher than, lower than, or equal in elevation to surrounding properties (for screening purposes)?
6. Could grading improve drainage or add interest to the lot?

The house
1. Is the home one story, two story, or split-level?
2. Is the design modern or dated?
3. Could changes in the house (painting, adding natural wood siding, extending windows) improve the effectiveness of the landscape?
4. Where are the windows which will look out on the landscape?
5. Are there unattractive permanent architectural features which should be concealed?

Weather and microclimates
1. In what USDA hardiness zone is the house located?
2. Are there frequent strong winds, and from what direction?
3. Where are the shady and sunny areas of the property?
4. Are there microclimate areas, and where? (See Chapter 1 for a discussion of microclimates.)
5. What is the pattern of sun and shade in the private living area from dawn to dusk?

Existing plantings
1. Are there large existing trees, and where are they located?
2. Do any trees need to be removed for shade, disease, or design reasons?
3. Does the existing lawn area need improvement?
4. Is there any native vegetation to be worked into the design?
5. Are there any plants sufficiently desirable to justify being saved (other than trees)?
6. What are the species of any existing plants which will be saved?

Existing construction
1. Are existing fences and plantings sufficient to provide privacy?
2. Should the drive be replaced or rerouted?
3. Is the entryway to the house sufficient, or should it be redesigned on a larger scale or in a more convenient location?
4. Are there utility meters, natural gas tanks, or air conditioners which require screening?
5. Is any existing construction in need of repair or replacement (fences, walks, retainer walls, and the like)?
6. Does the patio or deck area need enlargement?

be included, since modifying them would involve major changes.

Area Layout

After the basic drawing is complete, the parts of the property to be included in the public, private, and service areas should be chosen. In most established yards these areas will already have been defined through use, but unless there are no other alternatives, it is worthwhile to experiment with shifting the locations of the areas. A service area could be moved from the back to the side of the house, or a private living area from the back to an enclosed courtyard in front. Figure 10-4 shows how a property could be designed with the service area in three different locations.

Public Area Design

The public area is the first area a visitor sees when he approaches. It is this area that conveys an initial impression about the residents of the house. If poorly designed and ill-kept, it will give an impression of sloppiness; if sparsely planted or rigidly pruned, it will convey austerity. If walled as a courtyard, it will project a sense of seclusion.

The main functions of the public area landscape are to blend the house with its surroundings and to provide a pleasant and readily accessible entry to the house. To perform these functions, most landscapes include several basic elements: the entryway and walkway, the driveway, and the auxiliary plantings.

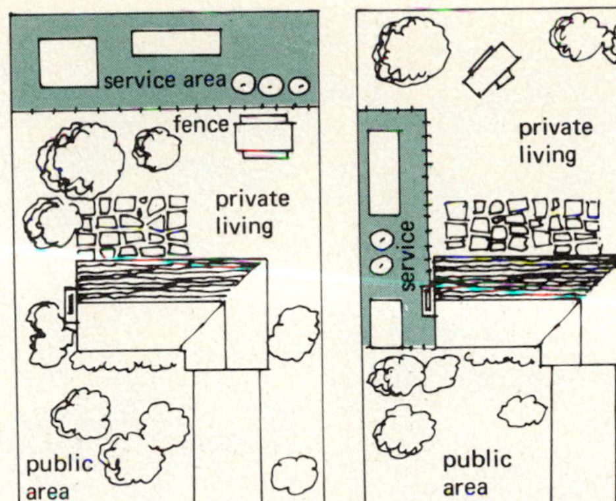

Figure 10-4 Three drawings of the same house, showing how the service area could be positioned in several locations.

ENTRYWAY AND WALKWAY AREAS

In most homes the builder includes a standard narrow walkway leading from the drive and an undersized concrete slab at the door to serve as a waiting area for guests.

Many homeowners consider these permanent features and are reluctant to use a design that involves removing and replacing them. Removing concrete is a major undertaking, but if the walk is too narrow and the waiting area at the door uncomfortably tiny, the labor of removing these features and replacing them with a gracious entry is justified.

At a minimum, any walk should be 4 feet (1.2 meters) wide, the least width that allows two people to walk side by side. Five or six feet (1.5 to 1.8 meters) wide is even better. At this width a person can walk without watching his feet to avoid stepping off the paving. A walk of this width may at first sound unattractive, but the monotony can be broken by small trees framing the walk, groundcover along the edge, or an interesting paving material.

Steps which need to be included should be wide and rise slowly, with threads at least 18 inches (45 centimeters) deep. Monotonous flights of evenly spaced steps should be avoided. Instead, several steps followed by a section of walkway and then more steps should be used to complete the level change.

One favored public area design is the entry patio. It combines the walk and entryway into one paved area interspersed with beds of landscape plants (Figure 10-5). Such patios require little maintenance and create a feeling of spaciousness.

A variation of the entry patio is the

Figure 10-5 An entry patio requires little maintenance and is visually pleasing.

Figure 10-6 An entry court.

closed entry courtyard (Figure 10-6). A wall or fence is used to delineate an area leading to the front door. The area is then planted with shrubs and small trees or paved and designed to include planting beds. An entry courtyard is useful when the front of the house is not overly attractive, when the house is situated close to the street, or when the home has large front windows which would otherwise have an unattractive street view. When the backyard is small, an entry court can also be the private living area.

Before constructing an entry court, it is wise to check the zoning ordinances in the community regarding the construction of walls or fences in the public domain. Occasionally, walls of any type will be prohibited, but more often there are height restrictions.

THE DRIVEWAY

Unless the driveway is a sweeping circular drive or leads through a large property, it will not contribute to the appearance of the public area and should be as inconspicuous as possible. Unnecessary curves should be eliminated, and the edges should never be outlined with flowers. Such treatment serves only to emphasize its presence.

In many homes the driveway doubles as a walkway to the point where the regular walk begins. This is completely acceptable provided the driveway is not blocked by cars; in fact it is preferable to bisecting the yard with a walk to the street. Another effective driveway treatment is to extend the driveway/walkway with a strip of paving (Figure 10-7) merging into a patio attached to the house. Visitors parking in the drive can then step onto a paved area instead of the yard.

AUXILIARY PLANTINGS

Auxiliary plantings are the third element of the public area design. They encompass the trees, shrubs, and groundcovers which are the living part of the landscape.

Trees form the backbone of the landscape, framing and forming a backdrop for the house. If a house has large, well-placed trees, it will always be inviting, regardless of what else is planted in the yard. The cool, restful feeling of a tree-shaded front yard is unequaled.

Figure 10-7 Extending a driveway with a strip of paving creates a graceful entrance and avoids bisecting the lawn.

If large trees are already present, their location must be worked around when designing the landscape. But in a treeless yard the selection and placement are the responsibility of the designer and are of utmost importance.

Evergreen trees such as pines are useful because they retain their foliage throughout the year and are a permanent screen from wind, sun, and street traffic. Deciduous trees, on the other hand, offer shade during the summer months but do not screen out welcome winter sunlight.

For easy maintenance and protection, it is helpful to locate trees in beds of shrubs or groundcover or to surround them with mulch. This eliminates the need for hand trimming grass and avoids the possibility of lawnmower knicks to the trunk.

Trees of moderate to large size should not be planted closer than 20 feet (6 meters) from the house, since their spread and height will increase as they mature. Small trees can be planted closer since their ultimate size is not as great, but they should never be planted under the eave of the house. In a similar manner, trees should never be planted where they will grow into overhead lines.

Auxiliary plantings of shrubs and groundcovers are used in most landscapes. Freestanding beds can be used to screen the house from the street (Figure 10-8), to balance other large plantings in the yard, or as accents. They can be used to incorporate shade trees since they are not usually close to the house.

Planting beds connected to the house can be either corner or foundation plantings. Corner plantings extend out from the front corners of a house, blending it into the site and eliminating the harsh right angle where the house meets the ground. They are among the most common landscape plantings, often including a small tree, several shrubs, and an attractive groundcover (Figure 10-9).

Foundation plantings were a necessity

Figure 10-8 A freestanding planting that screens the house from the street.

in the 1930s, when homes were built with about 18 inches (45 centimeters) of unattractive foundation showing aboveground level. The foundation had to be covered, and the foundation planting concept was formed. The idea persists today, and most homeowners still plant a line of shrubs along the front of their houses. A foundation planting should no longer be considered essential, and planting beds can be designed to cover none, all, or part of the front of the house.

Regardless of whether the bed is freestanding or connects to the house, most beds follow several simple rules. First, the shrubs are used in masses of three or more of the same species and are planted to grow together to form a mass of foliage (Figure 10-9). Second, the planting bed is always delineated by edging or paving, and all the included area is covered with groundcover or mulch. The clean lines and mulch- or groundcovered area create a finished, professional appearance. Third, sweeping lines are used to form large, curved, or geometric beds. The homeowner is frequently timid when designing planting beds, making them small and close to the house. A better effect can be achieved by large beds with bold shapes that jut out into the yard.

GUIDELINES FOR PUBLIC AREA DESIGN

There are many clues that separate the professionally designed from the obviously homeowner-designed landscape. By using a few techniques from professional landscapers, the finished design will be more pleasing.

First, do not try to save every plant in the yard and all the presently paved areas. If the design was poor initially, adding on will not improve it.

Second, if the lot is flat, consider construction such as terracing, mounding, or building fences, walls, or raised beds. The changes in level that construction creates add interest to the yard even before the plants grow and will eliminate the flat, suburban yard feeling.

Third, use only a few species of plants, but many of each of those species. For an average front yard, two to three species of shrubs, one species of shade tree, one to two

Figure 10-9 Corner plantings soften the edges of a house.

species of small trees, and one groundcover should be sufficient. Using individual plants of dozens of species creates an unharmonious and unattractive landscape.

Fourth, use groundcovers and mulches. Often the finishing touch can be put onto a landscape by underplanting trees and shrubs with groundcover or mulching with bark or stone (see Chapter 11 for mulch suggestions).

Fifth, avoid using ornaments. Birdbaths, gazing balls, wrought-iron chairs, and the like attract attention to themselves, deemphasizing the house and the remainder of the landscape. Ornaments should be restricted to the private living area or an enclosed courtyard and incorporated within decorative plantings.

Finally, do not be reluctant to use appealing design ideas from other people's landscapes or from books. Many do-it-yourself landscaping books give sample designs which, with slight altering, could be adapted to any house. The references at the end of the chapter include several idea books which are helpful in this regard.

Private Living Area Design

The private living area of your landscape should be just that: PRIVATE. Screening neighbors from both view and hearing is very important if the living area is to be used to full potential. Screening from wind and shading from sun become likewise important to overall comfort and will affect the usefulness of the area. A well-designed private living area can be pleasant for a Saturday morning breakfast, an afternoon nap in the shade, a dinner buffet at 6, and a party until 2. The elements of the private living area vary enormously with the family and climate, but almost everyone wants a deck or patio, screening for privacy, and shade.

PATIOS AND DECKS

A patio or deck is an outside room with a floor, walls, and a ceiling. The floor is the paving or decking. The walls can be fences, plants, or walls around it, and the ceiling can be the sky, a canopy of tree branches, or a constructed roof.

Most builders locate the patio area behind the house and adjacent to the living and dining areas. Here it is conveniently accessible from the house, an important consideration. However, other factors such as weather and privacy should also be considered when deciding on the location of this important area.

The orientation of the patio on the north, south, east, or west side of the house affects its livability and determines the amount of construction and shading which will be needed. A patio situated to the east of the house is in most cases best, since it is warmed early in the morning by the sun but shaded in the afternoon by the house. A west patio, by comparison, will remain cool through the morning and require shading in the afternoon for comfort.

South-facing patios receive direct sun all day. In warm climates shade will be needed almost all day. However, in cold climates the continual sunniness warms the patio and extends its usefulness into early spring and late fall. Conversely, a northern patio is shaded through most of the year by the house and remains relatively cool. This orientation could be ideal in the Southwest.

The size of a patio can greatly determine its usefulness. Homes are usually built with 12 × 12-foot (3.6 × 3.6-meter) patios, barely adequate for a table and two chairs and room to walk around them. Preferable minimum size is 15 × 15 feet (4.6 × 4.6 meters). The extra 3 feet (1 meter) per side will accommodate a table and four chairs or a picnic table easily, with room left for a lounge chair and/or two regular chairs. Although 15 × 15 feet (4.6 × 4.6 meters) is suggested as a minimum, building an even larger patio is desirable. A spacious patio enhanced by incorporated planting beds (Figure 10-10) will be well used and will become a focal point for gatherings of family and friends.

Figure 10-10 A private patio framed by raised planters.

While a main patio should be large and spacious, auxiliary patios or decks off a bedroom or bath need not be as large. Often they are designed for viewing from inside the house with only occasional usage. A 6 × 10-foot (1.8 × 3-meter) area will be adequate to provide room for a small table and two chairs.

VISUAL SCREENING

Most homes located in an urban or suburban area need screening, not only for privacy but also to keep animals and children in or out and sometimes for wind protection. The traditional fence along the property line does make maximum use of the available land but can create a "boxed in" feeling if not lavishly planted. In addition, local zoning laws may prohibit fences higher than 5 or 6 feet (1.5–1.8 meters) along the property line; if neighboring properties are higher, this may not be sufficient to ensure privacy. Alternatives to property line fencing include berms, raised beds, interior fences, and visual screening by plants.

Berms are mounds or ridges of soil 1 to several feet (0.3–1 meter) high (Figure 10-11). Although some properties have natural berms, usually they are created by grading. They are useful because they add interest to the landscape with elevation changes and, when planted with shrubs and trees or topped by a fence, can stay within zoning restrictions and still provide privacy.

Raised planters (Figure 10-10) of wood, stone, brick, or other materials holding tall shrubs and groundcover can also screen a patio. Almost any plant that will grow in the ground will grow in a raised planter. Moreover, a raised planter is ideal for growing plants with special soil requirements because the bed can be filled with the soil mixture of

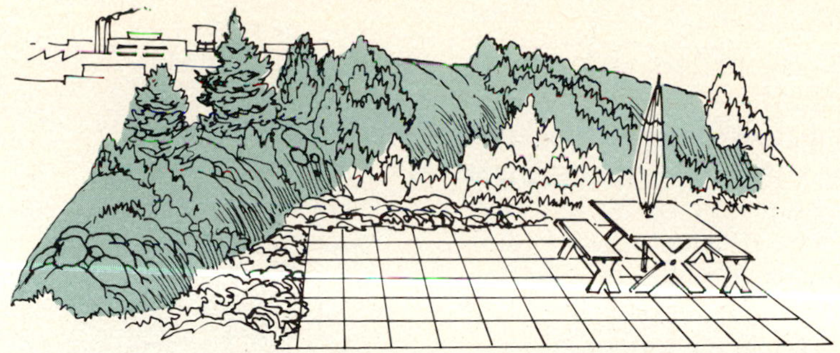

Figure 10-11 A raised berm used to screen a patio.

the owner's choice. In areas with alkaline soil, for example, a raised planter can be filled with an acidic mixture of soil and peat moss for growing azaleas and rhododendrons.

Interior fences are constructed within rather than along the boundary of the property (Figure 10-12). Where existing property line fences are too short, an interior fence can create a small private area and eliminate the necessity of replacing existing fencing. Interior fencing is also useful for screening the service/utility area from the view of the private living area, and it provides an attractive background for plants and garden ornaments.

Plants alone can be used to provide screening, but the effect is not immediate as it is with a fence. For the homeowner with patience, plants are among the best screening materials.

Many species of plants can be used for screening. Pines provide year-round evergreen screening; they can grow up to 100 feet (30 meters) high. A dense hedge of barberry plants will keep animals and trespassers out because of the sharp spines. When choosing plants for screening, ultimate height, denseness, growth rate, and whether the plants are evergreen or deciduous should be considered.

Figure 10-12 An interior fence used to delineate a service area.

SHADE

Afternoon shade is essential to outdoor comfort in most climates. Trees are one way to provide shade, and their location should be chosen carefully to ensure that they will cast their shade at the right time of day and over the desired area. The disadvantage of using trees for shade is that the effect is not immediate. The homeowner must wait several years until his trees reach an appreciable size.

Roofs offer immediate shade, although their cost is relatively high. They can create a rain-proof area for protecting patio furniture which trees cannot. Aluminum and fiberglass are two popular roofing materials. Wood roofs made of 2 × 6's or lath spaced in varying patterns (Figure 10-13) give a feeling of protection without complete shade, and they are architecturally pleasing in both public and private areas.

Figure 10-13 A wooden lath cover for a patio provides shade and creates a feeling of privacy.

OPTIONAL ELEMENTS OF THE PRIVATE LIVING AREA

For a family with youngsters, a play area is an essential feature of a landscape plan and can be included as part of the private living or utility area. The size of the area depends on the number of children who will use it and the total size of the property. For example, 12 feet by 12 feet (3.6 × 3.6 meters) is adequate for one or two children, although more room could be included if the yard was large. Surfacing can be grass, although daily use makes it difficult to maintain in good condition. Useful grass substitutes include smooth pea gravel, bark chips, and sand. A paved ring for tricycle riding might also be useful. When locating the play area in the landscape, it is practical to consider it as a temporary feature used only for a few years when the children are small. Accordingly, it should be placed so it can later be included into the landscape gracefully.

Game areas such as for croquet and badminton require large grassy areas, 24 × 54 feet (7.3 × 16.5 meters) for the latter and 30 × 60 feet (9.1 × 18.3 meters) for the former. This size area may be difficult to include in a small lot, however, a tetherball pole requires only a 20-foot (6-meter)-diameter circle and could be included in most landscapes. Paving material such as concrete or gravel is recommended for tetherball circle areas.

Emphasis of a view may be part of the landscape plan. With mountain or lake properties, plantings can be situated to frame the view (Figure 10-14), with taller plants to the sides and short ones in front.

The Service/Utility Area

A utility area is the part of a landscape that provides room for necessities such as clotheslines, garbage cans, and pet runs. The design inside the area and the space allocated for each element should be functional, and the area should have convenient access. The narrow strips along the sides of most houses are convenient service areas, since they tend to be difficult to landscape. If a vegetable garden

Figure 10-14 Landscaping to frame a view. Note how your attention is focused on the area between the two plantings.

is included, care should be taken that the site chosen will provide maximum sunshine.

A main consideration for the utility area is that it be screened from the private living area. Fences, beds of dense shrubs, and wire supports with climbing vines all perform this function. A baffle effect with sections of fence can substitute for a gate and give easy access while providing complete visual screening (Figure 10-15).

Choosing Plant Species for the Landscape

Selecting plants is the final step in designing a landscape. Establishing the basic design, with patios, walkways, and the like drawn in the plan, is preliminary; the plants to be included at this stage should be represented only by circles or similar symbols and labeled tree, shrub, groundcover, or whatever. The actual choice of plant species is made later to avoid the confusion of coordinating plant species and design elements at the same time. Regardless of whether the plant is a tree, shrub, or vine, the following factors should be taken into account before specifying plants in the design.

First, how adaptable is the plant to your climate? Winter hardiness is the most limiting

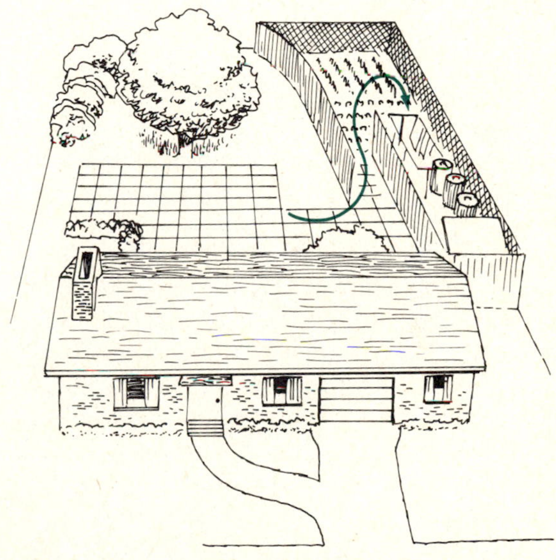

Figure 10-15 Sections of fence used as baffles to hide a service area.

climatic factor and the one over which there is least control. Choosing a plant that is not completely cold-hardy condemns the homeowner to extensive pruning of winter-killed wood every spring and may result in the complete death of the plant during a particularly cold year.

Rainfall is a second climatic characteristic to be considered. Natural rainfall in the eastern and southern parts of the country provides sufficient moisture for most plants. Irrigation is often necessary in the West, where water use for landscaping is fast becoming a luxury. Using drought-resistant plants to eliminate the majority of landscape water use works with, rather than against, the natural climate.

Smog tolerance can also be a limiting factor in the choice of plant species. Although smog will rarely kill a plant, different species including pines can be severely damaged. Most often leaf margins or areas between the veins brown, leaving the plants burnt-looking or unattractive. Since there is no protection against smog damage, choosing smog-tolerant plants is recommended where there are high levels of air pollution.

Native soil conditions are an environmental factor that can affect the growth of many plants. pH, the acidity or alkalinity of the soil, can be a serious hindrance. Plants native to acid soils (azalea and rhododendron, for example) planted in alkaline soils will turn yellow from inability to absorb nutrients at that pH range (see Chapter 13). These plants should be avoided in favor of more adaptable species or placed in raised planters in which the soil can be modified to the correct pH.

The relative amount of sun and shade the property receives should be considered when selecting plants. A house built in a wooded area will be largely shady, so shade-tolerant plants should be selected.

Resistance to pests and diseases is also important. Some species have a natural susceptibility to insect or disease problems, whereas others remain unaffected. To a large extent, disease and insect susceptibility is determined by the area in which the plant is being grown; species grown out of their native habitat are frequently prone to problems. A local nurseryman can offer advice regarding the adaptability of specific ornamentals to your area.

Not all landscape plants require the same amount of maintenance. Selecting the correct species can minimize the amount of time spent on maintenance chores like pruning and raking. High-maintenance plants are those which drop large quantities of dead flowers and fruits or which require frequent pruning of dead wood and unruly growth.

The mature height and spread of plants and their growth rate are often overlooked when homeowners choose plants. Only when the mature sizes of plants are known can they be intelligently placed in the design to blend without crowding. If plants are placed too close together, frequent pruning will be necessary to keep them within bounds; if they are placed too far apart, they will never grow large enough to create the intended effect.

Growth rate is a function of both the species and the amount of water and fertilizer received. If the plant has a specific function to fulfill, such as screening, fast-growing species will be desirable. Similarly, fast-maturing trees are frequently desirable because they provide shade quickly. However, ordinarily, moderate- or slow-growing plants are preferred over fast-growing ones because the latter tend to be weaker plants with short life spans.

A personal preference for the plant is, of course, important when selecting plant species to be included in the landscape. The attractiveness of the foliage, size of individual leaves (fine or coarse texture), and color should be noted, along with outstanding flowers, fruits, or fall color. An assortment of plants with varying foliage textures and colors is preferred to give interest to the landscape throughout the year.

CHOOSING TREE SPECIES

Trees can be divided into three sizes: large 45+ feet; 14+ meters), medium (30–45 feet; 9–14 meters), and small (less than 30 feet; 9 meters). Large and moderate-sized trees are used primarily for shade, although they can be highly ornamental and are then called "specimens." Trees in the small category can be up to 30 feet (9 meters) tall but are frequently in the 10- to 15-foot (3- to 4.5-meter) range. Small trees yield some shade when mature and are often used by patios and decks. "Patio trees" frequently have ornamental flowers or fruits and are pruned when young to be multitrunked (Figure 10-16).

Shade trees should be selected so that they will complement the scale of the house when fully grown. One-story houses are complemented by trees in the small and medium categories. Owners of two-story, split-level, and larger houses should choose shade trees of moderate or large size.

Figure 10-16 A multitrunked magnolia in a landscape setting.

Trees vary in the density of the shade they cast. Those with round heads of large leaves cast deep shade, whereas trees with an open branch structure and small leaves cast "filtered" shade. Trees should be selected which will provide the proper amount of shade for the climate; the denser the shade, the more difficult it will be to grow a lawn or landscape plants underneath.

Root systems are another consideration in tree selection. Some trees have "invasive roots" which grow vigorously and can clog a septic tank or sewer pipe. Planting such trees should be avoided within 30 feet (9 meters) of wells or sewer systems.

Other trees produce "surface roots" radiating from the trunk. The roots can crack driveways or walkways (Figure 10-17), and trees known to have them should not be planted near a paved area.

CHOOSING SHRUB SPECIES

Whereas trees provide height and shade, shrubs provide mass and bulk. They are divided into three sizes: small (less than 3 feet; 1 meter), medium (3 to 6 feet; 1-2 meters), and large (6 to 12 feet; 2-3.6 meters). Small and medium-sized shrubs are most frequently used in home landscapes. Those in the large category can be up to twice as wide as they are tall and are suited to commercial buildings and large properties. They can, however, be pruned to remove their bottom branches and used as small trees.

Along with small trees, shrubs are the main flowering plants in landscapes, and a design should include at least a few flowering species. Evergreen species are essential as well. Their foliage persists throughout the winter, making them the backbone of the landscape.

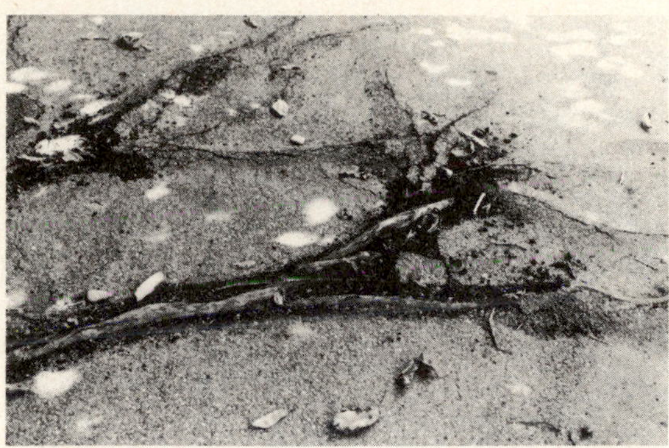

Figure 10-17 Tree roots which have cracked a paved area.

SELECTING GROUNDCOVER SPECIES

Groundcover plants are low-growing species which add variation in texture and unify groupings of larger plants. They give a finished appearance to a landscape, prevent weed establishment, and act as a living mulch. Although most groundcovers are less than 6 inches (15 centimeters) tall, they can be considerably larger: up to about 2 feet (0.6 meter). Shorter groundcovers usually are well suited to use on small properties; taller species are better suited to large properties.

A groundcover should be selected for disease and insect resistance and suitability to climate; it should also blend well with the other landscape plants. A groundcover should be unobtrusive—visually harmonizing, rather than competing, with other plants in the landscape. The choice of an evergreen versus a deciduous groundcover is a matter of personal preference. In an area with persistent snow cover, the groundcover will be hidden throughout winter anyway, but in other areas the evergreen foliage will be pleasing in the winter.

CHOOSING VINES

Vines are not often included in landscapes, although they can be a very attractive feature trailing along the ground, climbing up a wall, or supported by a trellis or arbor. Vines can form the floor, walls, or ceiling of the garden. They can provide screening and shade, supply colorful flowers or fruits, and soften a wall, among other uses.

The methods with which a vine climbs will determine whether it will require a supporting framework. The stems of some vines grow small "holdfasts" which look like small suction cups and anchor them to any flat surface. Others have adventitious roots along the stems. These vines adhere to a wall or climb a fence without additional support.

Other vines climb by twining. The entire vine may grow spirally or may send out tendrils which twine to support the rest of the vine. Vines in this category need a trellis or other means of support.

OBTAINING LANDSCAPE PLANTS

Most people purchase the plants they need for their landscape, but plants can also be propagated at home or transplanted from the wild or from friends' yards.

Home Propagation

Propagating woody plants by cuttings can be difficult for a home gardener. However, cuttings of some shrubs and most herbaceous groundcovers are relatively easy to root. If a source of cuttings can be found, it is practical to propagate the groundcover needed for the landscape at home

Transplanting from the Wild

Transplanting established trees and shrubs can be successful if the plants are small. Trees should be less than 10 feet (3 meters) tall and preferably growing in full sun, since a move from a shady woodland to a sunny yard can cause serious injury. Cool, rainy weather is best for transplanting (early spring in most cases, winter in warm climates where the ground does not freeze). If the tree is deciduous, it should be moved while leafless. Evergreens should be moved before a period of active growth.

The first step is to cut through the roots with a spade in a circle about 15 inches (40 centimeters) from the trunk. The tree can then be uprooted, keeping as much soil around the roots as possible. The soil ball should then be wrapped in plastic or wet burlap for transport to the new location. Replanting should be done immediately, following the suggestions given for balled and burlapped plants in this chapter. Spraying with an antidesiccant (available at nurseries) will slow transpiration in evergreens and aid their survival.

Purchasing Plants from Local and Mail-Order Nurseries

Both local and mail-order nurseries are good sources of plants, but each has advantages and disadvantages. Local nurseries may not have the selection of mail-order ones, but the buyer can see the plant and be assured of its quality. Often plants from a local nursery have been grown in containers and can be transplanted with very little chance of loss. Mail-order nurseries dig the plants from the field and ship them with the roots wrapped in damp wood shavings. The plants may be subjected to extremes of heat or cold during shipping and are less likely to survive transplanting.

As a compromise, many gardeners patronize local nurseries for more common plants but use mail-order nurseries for hard-to-find species. However, you should not be reluctant to ask for a species or variety which a local nursery does not normally carry. The salesperson will usually be willing to order it, particularly if you are willing to buy a sizable number (10 or more) of the species.

The nursery section of a discount store is another source of plants, and bargains are available there with careful shopping. These stores order rather than grow their own plants. You should ask the manager when the nursery shipment is due and inspect the plants a day or two after they arrive. Plants frequently receive poor care in discount stores, and what were originally healthy plants may deteriorate to poor quality within a couple of weeks after arrival.

PLANTING THE LANDSCAPE

Landscapes are planted in spring in most parts of the country. At this time the plants will be actively growing and will transplant with minimal shock. The ground will be thawed (in the North), and the temperature will be relatively cool, thereby slowing transpiration and preventing wilting. Fall is the second-best time for transplanting, again because of the cool temperatures. This prime planting period extends through winter in any climate where the ground does not freeze. Summer is the least preferable time for planting a landscape, since sun and high temperatures make the problems of wilting more acute. However, if container-grown plants are to be used exclusively, the roots will be practically undisturbed and will grow even in summer.

Transplanting Trees and Shrubs

Shrubs and trees are sold bare root, balled and burlapped, or container grown. The selling method determines the price, the chances of surviving transplanting, and the seasons for transplanting.

BARE ROOT

Only deciduous plants are sold bare root, and they are always dug and sold while dormant. Most bare-root plants are sold only in spring, however, occasionally they are available in fall or throughout winter in areas where the ground does not freeze. Bare root is an inexpensive way to buy plants, since they are grown in fields with minimal maintenance and simply dug up when the selling season arrives. Most mail-order nurseries utilize this selling method because it minimizes shipping weight.

Because bare-root plants are dug from the ground, many roots are lost, so a bare-root plant is more likely to die during transplanting than a container-grown one. It must be given particularly good care to assure its survival and establishment in the landscape. When selecting a bare-root plant from a nursery, you should pick the sturdiest plant with the largest root system. The roots should be covered with or bagged in wood shavings and be damp to the touch. If they have dried, it is likely that the plant may be dead. You should also scratch a tiny area of bark with your fingernail. If the plant is alive, a green layer will be visible beneath the surface.

A bare-root plant should be planted as soon as possible. If planting cannot take place for several days, the plant should be left in a shady area with the roots covered with moist soil. This "heeling in" prevents drying and the death of the roots.

At planting time the roots should be placed in a bucket of water and the tree taken to the planting site. A hole should be dug large enough to accommodate the roots easily without crowding (Figure 10-18). The topsoil should be piled on a piece of canvas or plastic next to the hole and the subsoil discarded. Next, the topsoil should be mixed with enough soil amendment to make a mixture to refill the hole. Part of the mixture should then be shoveled back into the hole so that when the plant is placed in the hole, it will be sitting at or slightly above its original growing level.

Root pruning should take place next, with any badly broken roots clipped off with

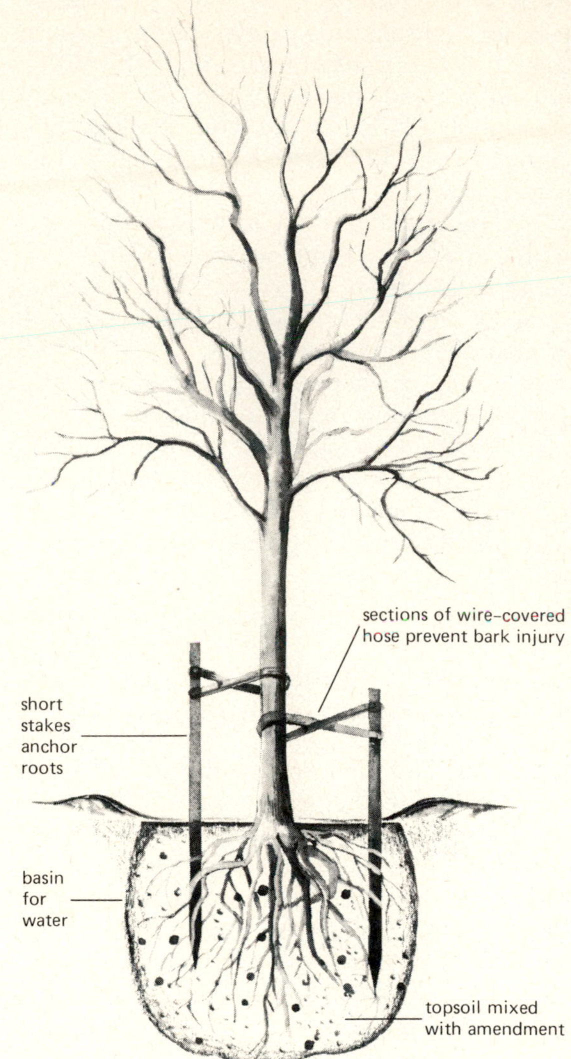

Figure 10-18 A bare-root tree after transplanting.

hand pruners. The remaining roots should be spread outward in the hole and covered with amended soil to ground level. A small ridge of soil should be mounded around the planting hole as a basin for water. Finally, the plant should be watered in well by filling the basin three or four times.

BALLED AND BURLAPPED PLANTS

Balled and burlapped (Figure 10-19) refers to trees and shrubs which are grown in a field but are dug to keep a ball of soil around the roots. Both evergreens and deciduous plants are balled and burlapped; since the process involves considerable labor, plants sold this way are expensive. Because the digging disturbs only part of the roots, balled and burlapped plants usually transplant successfully provided the ball remains moist and proper planting procedures are used.

Hole and soil preparation is basically the same for balled and burlapped plants as for bare-root plants, with the stipulation that the hole should be twice as wide as the ball and one to one and one-half times as deep. Balled and burlapped plants are heavy, and care must be taken when transplanting to prevent the soil ball from breaking. Once the plant is positioned in the hole, the strings can be cut and removed and the burlap peeled back from the trunk. If nails have been used to secure the burlap, they will rust in the soil and need not be removed.

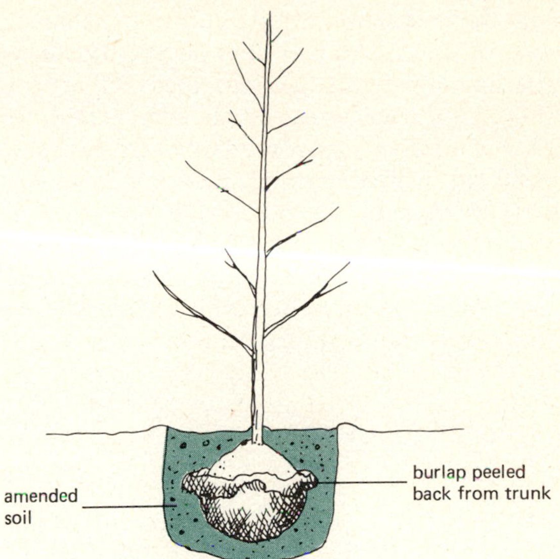

Figure 10-19 A balled and burlapped tree after planting.

CONTAINER-GROWN PLANTS

Many plants are grown and sold in containers. Because the root system develops within the container, the plant is only slightly disturbed at transplanting and can be moved whenever the soil is workable.

As with balled and burlapped plants, a planting hole at least twice as big as the container should be dug, then filled with amended soil to give the roots fertile, loose soil for reestablishment. Removal of the root ball from the container without breakage is the only difficult part of transplanting. With a plastic container, the plant can be turned upside down and will slide out easily with little shaking. For plants in metal cans, it is easiest to cut the container down opposite sides, peel it back, and remove the plant. Many nurseries cut the can at the time the plant is purchased, saving the inconvenience of cutting it at home.

BOXED TREES

Occasionally, you will see large trees for sale with their roots in boxes. These plants may have been container grown or field grown. With the latter, the boxing is in lieu of balling and burlapping.

Planting hole preparation should be the same as for balled and burlapped plants. Removal of the box must be done carefully to avoid breaking the soil ball. First, the tree should be turned on its side and the bottom of the box removed. Using a board as an inclined plane, the box should then be slid onto the planting hole. Once in place, the sides of the box can be removed and the hole filled with amended soil.

Pruning Newly Transplanted Trees and Shrubs

Pruning newly transplanted trees and shrubs encourages them to develop proper form (follow the instructions for pruning in Chapter 11). In the case of bare-root plants, it reduces the amount of top growth to a level which the sparse root system can support. Many times bare-root plants are sold prepruned and need no further pruning until the following season. However, if not, about one-third of the top should be removed at transplanting by removing the weaker branches and clipping the remaining branches back to one-half to two-thirds their original length.

Transplanting Groundcovers

Groundcovers are herbaceous perennials or woody shrubs. The following recommendations apply to herbaceous groundcovers. Shrubs should be planted as recommended in the previous section.

Because many herbaceous groundcovers spread by trailing along the surface of the soil, it is important that all soil in a groundcover area be conditioned. The area should be spaded deeply to at least 12 inches (30 centimeters), and amendment should be worked throughout. The groundcover is then transplanted with a trowel throughout the prepared bed and watered thoroughly.

Spacing of herbaceous groundcover plants is determined by economics and how quickly coverage is desired. Spacing plants 2 feet (60 centimeters) apart will cost less than spacing them 8 inches (20 centimeters) apart but will take up to a year longer to cover the area. As a general rule, spacing plants 1 foot (30 centimeters) apart gives coverage at a moderate rate and expense.

Staking and Wrapping Trees

Staking may be needed either to anchor the root system of the tree until it becomes established or to support the trunk in an upright position. Unless it can be determined that staking is fulfilling one of these functions, it is not necessary and can cause weakening of the trunk.

For root anchorage of bare-root trees, two short stakes (1–1½ feet; 30–45 centimeters) should be inserted on opposite sides of the trunk (Figure 10-19). The tree should then be secured to each with strips of rubber, cloth, or wire covered with a section of garden hose. Uncovered wire, rope, or twine is not recommended since friction will cause it to rub off the tender bark.

Trees which require trunk support should be staked as low as will allow them to stand upright under calm conditions. Again, stakes should be positioned on opposite sides of the tree and the trunk tied to the stakes at a single point near the top.

Wrapping (Figure 10-20) involves twining tree-wrapping paper or burlap secured with twine around a tree trunk from the base to the lowest branches. The wrapping protects the trunk from sun damage, a problem on trees grown in rows with their trunks shaded by neighboring trees. Wrapping should be left in place about a year, after which it will begin to decay and can be removed.

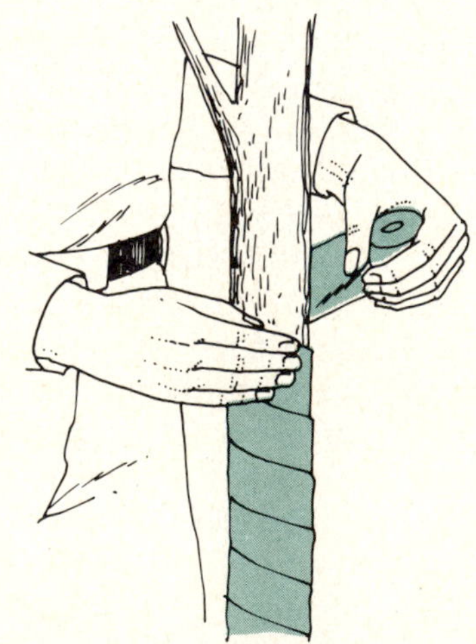

Figure 10-20 Wrapping a tree trunk with paper tape.

POSTPLANTING CARE OF THE LANDSCAPE

The most important part of postplanting care of a landscape is regular watering. During the first growing season plant roots occupy only a small volume of soil and need constant moisture to become established in the new location. Slow soaking with a hose is preferred for shrubs and trees, which should be watered every week to 10 days. Slow watering allows all the moisture to be absorbed into the soil and provides sufficient water to moisten the entire root area.

Groundcover watering is done by hand or sprinkler as frequently as required to keep the soil moist. The generous supply of moisture will enable the groundcover to fill in within a short period. It is advisable to check the depth of water penetration after each watering with a trowel, digging to the depth of the roots and examining the soil moisture. In this way there is no possibility of moistening the surface while the roots remain dry.

Mulching should be a part of postplanting care. It not only keeps in soil moisture but also prevents weed growth, a problem in newly planted groundcover areas. Mulching of landscape plants is discussed in Chapter 11.

Selected References for Additional Reading

Behme, R. L. *The Outdoor How-to-Build-It Book.* New York: Hawthorne Books, 1971.

Brimer, J. B. *Homeowner's Complete Outdoor Building Book.* New York: Harper and Row, 1971.

Harris, R. W., A. T. Leiser, and W. B. Davis. *Staking Landscape Trees.* University of California Agricultural Extension Service, Bulletin AXT 311, 1974.

Hoyt, R. S. *Check Lists for the Ornamental Plants of Subtropical Regions.* new rev. San Diego, CA: Livingstone, 1958.

Nelson, W. R. *Landscaping Your Home.* rev. ed. Urbana: University of Illinois Press, 1975.

Sunset Books Editors. *Ideas for Entryways and Front Gardens.* Menlo Park, CA: Lane, 1961.

———. *Outdoor Lighting.* Menlo Park, CA: Lane, 1971.

———. *Ideas for Landscaping.* Menlo Park, CA: Lane, 1972.

———. *Landscaping.* Menlo Park, CA: Lane, 1972.

———. *Sunset Western Garden Book.* 3d ed. Menlo Park, CA: Lane, 1976.

Wyman, D. *Shrubs and Vines for American Gardens.* rev. ed. Toronto: Macmillan, 1969.

———. *Trees for American Gardens.* New York: Macmillan, 1965.

Chapter 11

Landscape Maintenance

Regular maintenance of front- and backyard landscapes by pruning, fertilizing, watering, and weed control is a necessity if home grounds are to remain attractive and plants healthy. But maintenance does not need to be time-consuming. A landscape planned without a lawn and with a goal of low maintenance can be kept in condition with only about 2 to 3 hours of labor per month.

FEATURES OF THE LOW-MAINTENANCE LANDSCAPE

Landscapes designed to remain attractive with minimal maintenance include some or all of the following features:

1. Large paved or decked areas which require only occasional sweeping
2. Raised planters which require less bending for maintenance
3. Groundcover areas in lieu of lawns to eliminate mowing
4. Mulches or groundcovers around all trees and shrubs which do not require hand trimming (as grass does) and which suppress weed growth
5. An automatic irrigation system where natural rainfall is not sufficient
6. A limited number of carefully placed plants rather than large numbers of plants scattered throughout the landscape
7. Plants which do not require frequent pruning and which do not drop leaves, fruits, or dead flowers
8. Plants which, when mature, will not require extensive pruning to keep them the proper size
9. Flowering shrubs rather than high-maintenance annual and perennial plants
10. Plants which are naturally disease- and insect-resistant and which have been proven to grow well in the locale

TOOLS FOR GARDEN MAINTENANCE

Every gardener needs a selection of tools such as trowels, rakes, and pruners to maintain the landscape. The following sections describe the most frequently used and valuable gardening tools and equipment.

Soil-Working Tools

TROWEL

A trowel (Figure 11-1) is used for digging small holes for planting bulbs, setting out transplants, and other similar tasks. When pur-

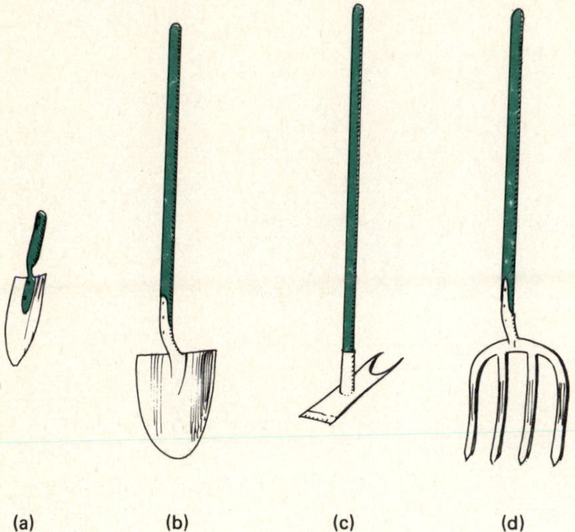

Figure 11-1 Most commonly used soil-working tools for the garden. (a) Hand trowel. (b) Round-point shovel. (c) Weeder hoe. (d) Spading fork.

chasing a trowel, check to make sure the handle attaches firmly to the blade and does not bend.

ROUND-POINT SHOVEL

A round-point shovel is used for digging larger holes such as for planting trees and shrubs and other general garden digging. The tip is pointed to make it easy to push into the soil (Figure 11-1).

HOE

A hoe is used for shallow cultivation of a previously prepared area and will penetrate the soil up to 4 inches (10 centimeters) deep. It is frequently used in vegetable gardening to prepare rows for seeding. Hoes are also used for weeding. For this purpose, they should be used with a scraping motion that severs the weeds at the soil line (Figure 11-1). If used in a cultivating manner, additional weed seeds will be exposed, germinate, and intensify the weed problem.

SPADING FORK

A spading fork is used to move piles of leaves or refuse, turn compost, and break up large clods in the soil. Although a shovel can be used for these purposes, a spading fork is more efficient and will move a greater volume of leaves or compost with each bite (Figure 11-1).

Watering Equipment

HOSE

Many grades of garden hoses are available. Inexpensive plastic kinds are stiff and develop permanent kinks which restrict water flow. Reinforced plastic hoses remain pliable, do not kink, and are lightweight. Rubber hoses are the most durable, but they are heavy and expensive.

PISTOL NOZZLE

A pistol nozzle provides on-off control and can be adjusted to give from a strong forceful stream to a mist. Unfortunately, when the stream is directed, it is too forceful for most watering, and when emitted at a lower volume, it sprays in a circle and cannot be accurately directed. However, some gardeners find the forceful stream useful for washing off insects and the mist ideal for watering seedlings (Figure 11-2).

FLARING ROSE NOZZLE (FAN NOZZLE)

This nozzle often does not provide on-off control. However, it applies a large volume of water without force and is good for general overhead watering (Figure 11-2).

OSCILLATING SPRINKLER

This sprinkler is the most adaptable to general yard use and can be adjusted to cover any size area by altering the water pressure and the sprinkler mechanism. Because the head moves back and forth, water is applied over the area

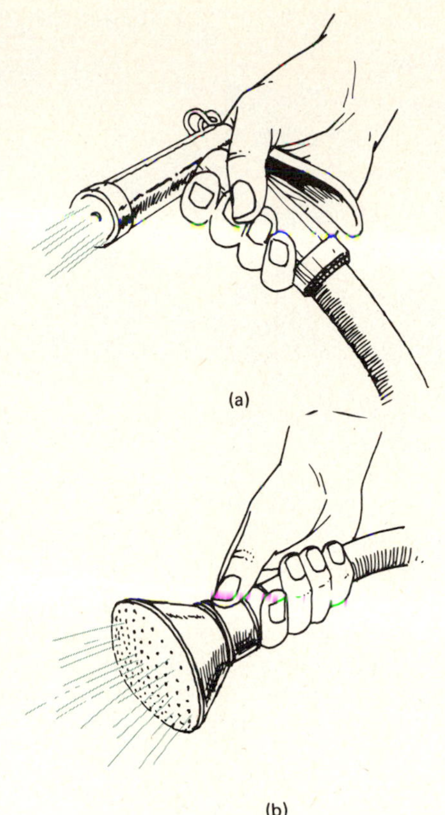

Figure 11-2 (a) Pistol nozzle. (b) Flaring rose nozzle.

at a slow rate, and runoff is less likely (Figure 11-3).

WATERING CAN

Either plastic or galvanized metal watering cans are acceptable. The can should contain a removable sprinkler head so that either a fast volume or a slow sprinkling of water can be applied. Although metal cans last longer, they are heavier and more expensive than plastic cans.

Rakes

LEAF RAKE

A leaf rake can be made of lightweight metal or bamboo and has long flexible teeth. It is used for raking leaves, grass, and other lightweight materials (Figure 11-4).

GARDEN RAKE (BOW RAKE)

A garden rake is generally metal with short, rigid, widely spaced teeth. It is used for smoothing the soil surface in preparation for seeding a lawn or garden, removing thatch from the lawn, and other purposes (Figure 11-4).

Fertilizing Equipment

HOSE-END SPRAYER

This apparatus attaches to a hose and automatically mixes any liquid concentrate into the water at dilute strength. It is used for fertilizing lawns, flowers, and other shallow-rooted plants and for applying certain insecticides (Figure 11-5). However, it is not accurate enough for most pesticide applications.

ROOT FEEDER

A root feeder works like a hose-end sprayer, proportioning liquid fertilizer into the irrigation water. It is equipped with a long, needle-

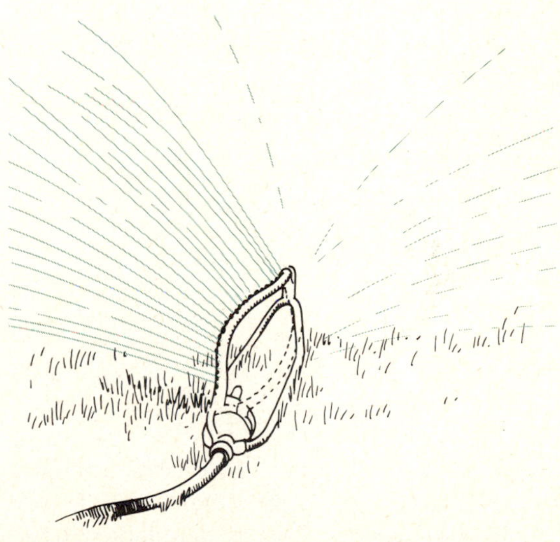

Figure 11-3 An oscillating sprinkler.

Figure 11-4 (a) Leaf rake. (b) Garden (bow) rake.

like tube for injecting the fertilizer into the root zone of shrubs and trees.

Pruning Tools

HAND PRUNERS

Hand pruners will cut branches up to ¾ inch (20 millimeters) in diameter. They are the most frequently used pruning tool (Figure 11-6).

LOPPING SHEARS

This tool will cleanly remove branches up to about 2 inches (5 centimeters) across in a single cut. "Loppers" are used on larger shrubs and trees(Figure 11-6).

Figure 11-5 Hose-end sprayer.

PRUNING SAW

A small, curved pruning saw can be used for removing larger branches and is specially designed to fit in tight places. It can substitute for lopping shears but will require more labor. Both lopping shears and pruning saws are available in pole-mounted models for pruning high in trees (Figure 11-6).

HEDGE SHEARS

These are needed only if formal hedges are to be maintained (Figure 11-6). Electric types are also available.

Care of Garden Tools

If properly cared for, garden tools will last many years. Regular removal of rust on rakes, trowels, shovels, and other tools is important. Wire brushing followed by rubbing with steel wool removes most corrosion. A light coating of oil will slow its return and is also useful as a prestorage treatment in areas where garden tools are not used in winter.

Wood handles should be inspected occasionally for cracks and splintering. Small cracks can be fixed with wood cement. Sanding followed by a coating of boiled linseed oil will cut down on splinters. As an alternative to oiling, some gardeners paint tool handles red. This seals the wood and makes misplaced tools easier to find in the garden.

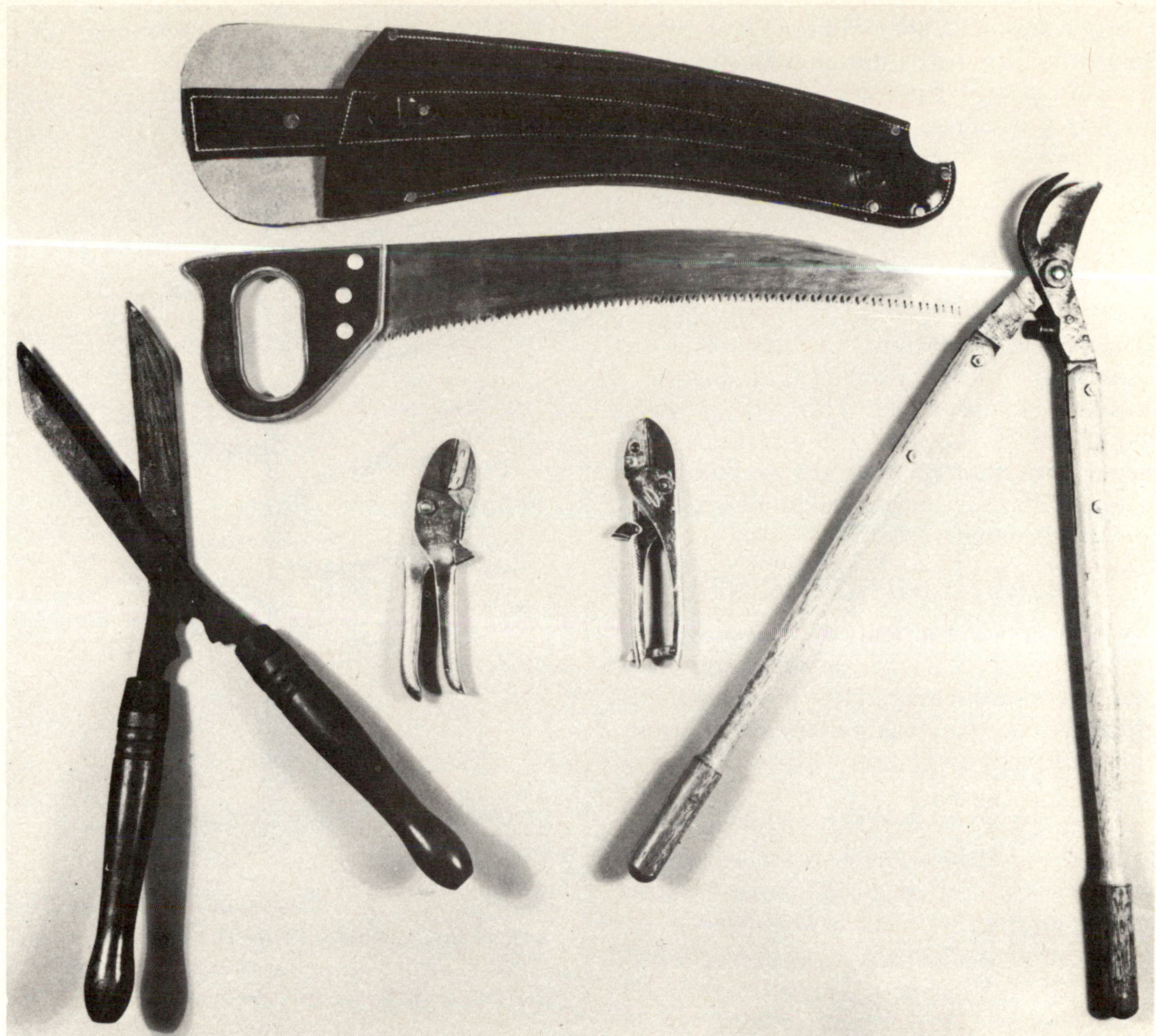

Figure 11-6 Pruning tools. Hedge shears (left), pruning saw (top), hand pruners (middle), and lopping shears (right).

Sharpening of pruning tools at least once per season is recommended so they will make clean cuts. A sharpening steel or whetstone can be used, and professional tool-sharpening services are also available. Spades, hoes, and shovels also require yearly sharpening and should be honed lightly with a hand file (but not filed razor sharp).

Hoses should be coiled after each use and stored in the shade. Sunlight will deteriorate a hose rapidly. In cold areas the hose should be stored over winter in a garage or basement.

PRUNING LANDSCAPE PLANTS

Pruning is probably the most misunderstood maintenance task. Considerable time can be spent on seasonal clipping of trees and shrubs, only to find that in a few months the plants are again in need of pruning. However, proper pruning encourages controlled growth, and time invested in pruning can be held to a minimum.

Purposes of Pruning

SIZE CONTROL

Most people use pruning for size control—to keep plants from growing over windows or a walkway, for example. However, pruning should not be necessary if the plants are selected with their mature size in mind and grown with forethought as to obstructions they might later create.

HEALTH IMPROVEMENT

Plant health is maintained or improved by such pruning practices as removing dead branches and diseased limbs. It also includes pruning trees when young to encourage strong branch structure.

APPEARANCE IMPROVEMENT

Pruning to remove scraggly branches, clip off dead flowers, and encourage bushy compact growth all improve plant appearance.

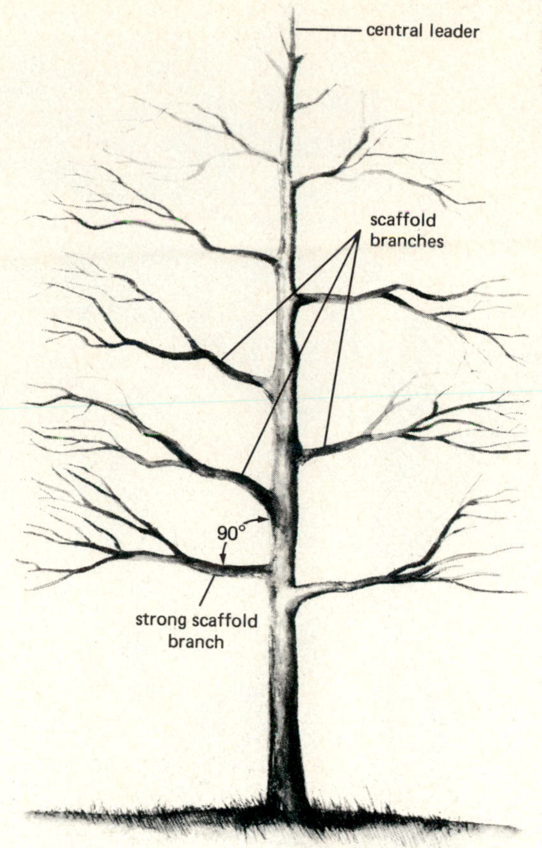

Figure 11-7 A well-shaped tree with a central leader form. Note the even spacing of the scaffold branches down the trunk and the wide branch crotches.

Pruning Trees

YOUNG TREES

Pruning trees at transplanting and in the first few years that follow will greatly affect their strength and health in later life. During this early period the tree develops the main branches which will form the support structure for the foliage. Pruning directs the growth of those main branches in order to create a strong tree.

Removal of Double Leaders. Most trees have a "central leader," that is, they grow with one main upright shoot (leader) and side branches called *scaffold branches* (Figure 11-7). Occasionally two leaders develop, causing a fork. The angle between the two leaders is a potential weak point of the tree, because as each leader increases in diameter, it exerts pressure against the other at the base. Eventually, conditions conducive to splitting of the trunk develop, and in a wind storm the weight of the leaders can pull the trunk apart. Accordingly, the weaker of the two leaders on a double-leader tree is removed while the tree is young. The remaining one will then develop normally to become a strong tree.

Selection of Scaffold Branches. Removal of a double leader is rare in tree pruning, but removal of excess scaffold branches is not. Ideally a young tree should have only a few branches spaced widely and evenly along the trunk. Numerous small branches are not desirable since they will enlarge and crowd each other as the tree grows. Scaffold branches should be selected so that they are evenly spaced up the trunk and not growing directly above one another. Branches growing from the trunk at a wide angle approaching 90° (Figure 11-8) are preferred since they are stronger, although many trees have narrow-angled branches when young.

When a double leader or extra branches are removed, the branches should always be cut to leave a wound area that is flush with the trunk. Stubs both are unsightly and will never heal. They can also be a source of rot entry into the trunk.

Topping. Topping is a practice used by some gardeners in the hope of forming a tree that is full-headed and dense. The central leader is removed, and the side branches below the cut then grow rapidly. The ultimate effect is a flat-headed tree with an unnatural appearance. Topping is not a recommended pruning method.

Limbing Up. Limbing up is the removal of the lower branches of a tree which are positioned too low on the trunk and thus create a hazard or inconvenience. Removal is necessary because the height at which a branch originates is the height at which it permanently remains. The trunk does not lengthen and elevate the branch, as is popularly believed. A young tree 4 feet (1.2 meters) tall with branches 2 feet (0.6 meter) off the ground would have large branches still 2 feet (0.6 meter) off the ground 20 years later if left unpruned.

Limbing up should be delayed for 3 to 4 years after planting, if possible, since removal of foliage from a young tree will slow its growth and weaken the trunk. Once the tree is established and growing rapidly, one or two branches can be removed each year until the desired trunk height is reached.

Desuckering. Desuckering is the removal of shoots which grow from the base or lower trunk of trees. If the tree is grafted, these branches often grow from the vigorous rootstock. In other cases, buds on the trunk are located far enough from auxin-manufacturing buds that they are not reached by the dormancy hormone. Such branches are called *suckers,* or *watersprouts,* and should be removed as they appear with hand pruners. If left in place, they give the tree an unkempt appearance and use up energy that should be diverted to top growth.

Although most watersprouts grow from the trunk at ground level or slightly above, they are sometimes found on scaffold branches. They are easy to locate because of their vigorous vertical growth and should be removed. If left, they will grow until they cross and rub against the scaffold branches.

Figure 11-8 Branches which begin growing inward toward the center of a tree should be cut out as shown.

MATURE TREES

A tree that has been properly pruned when young should need little, if any, pruning when mature. Desuckering may be necessary, along with the removal of branches broken by wind or growing toward the center of the tree (Figure 11-8). Small branches growing close together may rub one another (Figure 11-9); the smaller or more poorly positioned ones should be cut back to their parent branches to avoid injury to the bark.

Removal of Large Branches. Occasionally, large branches may need to be removed because of disease, incorrect pruning when young, or some other reason. Such branches are frequently heavy and, while being sawed off, can break and rip down the trunk. To avoid the chance of injury to the tree, the three-cut pruning method (Figure 11-10) is recommended for removing branches more than 2 inches (5 centimeters) in diameter. The first cut is made under the branch about 1 foot (30 centimeters) from the trunk and goes halfway through the branch. This cut provides insurance against tearing in case the branch should fall. The second cut is made farther out on the branch and cuts completely through and removes the branch. The third cut removes the stub flush with the trunk to promote quick healing. Tree wound paint, an asphaltlike sealer, is commonly used on wounds

Figure 11-9 The weaker or more poorly positioned of these two rubbing branches should be removed.

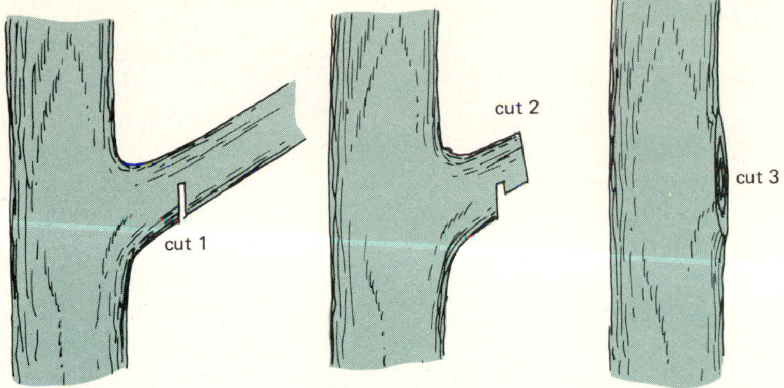

Figure 11-10 The three-cut pruning method for removing large branches. This method prevents the branch from tearing down the trunk.

larger than 1 inch (2.5 centimeters) in diameter to discourage the entry of microorganisms and insects.

Removal of Girdling Roots. A girdling root (Figure 11-11) usually results from the incorrect initial planting of a tree. If the roots are not spread out in the hole, they can enlarge and press against the trunk at or just below ground level. The resulting pressure halts the flow of water and nutrients within the trunk and will eventually kill that side of the tree.

The only cure for a girdling root is to cut it away with a chisel or hatchet. Signs of a girdling root include flattening of the trunk

Figure 11-11 A girdling root on a young tree.

on one side and, on deciduous species, dropping of the leaves on the affected side earlier in fall.

PRUNING VASE-SHAPED TREES

Many patio trees, weeping trees, and some larger shade trees like the elm do not naturally have a central leader, and it is a mistake to prune them to have one. Such trees may split into two leaders, forming a vase shape, or have all their scaffold branches arising at one point on the trunk (Figure 11-12). Pruning for these trees should include only desuckering and removal of crossing and inward-growing branches.

POLLARDING

Pollarding (Figure 11-13) is the formal training of deciduous trees and a common pruning technique in certain areas of the country. It restricts the height of the tree and creates a denser head of foliage than the species would normally have.

Pollarding is usually begun after the tree has attained the maximum height desired. When the tree is dormant it is cut back severely, the main branches sawed off deeply into the old wood. When the tree breaks dormancy, buds will sprout below each cut, producing vigorous watersprouts and creating the dense foliage mass. The next winter and each

Figure 11-12 A vase-shaped tree with all main branches arising from one point on the trunk.

Figure 11-13 (a) A pollarded tree in winter before annual pruning. (b) After pruning.

winter following, the recurring watersprouts are removed. Eventually the ends of the branches become gnarled, giving the plant a distinct appearance.

Although pollarding does give a tree a full, pleasing appearance during the growing season, the tree will be very unattractive when dormant. In addition, pollarding generally requires the yearly services of a tree maintenance person who has the equipment needed to reach high in the tree and remove the watersprouts. Consequently, pollarding cannot be recommended for home landscapes.

PRUNING NEEDLE EVERGREEN TREES

Pines are the most familiar needle evergreens, all of which are characterized by a pyramidal, or "Christmas tree," form with a central leader. Evergreens generally do not require pruning except to remove broken or dead branches and occasionally to eliminate a double leader. They may be limbed up above eye level, although this does not enhance their appearance.

Figure 11-14 Pruning a needle evergreen by breaking out the unexpanded candles.

When pruning needle evergreens, remember that unlike most other trees, they do not branch if cut back to old wood, so pruning into the woody portions has a permanent effect. The only pruning which can be used to limit size or make the foliage thicker is the removal of some of the new growth, or *candles* (Figure 11-14), while it is still immature. Breaking out a portion of the center candle and/or side candles will cause the growth to become shorter and fuller.

Pruning Shrubs

To preserve their natural appearance, most shrubs are pruned by thinning or "heading back."

THINNING

Many shrubs are composed of a clump of woody shrubs coming out of the ground and

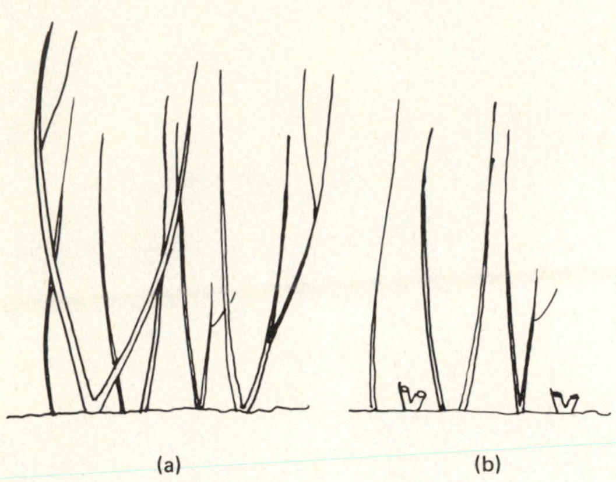

Figure 11-15 (a) Before pruning a shrub by thinning. (b) After pruning.

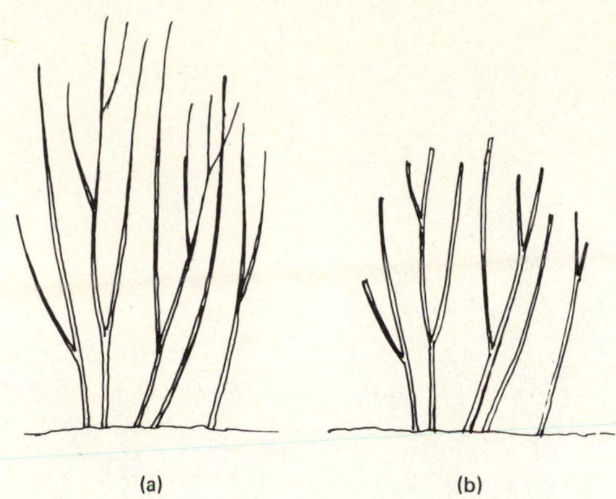

Figure 11-16 (a) Before pruning a shrub by heading back. (b) After pruning.

grow by the increase in height and number of these stems. As the oldest stems age and lengthen, they will have a tendency to shade out the bottom foliage, and the shrub will become bare at the base. Thinning consists of removing the oldest stems at ground level. This shortens the shrub and encourages new shoots to grow from the roots and refoliate the bottom. Thinning can be done with hand pruners, if the stems are small, or with loppers, if the shrub is heavily overgrown with large stems (Figure 11-15).

HEADING BACK

Heading back (Figure 11-16) is done with hand pruners to remove straggly growth or to limit size. It consists of cutting twigs or small branches back to a point where the cut is hidden by the remaining foliage. If possible, the stem should be cut back to just above an outward-pointing bud or branch (Figure 11-17). This will encourage new growth to develop outward and eliminate crossing branches.

RENEWAL PRUNING

Renewal pruning (Figure 11-18) is used to rejuvenate and shorten overgrown shrubs. It is always done in early spring before new growth starts and consists of cutting back all the branches to stubs 2 to 3 inches (5–8 centimeters) long. This forces new growth from dormant buds on the stubs, which can then be pruned as needed by heading back and

Figure 11-17 Pruning back to an outward-pointing bud.

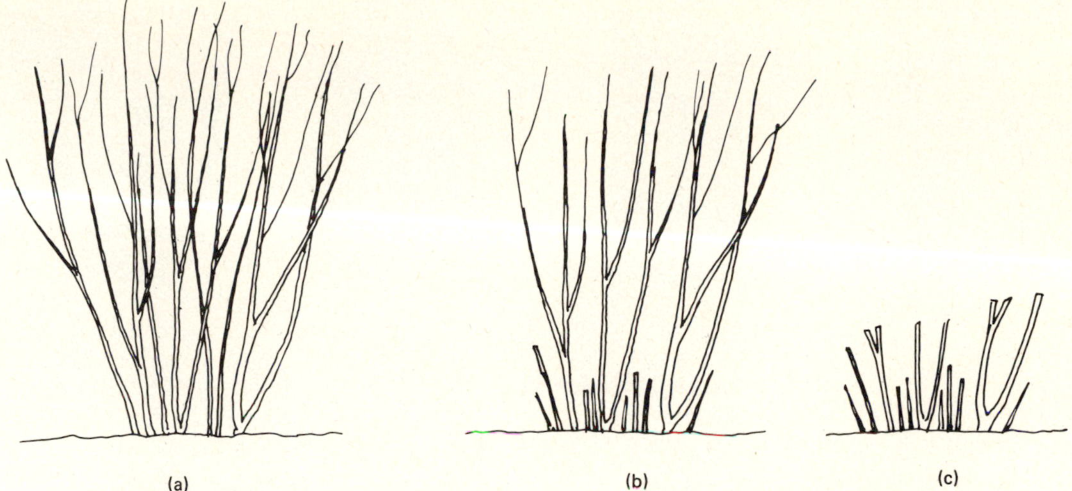

Figure 11-18 Renewal pruning a shrub over 2 years. (a) Before pruning. (b) After first pruning. (c) After second pruning. Note that only half the branches are removed the first year and the remaining branches the second year.

thinning to maintain the proper size.

Renewal pruning is a drastic measure and will make the plant unsightly for at least one season. As an alternative, half the branches can be removed the first year and the remainder the second year. In this way the plant will always keep some foliage and have a somewhat better appearance.

PRUNING HEDGES

Natural hedges are lines of shrubs planted close together to create a thick mass of foliage. Pruning natural hedges is basically the same as for other shrubs, that is, thinning and heading back once or twice a year.

Formal hedges require frequent pruning to maintain a neat appearance; shearing every 2 weeks during the growing season is typical.

Selection of the correct hedge shape is very important. Too often a hedge is pruned narrow at the base and wide at the top. This is incorrect because the wide top will shade the lower portions, causing them to drop their leaves. Pruning the top narrower than, or at the same width as, the bottom (Figure 11-19) will avoid shading of the bottom, and the plants will remain full and leafy to the base.

Pruning Groundcovers

Most groundcovers do not require regular pruning to improve their appearance or health, although shrub types such as junipers may require heading back to control spread (Figure 11-20). However, occasionally an older bed of groundcover will develop long unsightly stems and can be improved by renewal pruning. If the area is small, it can be clipped back severely with hand pruners. For large areas, a lawnmower set to cut as far from the ground as possible (and with a catch bag) can be used. Although the area will be unsightly at first, it will usually refoliate in 2 to 3 months.

Timing Pruning

The time of year at which maintenance pruning takes place will not affect the health of

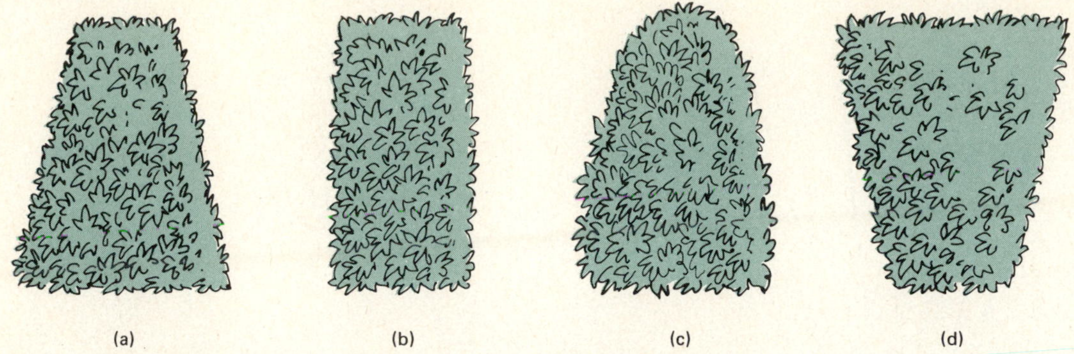

Figure 11-19 Good (a, b), acceptable (c), and poor (d) shapes for hedges based on the amount of sunlight that will reach the base.

a plant and need not be restricted to a particular season. However, early spring is a traditional pruning time for several reasons. First, it is easier to determine which parts of deciduous plants should be removed if the branch structure is not obscured by leaves. Second, by pruning before new growth starts, the growing direction of new shoots can be controlled. By contrast, summer or fall pruning removes poorly placed shoots produced that season.

There are exceptions to the early spring recommendation for pruning. If spring-blooming shrubs are pruned early in the year, some of the flower bulbs which have already formed will be cut off. Consequently it is best to prune these shrubs immediately after flowering and before they develop buds for the next year's bloom.

Deciduous trees which experience sap flow in early spring are the second exception. Pruning wounds will exude sap liberally in early spring. This loss of fluid will not seriously injure the tree, but it is unsightly and can be avoided by pruning later in the year.

In conclusion, light pruning twice or three times annually is preferable to severe pruning once a year to maintain an attractive landscape.

Special Pruning Techniques

ESPALIER

Espalier (Figure 11-21) is the stylized training of a shrub or tree flat against a wall or trellis, usually in a symmetrical branching pattern. Since not all plants can be adapted to this type of pruning, consult a knowledgeable

Figure 11-20 Heading back a groundcover juniper to control its spread. (a) Before pruning. (b) After pruning. Note how the branches are layered to prevent shading out of the lower branches.

Figure 11-21 Espalier (left) and topiary (right) pruning styles.

nurseryman or one of the references listed at the end of the chapter before espaliering a plant.

Espalier pruning must be started when a plant is small. The trunk is tied to the trellis, and the sites where future branches are desired are selected. The buds or branches closest to these points are then tied to the trellis, and all others are removed. This permits only these preselected branches to grow and achieves the desired shape in several years. Annual pruning will still be required to remove stray growth and keep the plant from outgrowing the trellis.

TOPIARY

Shrubs and small trees trained into pompoms or animal shapes are examples of topiary pruning (Figure 11-21), a novelty pruning style that originated in formal European gardens. Topiary training can be thought of as a combination of a formal hedge and an espalier. Branches are left at selected locations on the trunk and then sheared repeatedly to encourage fullness. As with espalier pruning, a topiary should be started while a plant is young, shaping it gradually to its final form over several years.

Topiary plants in the landscape should be considered specimens; one or two are sufficient to enhance any landscape. They can be used effectively in Oriental or modern landscapes and can be purchased pretrained at garden centers.

WATERING

The goal of watering is to replenish the moisture in the soil surrounding plant roots. A considerable volume of water is needed to accomplish this, and underestimating this volume is a common error. An inexperienced gardener may sprinkle a plant with a hose for a few minutes and consider it watered. In actuality, the sprinkling has only wet the top 2 to 3 inches (5–8 centimeters) of the soil, whereas the majority of the roots remains dry.

Correct watering of shrubs should give a slow flow of water at the base of the plant for 20 to 30 minutes. The water will penetrate evenly throughout the root area and wet the soil deeply. This will encourage the roots to grow downward into the moisture-retaining soil layers instead of along the surface. Trees should be watered deeply over the entire area covered by the branches, since roots will be growing throughout the area. Exception to the rule of deep soaking must be made for established plants accustomed to sprinkler irrigation. The root system of these plants will be shallow, and the plants will need to be maintained with sprinkler irrigation.

Water Frequency

The frequency with which a plant should be watered depends on its age, its drought tolerance, and the ability of the soil to retain moisture.

PLANT AGE

More newly transplanted trees and shrubs die from lack of water than from any other cause. When first transplanted, the root system of a plant occupies a limited volume of soil and must extract all its water and nutrients from that soil. During the first few months after transplanting, the roots grow into the conditioned soil of the planting hole and, after that, into the surrounding soil. But until they are established and the plant is able to absorb moisture through an extensive root system encompassing a considerable volume of soil, watering should be frequent. Watering every 7 to 10 days for the first year and every 2 weeks the second year is advisable. After that, irrigation will be needed only during periods of drought lasting more than 3 or 4 weeks.

DROUGHT TOLERANCE

Tolerance to drought varies immensely among plant species. Some are able to extract minute quantities of water present in the soil and remain healthy, whereas others under identical conditions would wilt to the point of death.

Cacti and succulents are the best-known examples of drought-tolerant plants, but they are not the only such plants. Trees with taproots and shrubs native to arid climates are similarly drought-tolerant and can be used to create a low-maintenance and ecologically sound landscape where natural rainfall is scarce.

MOISTURE-HOLDING CAPACITY

The proportions of sand and clay in a soil greatly affect its ability to retain moisture and consequently affect the watering frequency. Soils containing a large percentage of clay will be the most water-retaining, with loams intermediate and sandy soils least able to hold water. A sandy soil may need watering twice as frequently as a clay.

If a natural soil is sandy, its moisture-holding capacity can be greatly improved by adding organic matter, as described under "Water Conservation."

Water Techniques

SPRINKLERS

Sprinkler irrigation, whether by permanent or portable sprinkler heads, was the main method of irrigation used for many years in areas of the West lacking regular rainfall. It watered large landscaped areas at one time and, if left on long enough, would eventually saturate the ground to the desired depth. The main disadvantage of sprinklers is that they waste water; that is, they wet all the soil instead of only that which contains plant roots.

Today, watering by sprinklers can be recommended only for groundcover, lawns, and densely planted areas such as flower beds. However, sprinkler irrigation should not be discontinued on landscapes established by this watering method, since the plant root systems will be concentrated throughout the sprinkler pattern area.

HAND WATERING

In landscapes where supplemental watering is seldom required, hand watering with a hose is practical. The hose can be adjusted to low pressure and moved from plant to plant every half-hour, or the watering basin made at planting can be filled several times.

DRIP OR TRICKLE IRRIGATION

Drip or trickle irrigation (Figure 11-22) is a relatively new irrigation technique. Its advantages are low cost, ease of installation, and conservation of water. An additional advantage of drip irrigation is that, because it applies water only at one spot (instead of throughout an area, as a sprinkler), the germination and growth of weeds are inhibited.

A typical drip system would consist of four major parts: a filter, a pressure regulator, ½- to ¾-inch (12- to 20-millimeter)-diameter flexible tubing, and ⅛-inch (3-millimeter)-diameter microtubing. Water from an outside tap is first filtered to remove particulates and then passed through the regulator to lower the pressure to about 10 percent of normal. Next it passes through the ½-inch (12-millimeter) tubing located near the plants to be irrigated. Short lengths of the microtubing inserted into the main lines carry water to each plant, with all plants watered at once at a slow, even rate. The "drip" and "trickle" names come from the slow flow of water from each microtube, a total of about 1 gallon (3.8 liters) per hour.

Trickle irrigation is very easy to install, even for a person with no knowledge of plumbing. Because it operates at low pressure, no glue or threaded connections are needed. The splices between the main lines are held together with self-sealing plastic connectors. Likewise, the microtubes can be inserted into the main line by simply making a hole and pushing in the microtube until it fits snugly. Unlike a sprinkler, no digging is needed to install a drip irrigation. The main lines are laid on top of the ground and covered with mulch.

The main disadvantage of drip watering is the tendency of the microtubes to clog with mineral deposits from the water or with soil. These problems can usually be remedied by flushing the system periodically at high pressure.

Drip systems are most often used on tree

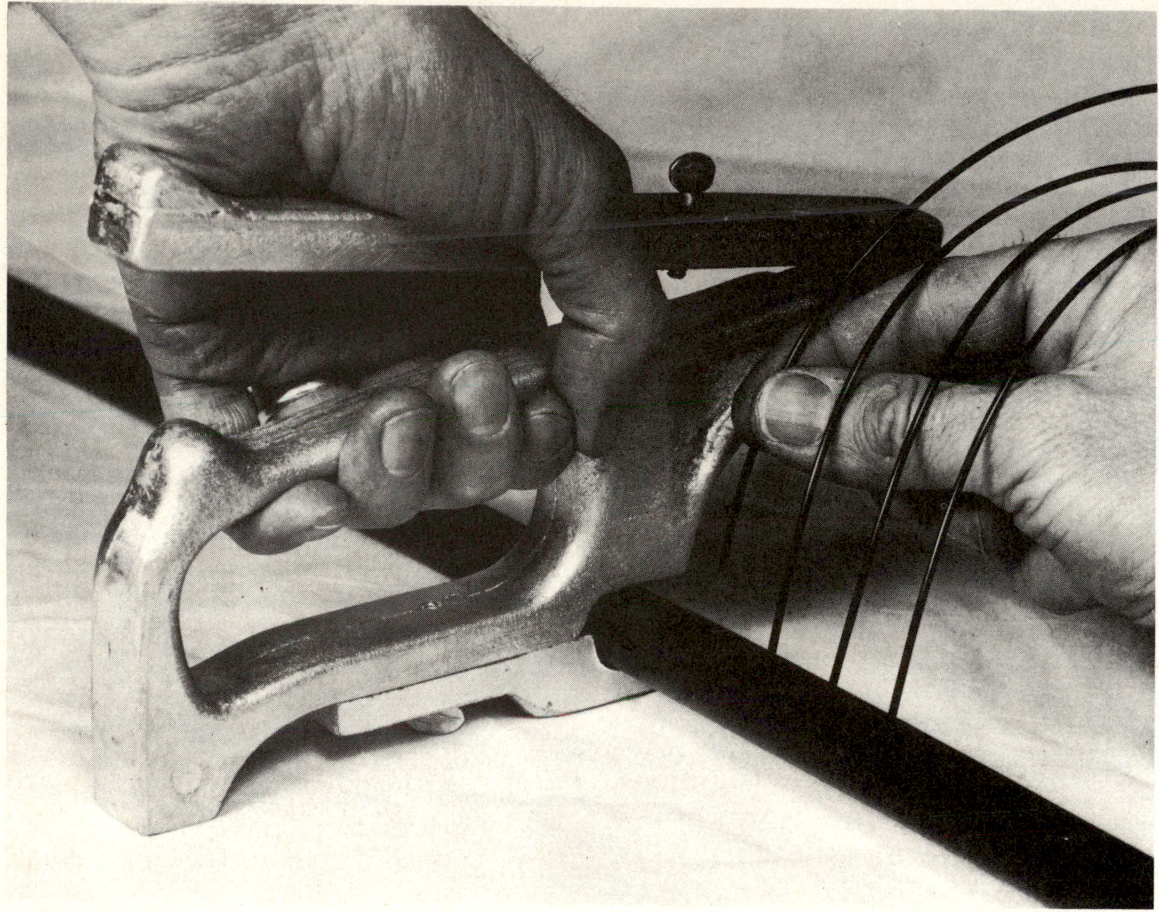

Figure 11-22 Putting together a trickle irrigation system. Note the small microtubes which will carry water to each plant.

and shrub plantings, since each individual plant can have its own microtube. However, the system can be adapted to water groundcover and grass areas with microsprinkler heads (Figure 11-23) which spray over an 8-inch (20-centimeter) radius.

Water Conservation

Even where water is abundant, conserving water used for irrigating landscape plants should be considered. There is never enough of this valuable natural resource to justify thoughtless use. The following techniques will reduce the amount of water used in a landscape planting.

IMPROVEMENT OF THE WATER-HOLDING CAPACITY

Though the natural soil in an area may have a poor water-holding capacity, it can be improved with the addition of organic matter. Compost, peat moss, shredded bark, coarse sawdust, and similar materials can be incorporated at planting. Herbaceous groundcover beds can be noticeably improved by mixing in organic matter to a 1-foot (0.3-meter) depth, since the systems of the plants are nat-

Figure 11-23 A microsprinkler head used with a drip irrigation system.

urally shallow and will remain within the improved layer.

PREVENTION OF EVAPORATION

Much of the water lost from landscape plantings leaves by evaporation from the soil. This evaporation rate can be slowed by the use of a mulch and by watering in the early morning when the temperature is lowest.

USE OF PROPER IRRIGATION TECHNIQUES

The proper irrigation equipment should be used: trickle for trees and shrubs, plus sprinkler for lawns and groundcover. The system should then be used wisely. Be cautious when using automatic timing devices, since they do not take weather conditions into account and may apply water in a rainstorm. Frequent checks for water runoff will prevent one of the most common means of wasting water.

FERTILIZING LANDSCAPE PLANTS

Many homeowners never fertilize their trees and shrubs, assuming that fertilizers are needed mainly for vegetables and lawns. But just as a fertilizer promotes the strong rapid growth of these plants, it will also speed the growth of landscape plants. A newly installed landscape which is fertilized regularly will grow much faster than an unfertilized one.

Timing of Fertilizer Applications

The time of year when fertilizer should be applied to landscape plants varies with the climate, but generally it should be restricted to just before or while a plant is actively growing.

In most parts of the country, fertilizer is applied annually in spring so the nutrients will be available for the new growth. However, for faster growth fertilizer could be applied two or three times during the growing season at 1- to 2-month intervals, for example, on April 1, May 5, and July 1.

In mild climates such as found in Florida and parts of California, plants grow continuously. Fertilizer will need to be applied every 2 to 3 months through the entire year to replenish the nutrients absorbed by the plants.

Rates of Fertilizer Application

TREES

In general, granular fertilizers approximately twice as high in nitrogen as in phosphorus or potassium are used on trees. The formulas 10-6-4 and 15-5-5 are ratios commonly found. Newly planted trees will benefit from annual fertilizing beginning no sooner than 1 month after they are planted and continuing the first two to three seasons. However, established trees need only be fertilized every 2 to 3 years. The drill hole or surface application (discussed in Chapter 5) can be used for established medium- or large-sized trees, but small and newly planted trees are normally fertilized only by surface application.

The yearly amount of fertilizer recommended for trees is about 6 pounds (2.7 kilograms) of actual nitrogen per 1000 square feet (92 square meters) of area under the branches. Actually, the amount of fertilizer applied is not critical, provided it falls within a reasonable range. In some cases it will not be practical to compute a fertilizer weight for a small tree [with, for example, a 4-foot (1.2-meter) branch span]. Estimating is completely satisfactory in most cases. About ½ cup (120 milliliters) of a fertilizer with 15 percent or less nitrogen is adequate for small trees less than 12 feet (3.6 meters) high. This allotment can be increased to 1 cup (240 milliliters) for 15-foot (4.7-meter) trees, 3 cups (700 milliliters) for 20-foot (6-meter) trees, and 4 cups (950 milliliters) annually for trees 30 feet (9 meters) or greater in height.

Shrubs

Shrubs, whether evergreen or deciduous, are fertilized in most circumstances at the rate of 2 to 4 pounds (0.9 – 1.8 kilograms) of nitrogen annually per 100 square feet (9.2 square meters) of bed area per year, using the surface application method. The fertilizer can be heavily nitrogen-based if the shrubs are grown strictly for their foliage, but a high-phosphorus fertilizer is better for flowering types.

The fertilizer can be split into two applications if desired: one in early spring and the second immediately after flowering. If acid-loving plants are being grown, an acid-forming fertilizer can be used (see Chapter 6).

If fertilizer calculations are not going to be made, a moderate amount of fertilizer can be estimated to equal ½ cup (120 milliliters) for shrubs less than 2 by 2 feet (0.6 × 0.6 meter), 1 cup (240 milliliters) for shrubs up to 4 feet (1.2 meters) tall and 3 feet (1 meter) across, and 1½ cups (350 milliliters) for larger shrubs. The fertilizer should be watered in after it is applied.

GROUNDCOVERS

Groundcover plants will benefit from yearly fertilization in spring using a surface application. Two to four pounds (0.9 – 1.8 kilograms) per 1000 square feet (92 square meters) watered thoroughly afterward is advisable, or ¾ cup (180 milliliters) can be estimated to be sufficient for a 10 × 10-foot (3 × 3)-meter area. Either a balanced or a high-nitrogen fertilizer is acceptable.

WEED CONTROL

Weeds growing in a landscape planting can seriously detract from its appearance. They also compete for the available light, water, and nutrition; in an establishing groundcover, weeds will thrive at the expense of the desirable plants. For these reasons weed control is essential, whether chemical or cultural.

Chemical Weed Control

With few exceptions, chemical weed control is impractical in landscape plantings. Many species are used in a typical landscape, and each will react to weed-controlling chemicals (herbicides) in a different way. An improperly applied weed killer may ruin not only your own plants but also those in neighboring yards.

Table 11-1 Herbicides for Ornamental Plantings

Chemical	Weeds controlled	Remarks
EPTC (Eptam)	Germinating annual grasses and some broadleaf weeds; will not control established weeds	For weed control in flower plantings; irrigate after application
Trifluralin (Treflan)	Same as above	For flower and shrub plantings; irrigate after application
DCPA (Dacthal)	Same as above	For use in herbaceous plants; most effective in sandy soils
Dichlobenil (Casoron)	Germinating annual broadleaf and some grassy weeds including quack grass; will not control already established weeds	For use around established deciduous trees and shrubs and around conifers
Dalapon (Dowpon or Basfapon)	Actively growing grasses, both annual and perennial	For preplant weed control in landscape and flower plantings; injury may occur on some plants by root uptake
Amitrol-T	Nonselective on all actively growing weeds	For spot treatment of localized weed problems
Diphenamid (Enide or Dymid)	Germinating annual grasses and some broadleaves	For use in groundcover, flower, and shrub plantings
Oryzalin (Surflan)	Germinating grasses and broadleaves	For use in woody ornamentals

Exceptions can be made. Herbicides of the "preemergence" type kill only germinating seedlings and not established plants. Such chemicals could be applied, for example, to a newly planted groundcover area to alleviate the inevitable sprouting of weed seeds which occurs when the soil is disturbed.

Table 11-1 on page 261 lists some of the herbicides used in ornamental plantings.

Cultural Weed Control

Cultural weed control consists of hand pulling and cultivating with a hoe, using ornamental plants competitively, and mulching.

Hand pulling and cultivation are effective for controlling annual weeds. But they are only temporarily effective in controlling perennial weeds, since the weeds often break off at the soil line and grow back from the roots. Cultivation exposes fresh soil and additional weed seeds to favorable germination conditions and necessitates repeating the cultivation several times each season. Although pulling and cultivation may be the only remedies for large weeds, mulching is more practical and labor-saving over a long period.

Using ornamental plants competitively can also reduce weed problems. By planting ornamentals thickly and using groundcovers on bare soil areas, weeds will not have an opportunity to become established. They will

Figure 11-24 A mulch of shredded bark.

be crowded out by the desirable plants.

Mulching is the third cultural weed control. A layer of mulch excludes light from the soil, checking the growth of both seeds and existing small weeds. A mulch can be any material used to cover the soil surface.

Mulching plants in the landscape serves other purposes besides weed control. It provides a decorative background for plants, conserves soil moisture, and maintains the soil at an even temperature.

ORGANIC MULCHES

The most popular organic (plant-derived) mulches used in landscape plantings are shredded bark and bark chips (Figures 11-24 and

Figure 11-25 A chunk bark mulch.

11-25). They are applied 2 to 3 inches (5–8 centimeters) thick throughout shrub beds and under trees and give these areas a neat, rustic appearance. Because they are organic, they decompose slowly over a number of years and will need to be renewed occasionally by adding a fresh layer to the existing mulch.

ROCK MULCHES

Rock mulches can be considered permanent since they do not decompose. Included in the inorganic mulches are stone materials such as marble chips, smooth stones, and coarse or smooth pebbles (Figure 11-26). Their only disadvantage is that, if replanting of the bed is required, they must be removed and replaced afterward. Organic mulches, on the other hand, could be simply worked into the soil.

Figure 11-26 A stone mulch.

Figure 11-27 Shredded bark applied over a clear plastic mulch layer. Note the uncovered plastic in the background.

PLASTIC MULCHES

Plastic (Figure 11-27) is one of the most effective mulches because it is completely impervious to weed growth. Although unattractive by itself for use in landscape plantings, it can be covered with a decorative layer of bark or stone. It will also be concealed when used with a groundcover shrub, which will spread and hide it soon after planting.

Plastic generally comes in rolls of varying widths and is rolled onto the area to be planted with the edges overlapping. Plants are then transplanted into slits cut through the plastic. Finally, the edges are covered with soil to hold the plastic firmly in place in case of wind. For decorative areas such as landscapes, a layer of bark mulch can be applied over the entire sheet of plastic to cover it. Water will run off the plastic during periods of rainfall and be diverted through the holes to the plants. If plastic wider than 3 feet (1 meter) is used, additional holes should be made for water penetration into the soil.

ADDITIONAL MULCHES

Mulches useful for the landscape are not limited to those just discussed. Rice hulls, cocoa bean hulls, and oak leaves are attractive mulches, but are available in only a few areas of the country. Likewise, corncobs, compost, and grass clippings are also effective but are generally considered somewhat unattractive for landscape use. They are discussed in Chapter 6 as vegetable garden mulches.

Selected References for Additional Reading

Bruning, W. F. *Home Garden Magazine's Minimum Maintenance Gardening Handbook.* New York: Harper and Row, 1970.

Cassiday, B. *Home Guide to Lawns and Landscaping.* New York: Times-Mirror Magazines, 1976.

Crockett, J. U. *Landscape Gardening.* New York: Time-Life Books, 1971.

Hudson, R. L. *The Pruning Handbook.* Englewood Cliffs, N J: Prentice-Hall, 1973.

Ishimoto, T., and K. Ishimoto. *The Art of Shaping Shrubs, Trees and Other Plants.* New York: Crown, 1966.

Pirone, P. P. *Tree Maintenance.* 4th ed. New York: Oxford University Press, 1972.

Sunset Books Editors. *Basic Gardening Illustrated.* 2d ed. Menlo Park, CA: Lane, 1975.

———. *Low Maintenance Gardening.* Menlo Park, CA: Lane, 1974.

Wallach, C. *The Reluctant Weekend Gardener.* New York: Macmillan, 1974.

Wyman, D. *The Saturday Morning Gardener: A Guide to Once-a-Week Maintenance.* rev. ed. Toronto: Macmillan, 1974.

Chapter 12

Lawn Establishment and Care

Lawns have long been a traditional part of the home landscape. Many landscape designers minimize or eliminate them altogether; others assert that a well-kept lawn is the most attractive setting for a home. Although it is true that care of the lawn is one of the most time-consuming maintenance activities, lawns are among the best surfaces for game and play areas.

ESTABLISHING A NEW LAWN

For owners of a new house, establishing a lawn is often the first landscape activity. Usually a lawn is seeded or sodded, but in some regions techniques called *plugging* and *stolonizing* are used. Each of these establishment methods is discussed in the following sections.

Establishment by Seeding

Establishing a lawn by seeding has advantages and disadvantages. First, seeding can require several months to make an attractive turf, whereas a sodded lawn is practically instant. More is also required to establish the seedlings, and there is a greater chance of failure. However, seeding is inexpensive, and when there is a large area involved, cost can be important.

TIME OF YEAR

The best time of year to seed a new lawn is dependent on the species of grass and the region in which it is being planted (Figure 12-1). In warm areas like the South and Southwest, warm-season turfgrasses (Table 12-1) are popular because they grow best in the temperature range of 80 – 95°F (27 – 35°C). Spring is best for establishing these grasses, since the hot summer temperatures to follow will supply ideal growing conditions.

Cool-season grasses such as bluegrass (Table 12-1) are the primary lawn grasses in areas with cold winters. These grasses grow best between 60 and 75°F (15 and 24°C). There, late summer is the preferred seeding period. Sown at this time, the grass will receive cool fall temperatures for establishment, and there will be less competition from annual weeds. Spring planting is also acceptable if fall is not practical.

SITE PREPARATION

Site preparation is the same whether seeding, sodding, plugging, or stolonizing is used, and it is a critical factor in assuring a quality lawn. Depending on the condition of the area, some or all of the following operations will be a part of site preparation.

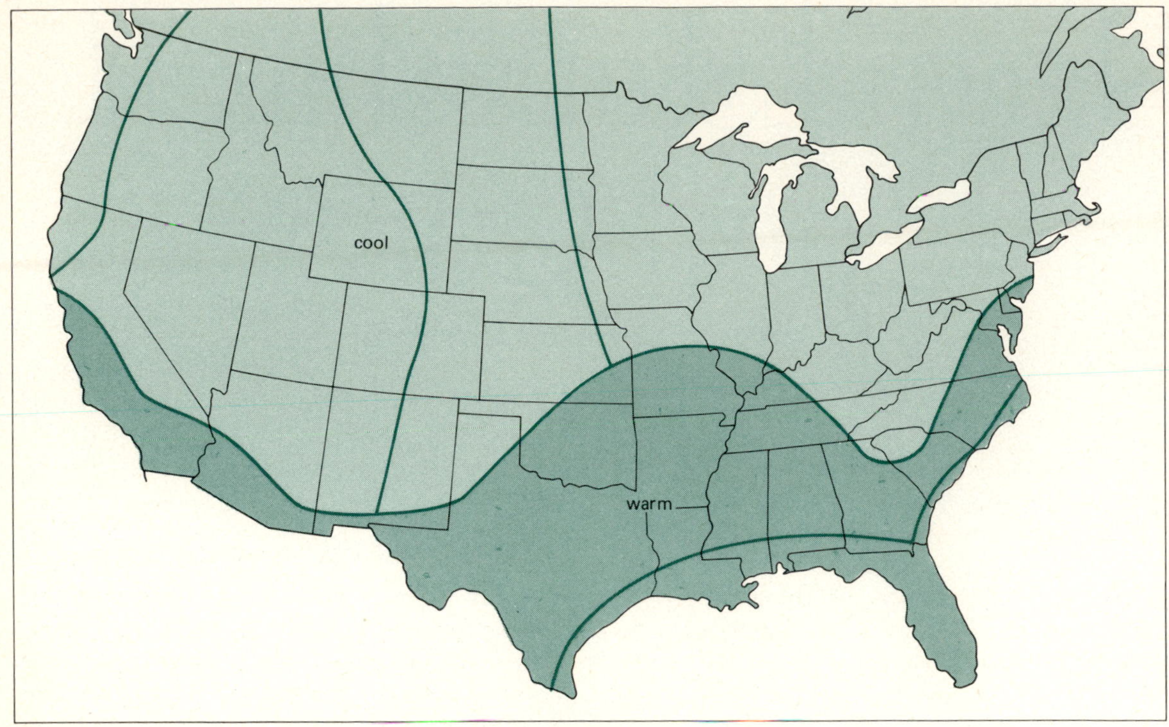

Figure 12-1 Areas for warm- and cool-season grasses.

Debris Removal. Debris removal is usually required around a newly constructed house. Pieces of wood, chunks of concrete, and other building materials will doubtless have been left at the construction site. Worse yet, they may have been buried and will later cause sunken areas or interfere with root growth. Debris removal should eliminate any foreign materials in the top 4 inches (10 centimeters) of the soil surface.

Elimination of Existing Vegetation and Preventative Weed Control. The best way to have a weed-free lawn is to prevent initial weed establishment. To accomplish this, the area should be evaluated in terms of the weeds already present and the potential for invasion from outside the lawn area. Noting whether the species are annual or perennial will pay dividends after the grass is planted. Perennial weeds can often be recognized by thick, fleshy roots or creeping underground stems. Generally speaking, you can assume that existing annual weeds will be killed during soil preparation, but that seedlings of the same species will soon reappear. Perennial weeds, on the other hand, will reestablish quickly from roots, tubers, and rhizomes after soil preparation is completed. Spading these weeds under or rototilling will spread them, resulting in a severe problem later. It is therefore important that perennial weeds be controlled prior to grading.

Perennial weeds may be controlled in several ways. If a small number of perennials is present or if time is short, they can be removed by digging. All portions of roots and stems must be removed, since even a small section will sprout a new plant. If the infes-

Table 12-1 Selected Turfgrass Species

Cool-season grasses
Velvet bentgrass *Agrostis canina*
Creeping bentgrass *Agrostis pelustris*
Colonial bentgrass *Agrostis tenuis*
Red fescue *Festuca rubra*
Perennial ryegrass *Lolium perenne*
Canada bluegrass *Poa compressa*
Kentucky bluegrass *Poa pratensis*
Warm-season grasses
Buffalo grass *Buchloe dactyloides*
Common Bermuda grass *Cynodon dactylon*
Bradley Bermuda grass *Cynodon incompletus* var. *hirsutus*
Magennis Bermuda grass *Cynodon × magennisii*
African Bermuda grass *Cynodon transvaalensis*
St. Augustine grass *Stenotaphrum secundatum*
Japanese lawn grass *Zoysia japonica*
Manila grass *Agrostis matrella*
Korean velvet grass *Zoysia tenuifolia*

tation is too widespread for hand digging, herbicides can be utilized. In selecting an herbicide, avoid those which kill only the tops, because the roots will survive and resprout. An ideal herbicide translocates through and kills the entire plant without leaving a harmful residue in the soil. Those available for homeowner use will often require several applications, and may delay the planting of the lawn for 4 to 8 weeks due to resprouting or soil residues.

Annual weeds are much more abundant than perennials, but they are much easier to control. One way to decrease the annual weed population in an area to be seeded is to allow the weeds to germinate prior to seeding, kill them, and then sow the grass. Using this weed-control procedure, called the stale seedbed technique, it is necessary to prepare the seedbed in the normal fashion until it is ready for seeding. At this point, irrigation is begun and continued for 7 to 10 days. If the weather is favorable, the weeds will sprout and can then be easily destroyed with a contact herbicide (one which kills the plant when sprayed on the leaves and stem). Be sure the herbicide selected does not leave a residue in the soil which will harm the grass. Care should be taken not to disturb the soil afterward, since this will cause new seeds to sprout. If the soil has become hard, a very light ¼-inch (6-millimeter)-deep raking may be used to break the surface crust. The grass can then be seeded, covered with a mulch, and irrigated. Resulting turf will generally contain 50 percent or less of the usual annual weed population. With some turf species, a preemergence herbicide (one which controls weeds as they germinate) can also be used at planting.

Some weeds will always be present after the grass has emerged, despite all precautions. These can be controlled by pulling or, after the lawn has been mowed once, by the application of a postemergence herbicide (one which controls a weed that has already germinated). Several postemergence herbicides will kill broadleaf (nongrass) weeds in turf; however, controlling emerged grassy weeds must still be done by hand.

Recommendations for specific herbicides are given in Table 12-1 or can be obtained from a local nurseryman, county Cooperative Extension Agent, or Farm Advisor. Like all pesticides, herbicides are dangerous to humans, pets, the environment, and desirable plants if used improperly. Read the lable thoroughly and follow all directions precisely. Particular care should be taken to prevent herbicides blowing to areas adjacent to the lawn (such as shrub beds), where damage to plants may occur.

Grading. Grading is generally done by the construction crew so that the ground slopes evenly away from the building at not less than a 1 percent grade. However, in some cases this has not been done or has been done improperly, or grade changes are called for in the landscape plan. In these cases care should be taken to remove the existing topsoil, grade the subsoil layer, and then replace the topsoil. When the topsoil is missing, it should be replaced with purchased topsoil of good quality mixed in to a depth of 4 inches (10 centimeters).

Soil Amending. Depending on the condition of the topsoil, soil amending can be needed to alter pH, improve fertility, or increase or lessen water-holding capacity with organic matter.

Ideally, a soil test should be made, and the proper pH- and fertility-correcting chemicals added as described in Chapter 5. However, it is usually sufficient to assume the pH is in an acceptable range and give a general fertilizer application of 2 pounds (1 kilogram) per 1000 square feet (92 square meters) each of nitrogen, phosophorus, and potassium. The fertilizer should be worked throughout the topsoil layer.

Amending the soil with organic matter need not follow any formula. Any of the amendments listed in Chapter 5 can be used, remembering the basic rule that a large amount of amendment is usually needed to cause any appreciable change in the soil.

Surface Preparation. The surface on which seed is to be sown should be free of clods and as smooth as possible. Clods can usually be broken up by rototilling or, in a small area, by hand. A light raking will remove remaining clods and provide the final surface.

SEEDING AND MULCHING

Seed Selection. Selecting the proper variety or mixture of grass for the area is important to growing an attractive lawn and should be done carefully. Selected grass species are listed in Table 12-1; because of the wide variety of grasses grown throughout the country, a discussion of the merits and drawbacks of all is not possible. The state Cooperative Extension Service or Farm Advisor's Office will be able to assist in the choice.

Buying seed is not simply a matter of finding the greatest amount of seed for the money. Information on both germination rate (percentage of the seeds which will germinate under ideal conditions) and purity (percentage of the seeds of the specific variety or mixture being purchased) should be noted. The label will give the germination test date, which should be within the current year. Otherwise the seed may be more than a year old and have a decreased germination rate.

The aim in buying seed is to obtain the highest germination rate and purity at the best price. In no circumstances should purity or germination rate be sacrificed in favor of low price. Problems of weeds and poor germination will far outstrip any small savings in seed cost.

Other factors to be considered include the amount of sunlight the area receives and the soil condition. Some seed mixtures are specially formulated for shade, poor soils, wet soils, and so on. Choosing a suitable seed mix can avoid problems later.

Seeding Rate. Generally speaking, package directions are the best source of information on seeding rate. They take into account variables of germination rate and ultimate plant density which would be very complicated for a homeowner to compute.

Several seeding rates may be listed on the package, depending on the speed of coverage wanted. Heavy seeding for quick results has some disadvantages, however. First, a heavily seeded lawn will sprout a large number of seedlings under overcrowded conditions. Eventually stronger seedlings dominate and form a stand of plants equal to that which would have been formed naturally at the lower rate. In the end, the sod may have taken longer to form at the high than at the low seeding rate.

Seeding Techniques. When seeding any area greater than 30 by 30 feet (9 × 9 meters), a mechanical seeder (Figures 12-2 and 12-3) should be used to assure even coverage. How-

Figure 12-2 Two types of mechanical seeders for distributing seed or fertilizer. Hand spreader (top), push type (bottom).

ever, for small areas hand seeding is satisfactory. Since grass seed is lightweight and easily blown by the wind, seed should be sprinkled on a windless day or during early morning or later afternoon, when the wind is likely to be at a minimum.

When seeding (or fertilizing with granular fertilizer), a pattern such as the one shown in Figure 12-3 will assure even coverage. To avoid missing strips between passes, the wheel of the spreader should be run just inside the previous wheel track. A steady walking pace is necessary to prevent double dosing and insufficient coverage problems.

Mulching. Mulches are applied to newly

Figure 12-3 A push-type mechanical seeder being run in a pattern to assure even coverage.

seeded areas to hold in moisture and to control erosion and washing away of the seed during rainfall or irrigation. Unlike mulches used for weed control, mulch over seed must be applied in a thin layer so the germinated seedlings can grow up through it. Coarse sawdust, shredded bark, peat moss, wood shavings, and straw are several mulches for this purpose; they should be applied so that the ground remains visible through the mulch (Figure 12-4).

POSTSEEDING CARE

Irrigation. Frequent irrigations to keep the soil surface moist are the key to obtaining good germination. Drying out even once during the germination and initial establishment stages can kill the majority of the seedlings and necessitate reseeding. Watering may be needed as often as three or four times a day to keep the soil surface moist. As the roots penetrate down into the ground, less frequent but longer waterings should be made; watering every 2 to 3 days will be sufficient.

Fertilizing. A light topdressing of about ½ pound (0.22 kilogram) of nitrogen per 1000 square feet (92 square meters) of lawn is recommended as the first fertilizing of turfgrass seedlings. The fertilizer should be applied only when the majority of the seedlings is 1½ to 2 inches (3.8–5 centimeters) high and should be immediately watered in.

Mowing. The first mowing should take place as soon as the seedlings exceed the normal cutting height. Afterward, mowing should be done regularly.

Figure 12-4 Straw mulch applied to a newly seeded lawn area.

Sodding

Sod consists of long strips of turf cut from the ground with the roots intact (Figure 12-5). Sodding can be done any time during the growing season, and it creates a high-quality lawn immediately.

BUYING SOD

Sod is expensive and should be selected carefully. An ideal sod is thick and uniform, free of weeds and diseased areas, and has only a small layer of thatch. You should always ask to see a sample of the sod you are purchasing. Specific arrangements for its delivery can then be made to allow adequate time for preparing the site in advance.

SITE PREPARATION

The same extent and methods of site preparation that precede seeding should also precede sodding. Negligence in site preparation may initially be covered up by a healthy sod, but the sod will not remain healthy if the soil into which it sends new roots has not been throughly prepared.

LAYING SOD

Sod must be laid as soon as possible after it is harvested. If it is stacked or left in rolls for more than a day or two, it will develop heat and disease problems and deteriorate quickly.

When laying sod, the strips should be staggered (Figure 12-5) so the ends do not form a line which will be noticeable later. Avoid stretching the sod, since it will shrink later and leave bare strips between the pieces.

After the sod is laid, the area should be rolled to ensure contact between the sod and the prepared soil beneath. Roll with a water-filled lawn roller in the opposite direction the sod was laid.

Figure 12-5 Laying sod.

CARE AFTER TRANSPLANTING

Immediately after being laid, sod should be watered thoroughly enough to wet both the sod itself and the first several inches (7–10 centimeters) of soil beneath. During the first 2 weeks it should be lightly irrigated once each day during the warmest period to minimize transpiration during reestablishment of the roots. Avoid walking, playing, or other heavy traffic for a couple of weeks.

Fertilization and weed control are usually not needed with sod during the first season. However, a light application of phosphorus can speed rooting of the turf.

Lawn Establishment by Stolonizing and Plugging

Stolonizing and plugging are used to establish new lawns of vigorous grasses such as Bermuda and creeping bentgrass. Stolonizing consists of spreading small stem sections or sprigs of the turf variety over a prepared area and then rolling or lightly covering with soil. The stolons root quickly and form a turf in several months.

In plugging, small 2- to 4-inch (5- to 10-centimeter)-diameter pieces of sod are transplanted into a prepared area 6 to 14 inches (15–35 centimeters) apart (Figure 12-6). The

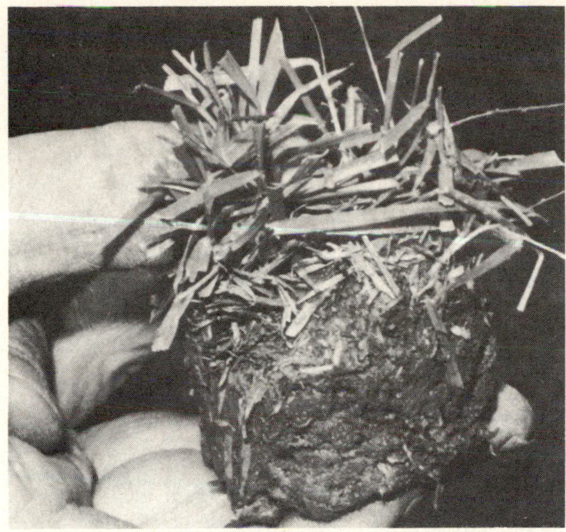

Figure 12-6 A plug of turfgrass ready to be planter in a lawn.

closer the plugs are spaced, the sooner the area will fill in. Warm-season grasses such as Bermuda grass, zoysia grass, St. Augustine grass, and buffalo grass are the species most commonly established by plugging.

LAWN RENOVATION

Renovation is used when a lawn is in poor condition but has not deteriorated to the point where it should be completely reestablished. The poor condition may be due to low soil fertility, weed invasion, thatch accumulation, soil compaction, or any combination of these and other causes. Although renovation can be started at any time during the growing season, late summer is best for cool-season grasses and late spring for warm-season grasses.

Determining the cause of the poor growth is the first step in renovation. If there are excessive weed growth, footpaths, and pale color, neglect may be suspected. However, if only isolated areas are dying, other causes should be suspected. Excess shade, poor drainage, diseases, and insects may be problems. Sending a soil sample for analysis will indicate nutritional or pH problems.

The next step should be weed control, by hand or with the chemicals recommended in Table 12-2. Excess thatch should be removed, if necessary, using a machine rented for this purpose.

Aeration of the soil can be done with a rented machine or a hand aerator. Finally, bare spots should be reseeded, followed by fertilizing as indicated by the soil test and a deep, thorough irrigation.

Given minimum care afterward, the turf should remain in good condition. If additional problems develop, or if renovation does not rejuvenate the lawn, consult your Cooperative Extension Service or Farm Advisor for advice.

LAWN MAINTENANCE

Mowing

Mowing is the most frequent maintenance chore in caring for a lawn, and it must be done correctly for the turf to look its best.

MOWING FREQUENCY

Mowing is essentially a pruning operation done on grass. It removes a portion of the photosynthesizing part of the plant, which in turn lessens the supply of carbohydrates going to the roots for growth and respiration. Thus roots are as affected by mowing as the blades of the grass.

Mowing should be as frequent as necessary to remove only 30 percent of the blade length at any one time. Cutting less often would remove more than 40 percent of the blade and be detrimental to the roots because of the abrupt drop in carbohydrate flow.

Table 12-2 Lawn Herbicides

Name	Time of application	Weeds controlled	Remarks
		Newly seeded	
Dalapon	Last application 3–6 weeks prior to planting	Annual and perennial grasses	Repeat applications required for perennials; soil residue will remain for 3–6 weeks; may leach into adjacent areas
Siduron	Immediately following seeding	Emerging annual grasses	Water immediately after treatment
Bromoxynil	After grass has emerged and weeds are in the 2- to 4-leaf stage	Many broadleaf weeds	Larger weeds require higher rates
		Established	
Benefin	Spring, before weeds germinate	Crabgrass and other annual grasses	Does not leach readily; do not use on bentgrass
Dacthal	Spring, before weeds germinate	Crabgrass and other annual grasses	Less effective on heavy soils
Bensulide	Spring, before weeds germinate	Crabgrass and other annual grasses, a few broadleaves	
2,4-D amine	When weeds are actively growing	Most annual and perennial broadleaves	Do not apply when windy; do not use on St. Augustine grass
2,4,5-TP (Silvex)	When weeds are actively growing	Most annual and perennial broadleaves, oxalis, chickweed	Do not apply when windy; often applied in combination with 2,4-D
DSMA	Postemergence	Young crabgrass, chickweed, oxalis, Dallis grass, sandbur	Do not use on St. Augustine, Bahia grass, or centipede grass; apply 1–3 times 5–7 days apart
Aminotriazole	When weeds are growing rapidly	Annual and perennial weeds	*Spot treatment only;* will kill turf if contacted; brush on with paintbrush

MOWING HEIGHT

The height at which grass should be maintained varies with the species and may range from ½ to 4 inches (1.3 – 10 centimeters). Table 12-3 lists suggested mowing heights for grasses grown throughout the United States.

Mowing at a height less than suitable for a particular grass constitutes "scalping" and leaves the turf an unattractive brown stubble. Occasional scalping will not permanently harm grass. But if repeated on a susceptible area (such as a berm top), it will eventually kill the turf.

RETURNING CLIPPINGS

Many people use grass catchers to remove clippings during mowing. Although this does give a neater surface, it is not essential to turf

Table 12-3 Preferred Cutting Heights of 19 Turfgrasses as Determined by the Resulting Turfgrass Quality and Vigor[a]

Relative cutting height	Cutting height (in.)	Turfgrass species
Very close	0.2–0.5	Creeping bentgrass Velvet bentgrass
Close	0.5–1	Colonial bentgrass Annual bluegrass Bermuda grass Zoysia grass
Medium	1–2	Buffalo grass Red fescue Centipede grass Carpet grass Kentucky bluegrass Perennial ryegrass Meadow fescue
High	1.5–3	Bahia grass Tall fescue St. Augustine grass Fairway wheat grass
Very high	3–4	Canada bluegrass Smooth brome

[a]Extracted from James B. Beard, *Turfgrass: Science and Culture*, Prentice-Hall, Englewood Cliffs, NJ, 1973.

maintenance. Returned clippings supply additional nutrients when they decay and reduce fertilization requirements. Clippings need only be removed if they are overly heavy and long and would inhibit passage of light to the grass underneath.

MOWER OPERATION

Mowing only when grass is dry is suggested for two reasons: first, because the mower is less likely to clog with soggy clippings and the clippings will distribute evenly rather than in clumps; second, because grass blades are more susceptible to disease invasion through the cut surface when wet.

The importance of sharpening mower blades so they cut grass cleanly is often overlooked. A clean cut minimizes the damaged area and reduces tip browning of the blades. It also decreases moisture loss and eliminates bruised blades, which are a site of disease entry.

Irrigation

As for irrigation of any plant, watering of grass should be done according to need rather than on a fixed schedule which does not take weather conditions into account. The aim is to apply water just before wilting, wet the top 6 to 8 inches (15–20 centimeters) of soil where the roots grow, and wait until wilting approaches before watering again.

Determining exactly when wilting is about to occur is difficult. One technique called "footprinting" is advocated by Dr. James Beard of Michigan State University. Walk across the grass, then observe how quickly the turf stands upright again. Turf with adequate water will recover rapidly, while turf approaching wilting will recover very slowly.

Where rainfall is insufficient, a permanent sprinkler system can be installed before surface preparation and seeding. Pop-up sprinkler heads, which are below cutting level normally but are elevated above the turf by water pressure during operation, are the most popular. They can be used in conjunction with plastic pipe to create a durable inexpensive system easily installed by a homeowner. To minimize water loss by evaporation, sprinklers should be used late in the afternoon or early in the morning. The former is preferable when disease problems are common because the foliage will not remain wet and susceptible to infection through the night.

Subsurface irrigation is a relatively new technique in use primarily on golf courses. Underground pipes spaced closely throughout the turf area ooze water at a slow rate. The water is moved by capillary action to the grass roots. The main advantages of a subirrigation system are the elimination of evaporation and runoff. Disadvantages include the possibility of clogging and buildup of salts. In addition, not all soils have sufficient capillary action to utilize the system.

Fertilization

CHOOSING THE FERTILIZER

Since the goal in fertilizing a lawn is the production of vegetative growth, the fertilizer chosen should be high in nitrogen. Fertilizers with two to four times as much nitrogen as either phosphorus or potassium are suggested, in ratios such as 4-1-2, 23-7-7, and 6-3-0. In addition, it is best to buy fertilizers with part of the nitrogen in a form that is immediately available and another part in a slow-release form.

Combinations of fertilizer plus herbicide, fungicide, or insecticide are widely sold for use on lawns. Their use eliminates separate applications of each chemical, but extra care must be taken to apply them at the proper rate.

TIMING AND RATE OF FERTILIZER APPLICATIONS

The frequency at which fertilizers should be applied to a lawn depends on whether the goal is to establish a young lawn or to maintain an older one, on the soil type, and on the weather conditions. The following recommendations are based on a loam soil with irrigation or precipitation every 1 to 2 weeks. In lighter soils or with more frequent irrigation, fertilizer should be applied more often.

Young Lawns. Establishing lawns require frequent fertilizer for rapid growth. As a general guide, a ½ to 1 pound (0.22 – 0.45 kilogram) per 1000 square feet (92 square meters) application of immediately available nitrogen should be applied when grass reaches 2 inches (5 centimeters). This should be followed by applications at the same rate every 3 to 4 weeks during the first growing season. In cold-winter climates applications should be stopped about 1 month before the onset of winter dormancy.

Established Lawns. Established lawns require less frequent fertilization to maintain an attractive appearance. Two fertilizer applications per year, one in early spring and one in fall, are normally sufficient. Two pounds (1 kilogram) of nitrogen per 1000 square feet (92 square meters) is recommended at each application.

In areas with mild winter temperatures grass frequently does not have a dormant period and will grow throughout the year. Accordingly, an additional fertilizer application will be needed in midwinter to supply nutrients for healthy growth.

FERTILIZER APPLICATION METHODS

Fertilizer is normally applied to lawns in granular form using a broadcast and water-in method. A fertilizer spreader is best to assure even coverage over large lawns. Hand broadcasting or applying liquid fertilizer with a hose-end applicator is practical for small areas. The latter method is used by some commercial lawn care companies and is valuable for very sandy soils with a poor nutrient-holding capacity. A diluted solution can be mixed and applied in place of plain water at recommended intervals.

Dethatching

Thatch is a layer of stems and roots (both dead and living) which accumulates between the turf surface and the soil (Figure 12-7). The accumulation occurs when the rate at which the turf sloughs off dead parts is greater than the rate at which they decompose.

Although a thin layer of thatch is not harmful, any layer greater than ½ inch (1.3 centimeters) can cause problems. The heavily thatched area may dry out fast, diseases and insects may invade, and there may be decreased tolerance to heat, cold, and drought. Vigorous grass varieties are the must susceptible to thatch accumulation because of their rapid growth rate.

Excess thatch is detected by examining a cross section of turf. It can then be removed mechanically by a "vertical mower" (Figure 12-8). This machine cuts and pulls the turf up in narrow strips. The material is then raked off and the lawn watered and fertilized. Timing of the mowing should allow at least 30 days of good growing conditions afterward for recovery of the turf. Early spring is suitable on most grasses.

For small lawn areas, renting a vertical mowing machine may be impractical. Hard raking with a stiff metal rake will loosen and remove the thatch.

Preventing future buildup of thatch involves altering the care of the turf to balance the accumulation and decomposition rates. If irrigation or fertilization has been excessive, it should be reduced to slow the growth of the turf. Liming may be needed since the pH of thatch is generally too low for the optimum decay rate. Finally, topdressing the turf with soil will hold in the moisture necessary for decomposition and provide a source of microorganisms for the breakdown process.

Figure 12-7 A thatch layer in turfgrass.

Figure 12-8 Dethatching with a vertical mowing machine.

Aeration

Aeration is needed when a lawn has been compacted by foot or vehicular traffic. Compaction is particularly prevalent in clay-based soils. Under compacted conditions, the movement of oxygen and other gases into and out of the soil is restricted. Water will soak in very slowly or run off. Turf roots will eventually begin to die.

Aeration removes small cores or slices of grass and soil from the turf area, penetrating to a depth of up to 4 inches (10 centimeters) from the soil surface. The remaining soil then can expand into the open area and relieve the compaction problem. Engine-driven aerators can be rented if a large area is to be treated. For spot treating or small lawns, a home aeration tool, which is used in a similar manner to a shovel, is more practical.

Weed Control

Weeds in a lawn are undesirable because they detract from its even appearance. Their varying textures and colors make them stand out, and their tendency to grow faster than the surrounding turf increases the mowing frequency. Seasonal weed species (crabgrass, for example) crowd out desirable turf and then die, leaving bare spots in the lawn. When weeds flower and set seed, they are even more conspicuous in a lawn and cause additional problems. Seed heads of common Bermuda grass, for example, generally will not be cut by reel-type mowers but must be removed by hand. Others have seeds that are spiny (burr clover and puncture vine) and will injure bare feet.

The objective of lawn weed control should be to keep weed populations within a tolerable range, not necessarily to have a flawless lawn. A few weeds seldom detract from the lawn enough to require control measures. In fact, many homeowners prefer a lawn containing a combination of plants including clover and ground ivy.

A healthy lawn will generally not have severe weed problems. However, due to disease, insects, or poor cultural practices, weeds sometimes become established in large numbers. Thus the first line of defense against weeds is to keep the grass healthy. Correct mowing height and frequency, fertilization, and insect and disease control will prevent weed seedlings from becoming established and even enable the grass to outcompete established weeds. For the few remaining, hand pulling is a sound control method. Pulling after a rain or irrigation will help remove plants with taproots.

Occasionally, weeds such as crabgrass become a severe problem in even the healthiest of lawns. Preemergence or postemergence herbicides can then be utilized. Crabgrass is most often controlled with a granular herbicide (often combined with a fertilizer) applied preventatively in spring before the crabgrass germinates. Weeds escaping preemergence treatment can be hand pulled later or sprayed with a "selective postemergence herbicide" that kills only the weeds without harming the grass. Several applications may be necessary for complete control. Table 12-2 lists herbicides used for weed control in lawns.

DICHONDRA

There are a number of plant species which will tolerate some foot traffic and can be planted as grass substitutes. Dichondra is the best known of these, although its use is limited to the warmest climates of the country.

Dichondra (Figure 12-9) is an extremely attractive grass substitute with a flat, matlike habit of growth. It is restricted to areas where the winter temperature does not drop below 25°F (−4°C) and is tolerant of the very warm summer temperatures such as those found in the Southwest.

A dichondra lawn must be established in spring or summer for soil temperatures to

Figure 12-9 Dichondra, a popular lawn in California and Florida.

be sufficient for seed germination and rapid growth. Soil preparation, sowing, and post-seeding care are essentially the same as for grass. Mulches of coarse sawdust, shredded peat moss, or other relatively fine material are preferred over straw because the creeping growth habit of the dichondra will not readily obscure a straw mulch.

Plug planting can be used to establish a dichondra lawn and is useful when only a small area is to be planted. In large areas the expense is prohibitive. As with plug planting of grasses, the area should be thoroughly prepared beforehand to facilitate rapid spreading. Flats of dichondra can be cut into plugs 2 inches (5 centimeters) square or larger and transplanted 6 to 12 inches (15–30 centimeters) apart.

The dichondra lawn requires regular watering and frequent fertilization to remain attractive. From this standpoint, it is not a low-maintenance plant. Fertilizing every 2 to 3 weeks is best, using a high-nitrogen fertilizer at 1 pound (0.45 kilogram) per 1000 square feet (92 square meters) of area.

Mowing frequency depends on growing conditions. Shade and abundant moisture increase dichondra lushness and height, and mowing may be required once per month. In full sun, growth will stay low, and mowing will never be required.

Primary weeds which invade dichondra are oxalis, chickweed, and miscellaneous lawn grasses such as Bermuda grass and crabgrass. Herbicides to control them are recommended in Table 12-4.

Table 12-4 Dichondra Herbicides

Name	Time of application	Weeds controlled	Remarks
		Newly seeded	
Diphenamid	Preemergence to weeds at seeding dichondra	Annual grasses, some broadleaves; weakens Bermuda grass but does not kill it	Follow label directions
Bensulide	Preemergence at seeding	Most annual weeds	Follow label directions
Napropamide	Preemergence at seeding	Most annual weeds	Follow label directions
		Established	
All of the above	Above	Above	Follow label directions
Monuron	Preemergence or when weeds are in the two-leaf stage	Most annual grass and broadleaf weeds, oxalis	Use exact rate; irrigate immediately after application
Neburon	Preemergence	Most annual grass and broadleaf weeds	Safer than monuron, lasts longer in the soil; available only in weed and feed mixtures
DSMA	When weeds are young	Crabgrass	Follow directions carefully; 1–3 applications 5–7 days apart are required
Dalapon	When Bermuda grass is growing actively (summer)	Bermuda grass	Use 2 oz/1000 ft^2 with no additional wetting agent; repeat each time regrowth occurs; use diphenamid treatments during winter to improve control
2,4-D	When weeds are growing actively	Broadleaf weeds	*Spot treatment only;* brush onto leaves of broadleaf weeds with a paintbrush; will kill dichondra if contacted

OTHER LAWN SUBSTITUTES

Dichondra is the only widely used lawn substitute for large areas. However, a number of uncommon plants will tolerate limited walking such as a front yard might receive. Most will be hard to find, and all are established by plugging. It is advisable to try a small area before using the species to cover an extensive lawn area.

Table 12-5 lists a number of lawn-substitute plants along with their temperature hardiness and ideal growing conditions. Many of the plants mentioned here are suited to growing in shade, where grass is particularly difficult to maintain.

Table 12-5 Lawn-Substitute Plants

Common name	Botanical name	Cold hardiness	Remarks
Corsican sandwort	*Arenaria balerica*	−5 to 5°F (−21 to −15°C)	Needlelike leaves; grows up to 3″ (7 cm) high; best in shade with moist soil; produces white flowers in May
Mountain sandwort	*Arenaria montana*	−20 to −10°F (−29 to −23°C)	Grows 1″ (2.5 cm) high with grasslike leaves; tolerant of poor soil and flowers profusely in spring with white blooms
Moss sandwort	*Arenaria verna*	−35 to −20°F (−37 to −29°C)	Bright green needlelike foliage; mat-forming, and develops lumps unless thinned occasionally
Camomile	*Chamaemelum nobile*	−20 to −10°F (−29 to −23°C)	Often sold under the name *Anthemis nobilis* and used commonly as a lawn plant in Europe; withstands drought and emits a pleasant odor when stepped on; fernlike, bright green leaves and good traffic tolerance; daisylike white flowers
Mock strawberry	*Duchesnea indica*	−10 to −5°F (−23 to −21°C)	2″ (5 cm) high with yellow flowers in spring; strawberrylike leaves
Beach strawberry	*Fragaria chiloensis*	−20 to −10°F (−29 to −23°C)	Native plant good for sandy soils and spreading by runners; white flowers and small edible fruits in spring
Colchis ivy	*Hedera colchica*	−10 to −5°F (−23 to −21°C)	Relatively rare ivy with heart-shaped leaves which tightly overlap over the ground; shade-tolerant
Mazus	*Mazus reptans*	−35 to −20°F (−37 to −29°C)	Deciduous plant, although leaves hang on long in fall; plant grows 1″ (2.5 cm) tall with 1″ (2.5 cm) leaves; purple flowers in May; prefers moist soil and will grow in partial shade
Dwarf lily turf	*Ophiopogon japonicus*	5–10°F (−15 to −12°C)	Dark evergreen foliage is grasslike, to 6″ (15 cm) tall; covers an area rapidly, grows in shade or sun, and produces lilac to white flowers in summer; good in sandy soil.
Garden lippia	*Phyla nodiflora*	0°F (−18°C)	Sometimes sold under the name *Lippia repens* or *Lippia canescens;* has a matlike growth habit with gray-

Table 12-5 (continued)

Common name	Botanical name	Cold hardiness	Remarks
			green leaves and is good in heat and sun; mow occasionally to maintain an even surface; very drought- and traffic-tolerant
Rusty cinquefoil	*Potentilla cinerea*	−35 to −20°F (−37 to −29°C)	Strawberrylike leaves and yellow flowers on a plant 2 to 4″ (5–10 cm) high
Self-heal	*Prunella vulgaris*	−35 to −20°F (−37 to −29°C)	Native to temperate Northern Hemisphere; leaves are semievergreen, lying flat on the ground; purple flower spikes in summer; best in damp soil and spreads rapidly
Pearlwort	*Sagina subulata*	−20 to −10°F (−29 to −23°C)	Mat-forming with needlelike leaves, similar to *Arenaria;* grows up to 4″ (10 cm) tall and is good in shady areas
Creeping thyme	*Thymus serphyllum*	−35 to −20°F (−37 to −29°C)	Very flat growing (less than 1″; 2.5 cm); evergreen will withstand poor soil, shade, drought; small purple flowers in summer; many species of thyme are sold under this name, including *Thymus lanuginosus* and *Thymus lanicaulis,* and since these grow taller, they are less tolerant of foot traffic
Creeping speedwell	*Veronica repens*	−10 to −5°F (−23 to −21°C)	Grows to 4″ (10 cm) tall, with a mosslike appearance and shiny foliage; bluish flowers in spring

Selected References for Additional Reading

American Society of Agronomy. *Turfgrass Science.* Ed. A. A. Hanson and F. V. Juska. Madison, WI: American Society of Agronomy, 1969.

Atkinson, R. E. *The Complete Book of Groundcovers: Lawns You Don't Have to Mow.* New York: David McKay, 1970.

Beard, J. B. *Turfgrass: Science and Culture.* Englewood Cliffs, N J: Prentice-Hall, 1973.

Organic Gardening and Farming Magazine Editors. *Lawn Beauty the Organic Way.* Emmaus, PA: Rodale Books, 1970.

Sunset Books Editors. *Lawns and Groundcovers.* 3d ed. Menlo Park, CA: Lane, 1964.

Chapter 13

Diagnosing and Treating Outdoor Plant Disorders

The diagnosing and treating of some plant disorders are easy; for example, wilting usually indicates lack of water, and caterpillars found chewing through a leaf are proof of an insect infestation. But as often as not, a problem is hard to diagnose, and a systematic approach to solving the problem should be used.

IDENTIFYING THE AFFECTED PLANT

The first step in diagnosing a plant problem is identifying the plant that is affected. A common name is sufficient for fruits, nuts, or vegetables, since those names are standard throughout the country. However, with ornamentals, the botanical name is often essential.

Identifying the plant will narrow the list of possible causes considerably, since species vary in their susceptibility to different maladies. A species will be prey to the same problem repeatedly, and a knowledgeable nurseryman can often diagnose a problem given only the plant name and the symptoms.

DETERMINING THE CAUSE OF THE PROBLEM

Plant problems are almost always attributable to either a parasitic organism or an environmental condition. Parasitic organisms include insects, rodents, and microorganisms such as bacteria and fungi. Environmental conditions include heat, drought, cold, and the like. Each of these causes of plant injury will be discussed along with its typical symptoms in the following sections.

Insects and Related Pests

Insects pass through several stages in their maturation process, and in one or more than one of these stages, they may damage plants. Butterflies are the best-known example of this phenomenon. They begin life as eggs laid on a host plant. The eggs hatch into a larva or caterpillar form, and during this stage the insect feeds on and causes damage to plants. After reaching full size, the caterpillar spins a cocoon (pupates), later emerging as the adult butterfly, a nectar-feeding form. However, although the butterfly itself is harmless, it lays the eggs for the next generation of caterpillars.

This egg-larva-pupa-adult sequence (Figure 13-1) is common to many insects, with the larva stage often referred to as a worm or grub. The adult will vary, with adult beetles, flies, or moths emerging as commonly as butterflies.

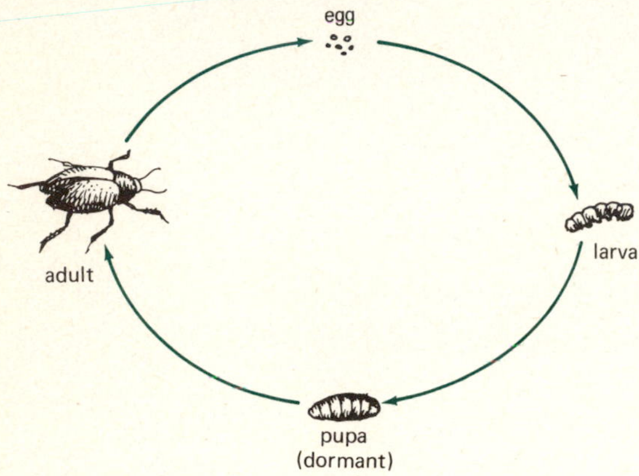

Figure 13-1 The typical maturation sequence of a metamorphosing insect.

Damage to plants by insects is almost always from feeding of two types: chewing and sucking. Chewing insects include many larvae (such as caterpillars and grubs) as well as grasshoppers, boring insects, snails, and beetles. These insects eat through plant parts rapidly, becoming larger and more damaging as they mature. Sucking insects, on the other hand, do not chew holes through a plant. They feed on the sap either by inserting their mouthparts into the phloem or by rasping a small area and feeding on sap oozing from the wound. Sucking insects include pests such as aphids, scales, and mites. Their feeding damage is harder to diagnose.

DIAGNOSING INSECT DAMAGE TO PLANTS

Damage by chewing insects is relatively easy to identify. Irregular holes (Figure 13-2) appear on leaves, or a stem may be severed completely. The insects may still be present and eating or may have left droppings on the plant.

If insects are suspected but not present, an effort should be made to determine what insect is causing the damage, for this affects the control to be applied. The plant should first be inspected under the leaves and near buds and feeding sites. If no insects are present, the pests may be in the soil at the base of the plant. Many insects hide there during the day, coming out to feed at night. For the same reason, checking the plants after dark will often reveal the insect.

Sucking insects present a different detection problem. They are usually small (some microscopic), and their damage varies. Typical symptoms include puckering and bleaching of the leaves (Figure 13-3). A hand lens is helpful in detecting these pests.

Recognizing damage by insects feeding on roots is not as easy as recognizing aboveground infestations. Grubs and other larvae are most often the culprits, chewing the roots of a plant until the root system is insufficient to sustain the foliage. The resulting symptoms are the same as for any condition harming the roots: yellowing, wilting, and eventual death of the top portions. When digging up the plant, signs of insect damage (and often the insects themselves) will be apparent.

The following sections include most of the chewing and sucking pests found in home gardens.

CATERPILLARS

These are chewing insects, the larvae of beetles, butterflies, and moths. Just before pupation they consume amounts of leaves equaling several times their body weight each day.

Figure 13-2 Damage due to a chewing insect, in this case a caterpillar. Similar damage is caused by slugs and snails.

Figure 13-3 The leaf on the bottom is puckered from a sucking insect feeding on it. The leaf on top is normal.

GRUBS AND BORERS

Grubs and borers are usually the larvae of beetles. They live underground eating plant roots (grubs) or burrow into stems and fruits (borers). Some, such as the Japanese beetles, are destructive in both larval and adult stages.

LEAF MINERS

Leaf miners are miniscule worms which tunnel between the upper and lower leaf surfaces. Their feeding develops scribblelike white patterns on affected leaves.

GRASSHOPPERS

Grasshoppers are chewing, defoliating insects throughout their life cycle. They have no distinctly different larval, pupal, and adult stages; young grasshoppers resemble the adults.

BEETLES

Beetles damage large numbers of plants by chewing and are particularly troublesome in late summer.

APHIDS

Aphids are small, sucking insects generally found clustered on the growing tips of plants. Their feeding causes disfigurement of the leaves and buds (Figure 13-4).

SNAILS AND SLUGS

These pests feed at night, chewing holes in leaves, flowers, and fruits. Because they are not insects, they cannot be controlled with most insecticides.

SCALE AND MEALYBUGS

These sucking insects spend their lives in colonies attached to leaves or stems. Scales (Figure 13-5) look much like raised spots, and mealybugs (see Figure 18-2) like dots of cotton. Their unique appearance stems from the waxy secretion which oozes from the back of the insect after it has settled and begun feeding. Once protected in this manner, the pests are unaffected by most insecticide sprays.

Figure 13-4 Typical symptoms of aphid damage to a delphinium.

MITES

Mites cluster on the undersides of plant leaves, sucking juices and giving a pinprick pattern to the top surfaces. In heavy infestations, they spin fine webs. Since the insects are nearly microscopic, a magnifying glass and strong light are helpful in diagnosing a mite problem (Figure 13-6).

THRIPS

Thrips are sucking pests only slightly larger than mites. They attack leaves or flowers and cause damage in the form of streaks or small bleached spots (Figure 13-7).

GALL-FORMING INSECTS

Both mites and aphids can cause the abnormal swellings known as galls. Although unsightly,

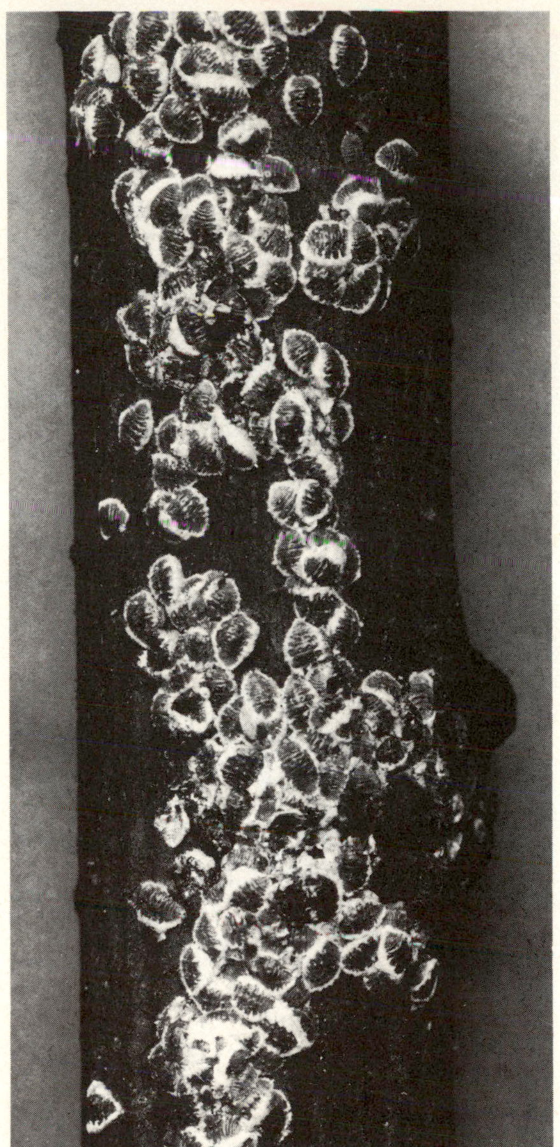

Figure 13-5 European elm bark scale.

most are not seriously deleterious to plant health.

NEMATODS (EELWORMS)

Nematodes are microscopic or nearly microscopic worms. Most are soil dwelling, burrowing into roots and decreasing their ability to translocate nutrients. They cause yellowing and general lack of vigor. Root nematodes can sometimes be seen with a magnifying glass, but one of the most common types, the root-knot nematode, is identified by the characteristic swellings formed on roots by its feeding (Figure 13-8).

Damage from foliar nematodes is less common. But some species emerge from the ground and crawl up to the leaves on a film of water such as from dew or rain. They then enter the leaves through the stomata and begin feeding. Typical symptoms include death of irregular sections of leaves and spotting.

Microorganisms Damaging to Plants

Several types of microorganisms damage plants, causing diseases which kill the foliage or rot the roots. These microorganisms can be divided into four types—fungi, bacteria, viruses, and mycoplasmas. Each is discussed in the following sections.

FUNGI

Fungi are single- or multicelled plants which lack chlorophyll. Therefore they rely on other sources for carbohydrates and parasitize green plants. A typical fungus has an extended threadlike body which increases in size by cell division. At maturity, a reproductive system is formed that produces dustlike spores which spread the fungus.

Fungi are responsible for more plant diseases than any other microorganism, but they are also the most easily treated. The diseases vary widely, but most can be classified into one of the types listed in Table 13-1.

BACTERIA

Bacteria are single-celled, lack chlorophyll, and frequently grow in colonies. They are the cause of relatively few plant diseases, but plants infected by them are difficult to treat.

Figure 13-6 Spider mites.

The diseases caused by bacteria include leaf spots, rots, cankers, blights, and wilts. In addition, bacteria can cause galls.

VIRUSES

Viruses (Figure 13-13) are among the smallest microorganisms, and only a few cause plant diseases of importance to the home gardener (Table 13-2).

They are spread between plants by insects. The insect feeds on a virus-infected plant, contracts contaminated mouthparts, moves to an uninfected plant, and spreads the infection. Aphids, mites, leafhoppers, and nematodes can spread plant viruses. The virus will eventually be fatal to the plant, although death may be quite slow. To date, virus diseases are untreatable.

MYCOPLASMAS

Mycoplasmas are very similar to bacteria and were long thought to be a small bacteria type. Of the 30 types that are known, only a few cause plant diseases. Symptoms are similar to those of virus disease and are difficult to diagnose.

Figure 13-7 Cuban laurel thrips on an ornamental fig (*Ficus* spp.).

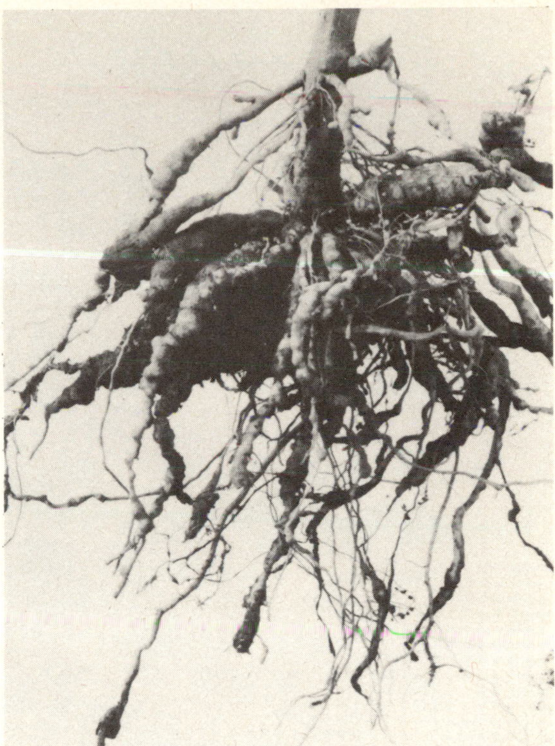

Figure 13-8 Root-knot nematode damage to a tomato.

Large Animal Pests Damaging to Plants

A number of rodents and similar animals cause damage in the garden. The more common ones and their symptoms are discussed in the following sections.

RABBITS

Rabbits are troublesome in many areas of the country. They nibble vegetable gardens and the tender spring growth of many ornamentals including peony and dahlia.

MICE

Mice are a problem chiefly in the winter when food is scarce, burrowing under mulch or snow layers to feed on the bark of fruit trees. The feeding can girdle the tree and cause its death.

POCKET GOPHERS

These burrowing rodents create a nuisance by digging holes and tunnels throughout the garden (Figure 13-14). They also feed on roots and can kill a large plant in a day or two. Pocket gopher damage can be easily diagnosed; a plant will wilt suddenly and, when lifted, will come out of the ground easily and have no roots.

MOLES

Moles also burrow, but feed mainly on insects they find in the soil. The damage results from their burrows, which leave a raised pattern throughout an area with mounded exit holes.

Table 13-1 Fungus Diseases of Cultivated Plants

Disease	Symptoms and parasitizing pattern
Mildew (Figure 13-9)	The fungus grows on the surface of plant leaves and stems. The threadlike body can be easily seen with the naked eye as a powdery white covering. Infected leaves eventually die and drop, and the whole plant dies.
Wilt	Fungi invade and multiply in the phloem and xylem of the plants, clogging them and restricting translocation. The entire plant or only a section will wilt and not recover.
Rot	Roots, underground storage organs, and the lower portions of stems are attacked by these soil-inhabiting fungi. Cells in infected areas die and decay, disrupting absorption and translocation and causing the death of the plant.
Leaf spot (Figure 13-10)	Blotches or spots appear on the foliage and, sometimes, stems or flowers of plants, spreading rapidly to cover the plant. The spots are usually very distinctive in appearance, having, for example, a dark center surrounded by rings of different-colored tissue. The spots gradually reduce the photosynthetic area of the leaves and cause the death of the plant.
Rust (Figure 13-11)	The fungi which cause rusts are distinct by virtue of requiring two different species of host plants in order to complete the life cycle. Often one host will be a cultivated plant and the other a native plant or weed. Symptoms of rusts are reddish-brown spots on the foliage and stems of susceptible plants. On some species, the fruiting body which bears the spores is a 1-in. mass of orange gelatinous horns with a bizarre appearance.
Canker (Figure 13-12)	Relatively uncommon disease affecting woody plants. Fungi cause a sunken bark area, which girdles the branch or trunk and eventually kills that portion of the plant above the infected area.
Smut	Uncommon disease type most often encountered on corn. Invaded portions swell drastically, eventually bursting open and dispersing a multitude of dusty black spores.
Witches'-broom	Mainly a disease of woody plants and not frequently encountered. Fungi infect the entire plant, but symptoms appear on small branches as unrestrained twiggy growth resembling a bird nest. The disease is sometimes fatal.
Blight	Usually affects woody plants and causes quick wilting and death of young and growing tissues such as flowers and twigs. Can spread downward to kill more mature portions of the plant.

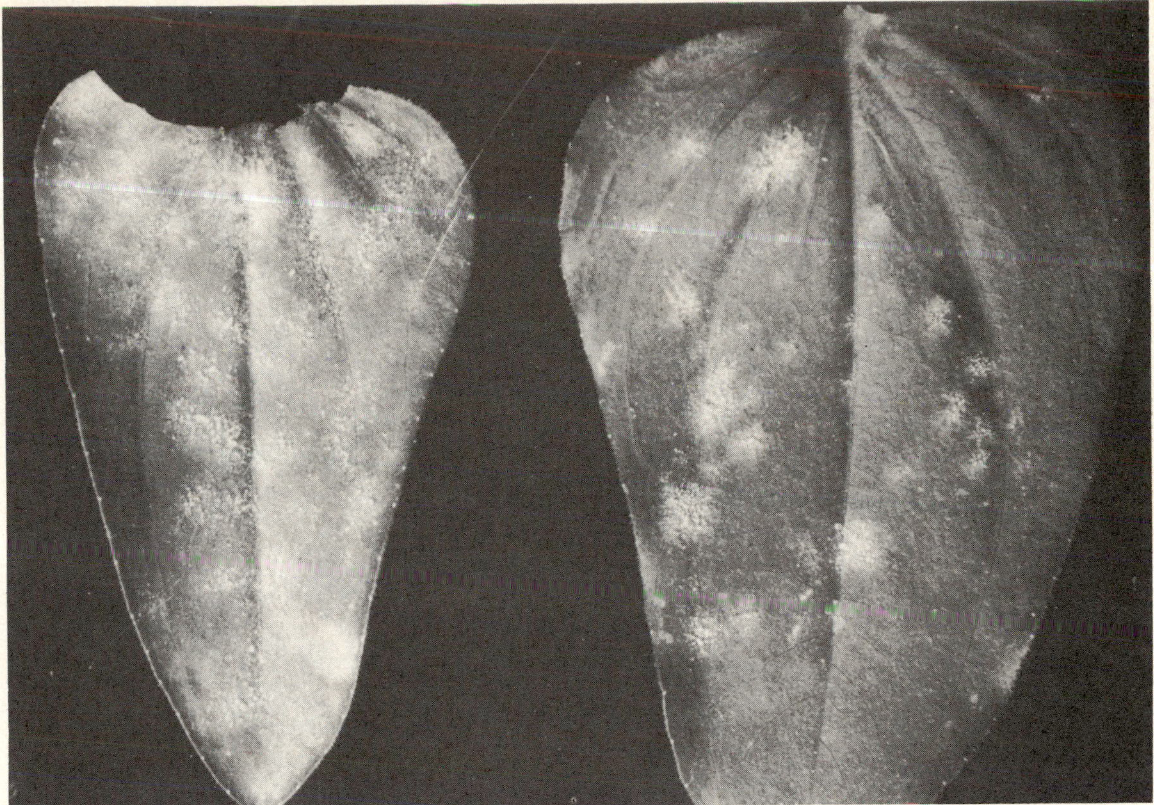

Figure 13-9 Mildew on zinnia leaves.

Moles are particularly damaging to lawns because they burrow through grass roots, causing the death of the turf above.

DEER

Deer are an increasing problem in rural areas and the West. They eat twigs from shrubs or trees and will also attack vegetable plantings. They usually feed late at night.

BIRDS

Birds are a problem where fruit plantings are grown. They will eat cherries, raspberries, and other species.

DOGS

Dog urine is toxic to plant foliage, causing browning or yellowing. Male dogs frequently damage the lower leaves of shrubs, and females leave circular spots on turf.

Environmental and Other Nonparasitic Causes of Plant Injury

FROST DAMAGE

Plants differ in their susceptibility to frost according to age and species: young tissues are more tender than mature ones; woody plants are more frost-resistant than annuals. Accordingly, a light frost may strike sporadically, damaging one plant and not the next. A key symptom is the sudden overnight wilting of tender plants, particularly on the uppermost leaves and buds.

Figure 13-10 Leaf spot disease on strawberry leaves.

WINTERKILL

Winterkill is the partial or total death of perennial plants due to cold temperatures. In minor cases a few branches will die back, and the plant will resprout from buds on the lower stems or crown. With complete winterkill, both tops and roots die. This occurs during very cold winters when the critical temperature for root survival is exceeded.

Table 13-2 Typical Virus- and Mycoplasma-caused Plant Diseases

Disease	Causal organism	Symptoms
Stunt	Virus or mycoplasma	Slowed or stopped growth under favorable growing conditions and with no other apparent cause
Curly top	Virus	Causes twisting and distortion of foliage; symptoms are similar to weed-killer injury
Mosaic	Virus	Mottled light and dark areas on plant foliage
Yellows	Mycoplasma	Yellowing of foliage, stunting, and witches'-broom-type shoot growth under good growing conditions and due to no other apparent cause

Winterkill or "burning" of evergreens is a common occurrence in cold climates. Leaf tissues become brown, and since they do not refoliate yearly, the appearance of the plants can be permanently marred. Burning is usually due to lack of soil moisture during the winter combined with continuing water need and drying of the needles by winter winds. Watering and mulching are therefore recommended during the winter.

DROUGHT

Wilting is the main symptom of drought, and watering revives a plant in a drought condition. Drought-struck plants will usually revive completely unless their leaves are already leathery or crispy. Even beyond that stage, many species will regenerate from the roots after complete death of the aboveground parts.

POOR DRAINAGE AND FLOODING

Poor drainage, in which roots remain in water-saturated soil, will kill most plants. Those native to bog and other lowland areas are exceptions. Death is due to root rot caused by lack of oxygen. The length of time a plant can remain in standing water without injury varies, but most will tolerate several days to a week of occasional flooding.

Figure 13-12 A bark canker along upper part of bark. Normal, healthy bark is at bottom.

LACK OF SOIL NUTRIENTS

Lack of adequate soil nutrients will stunt and bleach plants but will not usually kill them.

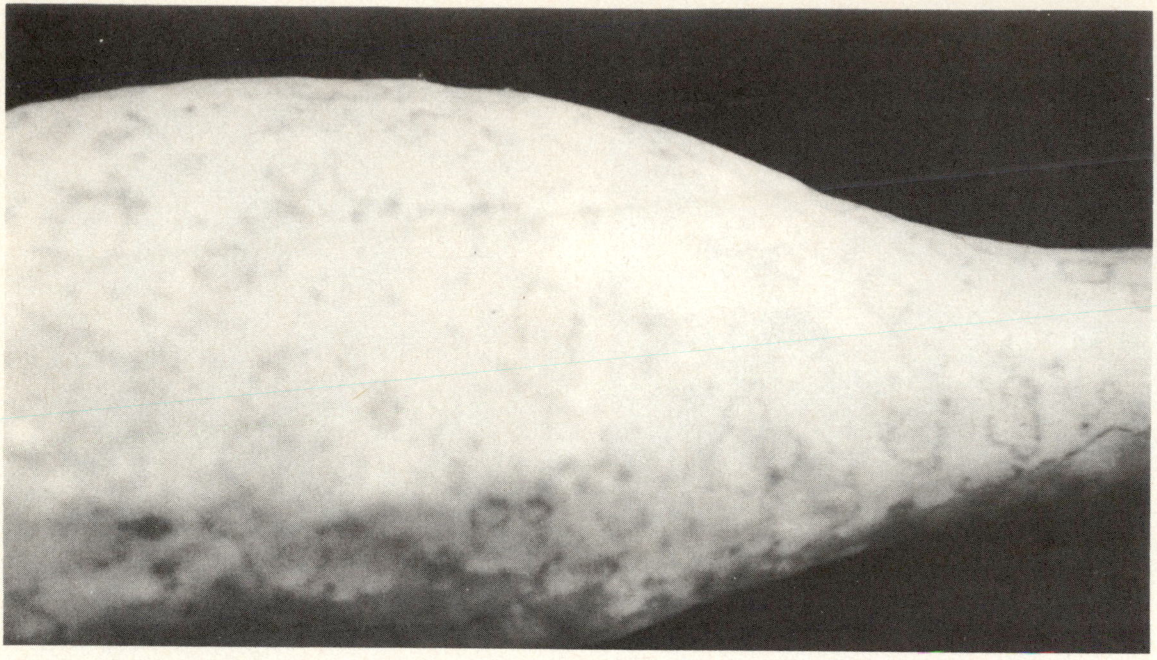

Figure 13-13 Watermelon mosaic virus on squash.

Figure 13-14 Gopher damage.

Lack of a required element will cause different symptoms on different species, and this makes nutrient deficiencies difficult to diagnose. However, probably 80 percent of all deficiencies are due to a lack of nitrogen, phosphorus, potassium, magnesium, or iron. Descriptions of deficiency symptoms for these elements are given in Table 13-3.

UNAVAILABILITY OF SOIL NUTRIENTS DUE TO pH

Even if an essential element is present, a plant may be unable to absorb it due to an excessively high or low soil pH. Iron and manganese are most commonly affected by pH, generally being tied up in soils. Symptoms will be the same as for normal deficiency.

EXCESS SOIL SALTS

A nutrient excess is frequently called a *salt buildup*. Nitrogen can be the primary damaging element if the toxic buildup stems from

Table 13-3 Deficiency Symptoms for Commonly Lacking Nutrients

Element	Symptoms
Nitrogen	General leaf yellowing, often showing up first on young leaves as new growth begins; decidious trees defoliate earlier in fall
Phosphorus	General yellowing of foliage, with young expanding leaves and shoots frequently showing purple discoloration
Potassium	Older leaves die at the margins, and there is reduced shoot growth and leaf size; the marginal burn may be preceded by yellowing and is most severe late in the growing season
Iron	Yellowing of areas between the veins on young leaves
Magnesium	Retarded growth followed by yellowing of areas between veins on old leaves; yellowing spreads later to young leaves

overfertilization, but compounds present in normal irrigation water and alkali soils can be as damaging.

Symptoms of excess salts are wilting and browning of leaf margins. This is due to death of the roots from dehydration by salts, and the effect is often indistinguishable from that of drought. The cure is by leaching, as described in Chapter 11.

EXCESS HEAT (HEAT SCORCH)

Excess heat damage (see photo in Chapter 1) frequently occurs on plants growing in paved areas or on south-facing walls; it is due to the reflection of sun-generated heat from these surfaces. The leaf margins and areas between the veins brown, and succulent new growth may die. Tolerance to excess or reflected heat varies between species.

SOIL COMPACTION

Constant foot or equipment traffic causes soil compaction around plants. The air spaces normally present in the soil are compressed, and water absorption and air penetration are inhibited. Plants in compacted soil grow slowly, and portions may die back. In severe cases, the entire plant may die.

IMPROPERLY APPLIED CHEMICALS

Pesticide sprays can injure plants if applied improperly, in excess concentration, or to plants for which they are not intended. The lawn weed killer 2,4-D is among the most frequently damaging. Spray drift can carry the chemical to neighboring areas, and curling and distorting of plant foliage (Figure 13-15) will result. Similar damage can appear when one sprayer is used for both weed killer and insect or disease sprays. Only a small amount of weed killer residue is needed to cause injury.

Sudden bleaching of plant foliage is a symptom of weed killer injury by chemicals such as aminotriazole (Amitro-T®). This weed killer injury can be distinguished from nutrient deficiency by the suddenness of the symptoms: weed killer injury will appear in 2 to 3 days, whereas nutrient deficiency will progress over several months. There are no remedies for weed killer injury, and the chances of its killing the plant depend on how much chemical was absorbed. Often the foliage will be marred for one or two seasons, but the plant will grow out of the damage and regain its former appearance.

Insect and disease sprays applied in excess concentrations or on plants for which they are not approved show different symptoms. Sudden browning of foliage to which the chemical was applied is the most common and may be followed by defoliation. A single defoliation from chemicals, disease, or insects is seldom fatal, but if repeated annually, the plant will die in 2 to 3 years.

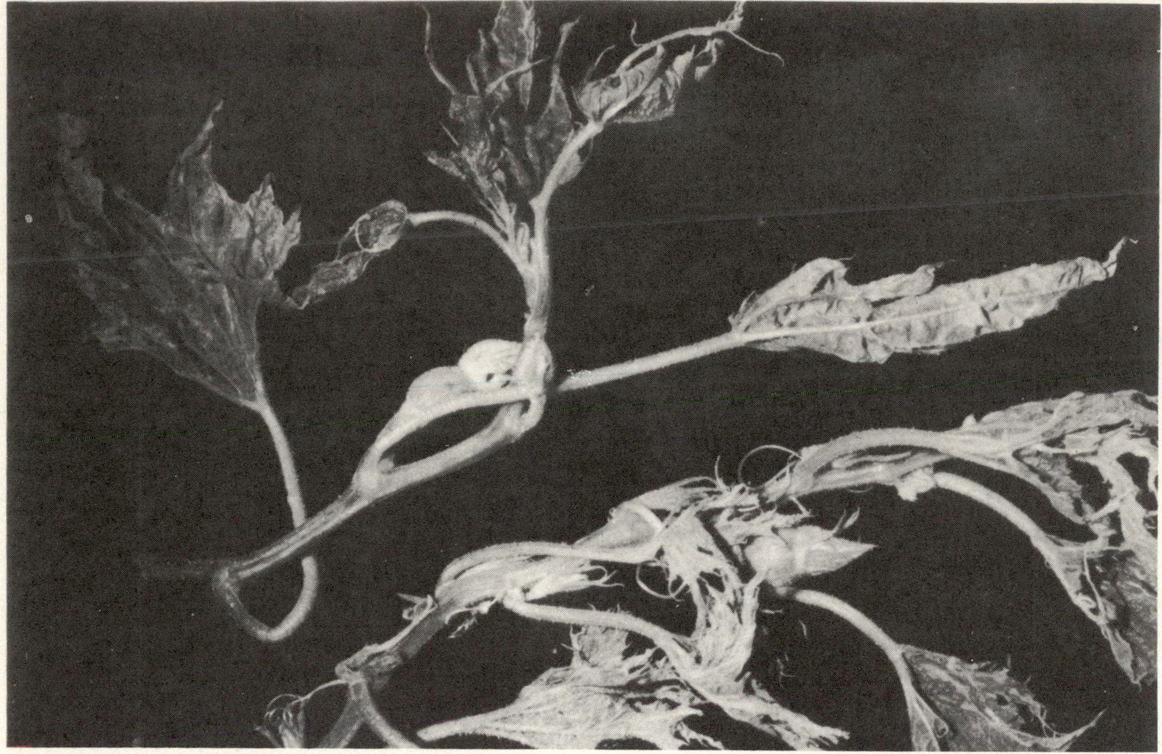

Figure 13-15 2, 4-D damage on squash leaf.

GRADE CHANGES

Raising the soil level over the roots of plants more than a few inches (5 centimeters) can be fatal because water and oxygen flow will be restricted from the roots. For this reason, tree wells are necessary when grade changes are planned. They should be made as large as practical, extending to the branch tips if possible. Supplemental drain tiles should be installed (Figure 13-16).

AIR POLLUTION

Air pollution damage is a problem only in a few areas of the country. It affects only susceptible plants, such as conebearing evergreens. The damage takes many forms, but usually involves some browning and death of the foliage. Pollution damage is difficult to diagnose and should only be suspected if the area frequently accumulates smog.

ALLELOPATHY

Allelopathy is the injury of one plant by another via the excretion of toxic substances by the roots. Both black walnut and eucalyptus trees are allelopathic to some other species.

TREATING PLANT DISORDERS

The steps in restoring a sick plant to health relate directly to the cause of the disorder. For environmental problems, there are two choices: correct the condition harming the plant, or find a plant that will tolerate the situation.

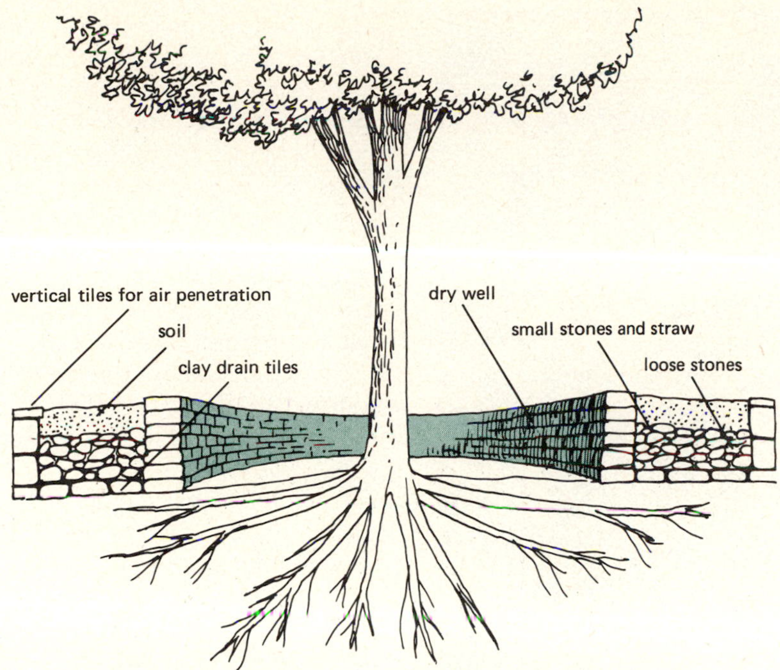

Figure 13-16 A tree well with drainage tiles.

Excess salts, drought, soil compaction, and pH problems, for example, can all be corrected. Air pollution, winterkill, and reflected heat problems, on the other hand, cannot be eliminated, but plant selection can be changed.

Alleviation of animal damage to plants can be aimed at either preventing the damage or killing the animal. Fences will keep out rabbits, dogs, and deer, although an 8-foot (2.4-meter) fence will be needed to prevent intrusions by deer. Blood meal may also work as a deer and rabbit repellent. Netting or noise-making devices are most successful with bird problems. Girdling of trees by mice and rabbits can be avoided by pulling away mulch layers and enclosing the trunk with a tall collar of tin or hardware cloth (Figure 13-17).

Damage by moles, gophers, and ground squirrels is harder to prevent. Poison baits are the best recourse against pocket gophers and ground squirrels, but moles are less likely to accept baits. Trapping with specially designed mole traps is frequently successful, but to be effective the trap must be placed in a main burrow used repeatedly by the animal. When moles infest a lawn, it is often because the lawn contains grubs, a favored food. Applying an insecticide drench will destroy the food source, and the moles will retreat.

Treating Microorganism-caused Diseases

Plant disease control is both cultural and chemical, and an understanding of both is essential to restoring and maintaining plant health.

CULTURAL DISEASE CONTROL

Cultural control involves using plant care techniques which minimize disease infection and spread.

Figure 13-17 A hardware cloth collar to protect the base of a tree from rodent damage.

Watering. Water is a key factor in plant disease attack and spread. Many fungi spores lie dormant until they are in the presence of water, germinating only then to invade the host plant. Accordingly, keeping water off plants susceptible to foliage spots and mildew lessens the chances of infection. Excessive soil water similarly favors the growth of disease organisms causing root rot, and overwatering will trigger that disease.

Splashing water is often a factor in foliage disease spread, carrying the microorganisms from infected to uninfected plants. Careful watering, with the water applied only to the soil, can decrease disease problems on susceptible species. The timing of watering can also reduce or enhance disease spread. The foliage of plants watered late in the day stays wet overnight and is susceptible to disease. Plants watered from morning to midday dry quickly and are less vulnerable.

Sanitation. Sanitation involves removing all disease-harboring materials to reduce the population of microorganisms. Many disease organisms have a seasonal cycle with dormant and active periods. The overwintering site for many foliage disease organisms is the fallen leaves which were infected during the summer. Accordingly, leaf and twig litter under disease-infested plants should be cleared away promptly in fall to prevent it from being a source of infection next season. These infested plant parts should be burned or hauled away, not composted.

Weed removal is another sanitation measure. One disease will frequently attack several related plants. If one of those plants is a weed, it can be a source of infection for the other plants. As another example, rust fungi require two plants to complete their life cycle, one of which is often a weed. Without a weed host nearby, the rust will not be able to attack the cultivated species.

Pruning. Pruning can be a valuable control technique for localized diseases such as gall, wilt, and canker. The infected portion of the plant should be pruned out and burned as soon as it is detected, cutting back several inches (5–10 centimeters) into healthy tissue. Between cuts and after pruning, the pruners should be dipped in alcohol to avoid spreading the microorganisms by the blades.

Roguing. Roguing involves pulling out diseased plants to stop the spread to others. It is valuable for annual plants or when only a few plants are infected. In addition, it is the only control for virus diseases and the most practical for mycoplasmas and most bacteria.

Crop Rotation. Crop rotation is usually associated with vegetable plants, but it can apply to flowers as well. It involves planting a different species in an area each year instead of the same one repeatedly. Rotation is a valuable control for soilborne diseases, most of which are hard to control by chemical means.

By using crop rotation, the parasitic microorganisms growing in conjunction with a species will be depleted yearly because they are not able to prey on the new crop.

Resistant Varieties and Species. Planting species and varieties resistant to disease is one of the soundest disease control techniques. Many vegetables and flowers have resistance to diseases imparted by breeding, and the small extra amount these varieties cost is well spent. Landscape plants are similarly bred for disease resistance, although some have natural disease and insect resistance. Planting these ornamentals in lieu of disease-susceptible ones can greatly lessen maintenance.

CHEMICAL DISEASE CONTROL

Chemicals are very helpful with some disease problems and nearly useless with others. Leaf spots, rust, mildew, smut, and blight diseases can usually be chemically controlled, but wilts, rots, cankers, witches' brooms, and all virus diseases are only marginally controllable or uncontrollable by chemicals. Fortunately, most commonly encountered diseases fall in the controllable group.

Fungicides are the most common disease-controlling chemicals, and most can be used on a variety of plants for numerous diseases. The container label and Cooperative Extension Service publications will give information regarding which diseases the product controls.

Although most fungicides are used as foliage sprays, a few combat soil-borne diseases and are applied to the root zones of plants. These are called *fungicidal soil drenches*. Still fewer are applied by one of the methods just described but are absorbed into the plants instead of simply forming a protective covering. These are called *systemics* and are effective on wilts when few other chemicals are.

Bactericides are generally antibiotics such as streptomycin and Terramycin. They are less common than fungicides and are used only on a few bacteria- and mycoplasma-caused diseases including blights.

Treating Nematode and Insect Problems

CULTURAL CONTROLS

The cultural controls for combating insect problems are less successful than those for combating diseases but can sometimes be effective. The objective is to remove the insect hiding places. Accordingly, weed control and removing litter at the base of a plant are two controls. Pruning off of an insect-infested portion of a plant is also used.

Supplying plants with adequate nutrition and other requirements for healthy growth is another cultural control technique. Healthy plants are less susceptible to both insect and disease problems than those under stress.

Physically removing insects by hand picking or with a strong stream of water (for aphids) is effective in halting insect damage.

CHEMICAL CONTROLS

Chemicals used to control insect pests are most often applied to plants as sprays or soil drenches and protect the plant by killing the predators. However, baits, which attract insects and kill them, and repellants are two other chemical controls.

Sprays. Surface sprays either kill on contact or are absorbed by insects as they walk across a sprayed leaf. These sprays are effective on most pests but are not useful for those with protective coverings such as mature scales and mealybugs.

Stomach poisons applied as foliage sprays are useful on chewing insects, which ingest the poison as they feed on the plant. They are less effective on sucking insects, which feed on the untreated sap.

Dormant oils are the third type of spray and are used on dormant deciduous plants. Applied in early spring, they are the primary means of combating scales and mealybugs. The oil coats trunks, branches, and twigs and excludes air, suffocating insects and their eggs. Later in the season, the oils can be diluted to a lower concentration safe for foliage.

Soil Drenches. Soil drenches of insecticides are used to combat underground pests. They are contact killers and are usually the same chemicals as surface sprays but in lesser concentrations.

Systemics. Systemic insecticides are absorbed into a plant and kill only those pests which feed on the plant. They can be applied as foliage sprays or soil drenches (penetrating by root uptake) and, occasionally, by injection into the trunk (for trees). Some insecticides have both contact and systemic action.

Baits. Baits are the most commonly used control material for snails, slugs, sowbugs, centipedes, and sometimes rodents. They act as stomach poisons, and careful attention should always be paid to placing them where other animals will not eat them.

Repellents. Few repellents are available commercially except for rodent repellents. However, homemade solutions of varying effectiveness can be made by following the instructions in organic gardening books.

Choosing Pesticides

With the numerous chemicals sold for controlling insect and disease pests, it is difficult for a gardener to know which to apply for a particular pest problem. The criteria for selecting a chemical should be that it controls the pest satisfactorily and is as nontoxic as possible.

The best sources of the toxicity and uses of pesticides are government publications (Cooperative Extension, Farm Advisor, United States Department of Agriculture) and the label of the material. The label will state "Danger," "Warning," or "Caution" in bold letters. "Danger" indicates that the chemical is highly toxic to humans, "Warning" that it is moderately toxic, and "Caution" that it is of a low toxicity.

Table 13-4 lists the LD_{50} of many common insecticides, fungicides, and bactericides. LD_{50} is an abbreviation for the phrase "lethal dose killing 50 percent of a population." It

Table 13-4 Oral LD_{50} Values of Selected Garden Chemicals[a]

Chemical and associated trade names	Use	LD_{50}[b]
Aspirin	—	750
Bacillus thuringensis (Biotrol®, Thuricide®, Dipel®)	Insecticide	No toxicity at tested dosages
Baygon	Insecticide	95–104
Benomyl (Benlate®)	Fungicide	9590
Caffein	—	200
Captan (Orthocide®)	Fungicide	9000–15,000
Carbaryl (Sevin®)	Insecticide	500–850
Chlorodane	Insecticide	335–430 (persistent in the environment)
Cyclohexamide (Actidione PM®)	Fungicide	1.8–2.5

[a]Extracted from Extension Bulletin E-751, Farm Science Series, Michigan State University Cooperative Extension Service.

[b]Class 1: highly toxic (1–50); class 2: moderately toxic (50–500); class 3: low order of toxicity (500–5000); class 4: very low toxicity (5000+).

Chemical and associated trade names	Use	LF_{50}
Daconil	Fungicide	Less than 3750
Dexon	Fungicide	60–150
Diazinon	Insecticide	300–400
Dibromochloropropane (Nemagon®)	Nematicide	172
Dimethoate (Cygon®)	Insecticide	215
Dipterex	Insecticide	560–630
Di-Syston	Insecticide	2–12
Dursban (Lorsban®)	Insecticide	97–276
Dylox	Insecticide	450
Ethion	Insecticide	65–119
Ferbam	Fungicide	17,000
Gasoline	—	150
Guthion	Insecticide	13–76
Karathane	Insecticide, fungicide	980–1190
Kelthane	Insecticide	100–1000
Malathion	Insecticide	1000–1375
Maneb (Dithane M–22®)(Manzate®, Manzate D®)	Fungicide	6750–7500
Methaldehyde	Insecticide	600–1000
Meta-Systox R	Insecticide	65–75
Methoxychlor	Insecticide	5000
Nicotine sulfate	Insecticide	83
Omite	Insecticide	2500
Parathion	Insecticide	4–13
PCNB (Terrachlor®)	Fungicide	1650–2000
Phosdrin	Insecticide	4–6
Plictran	Insecticide	540
Pyrethrum	Insecticide	820–1870
Rabon	Insecticide	4000–5000
Rotenone	Insecticide	50–75
Ryania	Insecticide	750–1200
Salt	—	3320
Streptomycin	Fungicide/ bactericide	9000
Sulfur	Fungicide	17,000
Tetradifon (Tedion®)	Insecticide	Less than 14,700
Thiram (Arasan®)	Fungicide	780
Zectran	Snail and slug killer	25–37
Zineb (Dithane Z–78®)	Fungicide	More than 8000

indicates how much of a given pesticide is required to kill 50 percent of a population of laboratory animals and is expressed in milligrams per kilogram of total body weight of the animal. Therefore the higher the number, the more of the material was needed to kill the animals and the *less* toxic the material. For comparison purposes, the LD_{50} values of salt and aspirin are included in the table, but this comparison in no way implies that pesticides less toxic than these materials should be used injudiciously.

ORGANIC GARDENING

Organic gardening is a popular way of gardening that does not permit the use of any synthetically manufactured granular fertilizers, insecticides, disease sprays, and the like. Organic gardeners contend that such materials work against rather than with nature, upsetting the natural balance of organisms in the environment and creating more problems than they solve. In lieu of conventional manufactured chemicals, they advocate mainly cultural solutions to plant disorders and, when necessary, pesticides, repellents, and baits which occur naturally in the environment. They contend that by virtue of being naturally occurring, chemicals do less harm to the environment and at the same time pose little threat to humans.

The cultural controls discussed for diseases and insects are essentially the same as those used by organic gardeners. In addition, the following pest control methods are advocated by organic gardeners, although the effectiveness of several has not been proven experimentally.

COMPANION AND REPELLENT PLANTING

In companion planting, relatively insect-resistant plants are grown next to susceptible ones to repel potential pests from the susceptible plant. Marigold roots, for example, exude a substance which repels nematodes, and onions and garlic are thought to ward off a number of pests. Garlic has been used successfully on some bacterial and fungal diseases.

TRAP CROPS

Trap planting is believed to work in the opposite manner as repellent planting. Insects often have definite plant preferences. In theory, planting a more attractive plant nearby will lure potential pests away from the principle crop. Dill can attract hornworms away from tomatoes, and Japanese beetles are reportedly lured by zinnias.

The effectiveness of trap crops is questionable. It is possible that the presence of two attractive crops compounds rather than alleviates a pest problem.

TRAPS

Traps can attract insects by providing food or shelter. Molasses, beer, and rotting fruit are reputed feeding lures which attract insects and then drown or stick them in the trap as they feed. Since many insects spend the day under plant refuse, a substitute hiding area can be made by laying a board on the ground. Slugs, sowbugs, and so on will congregate there and can be destroyed. More sophisticated light and electrocuting traps can also be purchased.

BARRIERS

Barriers prevent insects from reaching a plant. Flypaper, for example, can be placed around a tree trunk to stop insects crawling up from the ground. Wood ashes scattered around the base of a plant will repel snails and slugs. Vegetable garden transplants can be protected from cutworms with a paper collar held together with a paper clip.

BIOLOGICAL CONTROL

Biological control uses beneficial insects and microorganisms to control harmful insects. A number of insects use other insects as food or for part of their reproductive cycle. These beneficial insects can reduce the population of pests to a level at which their damage does not justify additional control. Among these are ladybugs, praying mantises, fireflies, lacewings, spiders, and many wasps.

Beneficial insects can be encouraged in the garden by minimizing the use of broad-spectrum chemicals which kill both pests and beneficials. Buying and releasing predators is only slightly effective, since the beneficials disperse over a wide area instead of remaining congregated on one property.

Bacteria, fungi, and viruses infect and kill insects, but research in this area has been slow, and only a few insect diseases are available for home use. Ultimately, diseases may prove to be one of the most valuable forms of insect control.

One disease of insects is *Bacillus thuringiensis,* a bacterium that kills the larvae of moths and butterflies. Cabbageworms, cutworms, corn borers, and tent caterpillars are among the insects controlled. The bacteria are applied as a spray as soon as worms are detected. After ingesting the bacteria, the worms stop feeding and are dead 2 to 4 days later. The dead insects then serve as sources of infection for future worms. *Bacillus thuringiensis* is completely harmless to plants and to animals other than the larvae.

A relatively new insect growth-regulating chemical, Enstar®, is being used to combat infestations of whiteflies and aphids. In addition to killing eggs and preventing maturation of developing insects, it also sterilizes adults. At stronger concentrations, it will kill adults outright.

BOTANICAL INSECTICIDES AND REPELLENTS

The two most popular botanical insecticides (those derived from plants) are:

Pyrethrum. A substance derived from a daisylike plant in the chrysanthemum family, pyrethrum kills many insects and mites. It is a common ingredient of many combination insecticides.

Rotenone. This insecticide can be extracted from several plants. It can be used on ornamentals and edibles and has little residual life.

Among the other botanical insecticides are ryania, sabadilla, and nicotine. They are not as readily available as pyrethrum and rotenone, and nicotine is considered dangerously toxic to humans (see Table 13-4).

HOMEMADE INSECT SPRAYS

Recipes using common household substances for insect control are also available. Buttermilk, soap and water, turpentine on rags, and various oils are suggested by organic gardeners as insecticides and repellents.

Botanical repellents can be made at home from a number of common plants. Their effectiveness has never been thoroughly investigated. Onions, garlic, hot peppers, tomato leaves, nettles, and horseradish are all reputed to have excellent qualities when mixed in water and applied as a spray.

INTEGRATED PEST MANAGEMENT

Integrated pest management is intermediate between spraying indiscriminately with insecticides and rejecting them completely, as organic gardeners do. It combines many controls—cultural, biological, and chemical— to prevent and treat plant problems. Emphasis is placed on learning as much as possible about the pest or disease problem—the life cycle, host plants, most favorable environment, and so on—and using the information to formulate a control plan. Integrated pest management also involves establishing an amount of damage to plants which can be tolerated and accepted, as opposed to assuming that all garden plants be completely pest and disease free. In an integrated program insecticides are used only when cultural and biological controls are ineffective.

When an insecticide is necessary, it is chosen carefully with regard to effectiveness, toxicity to beneficial insects, persistence in the environment, and danger to humans. It is unnecessary, for example, to spray a long-lasting, toxic insecticide for pest control when a less toxic and less persistent one will work as well. Similarly, one should not use a multipurpose spray which will kill both harmful and beneficial insects when a systemic would kill only insects damaging the plant.

While it would be difficult for a gardener to be familiar with the life cycles of all the insects and diseases of plants, he or she can use integrated pest management to a limited extent by following these steps.

1. *Determine the cause of the problem.* Decide whether it is due to an insect, disease, rodent, nutritional deficiency, or other source by analyzing symptoms, inspecting for pests, and using a reference book for additional information where necessary.
2. *Determine whether the harm being done is sufficient to justify control.* Set a level of tolerable damage. Mildew attacking annuals in late fall should, for example, be more tolerable than mildew on plants in spring, since the plants will soon be frosted anyway. Treatment might not be justified at all. In the same way, if lettuce has already been severely chewed by insects, spraying may be superfluous because the lettuce will not recover.
3. *Use cultural and biological controls, if practical and available, to treat the problem.* Hand picking a few cabbageworms or trapping slugs in beer is more ecologically sound than spraying and can be nearly as effective. Similar cases can be made for using *Bacillus thuringiensis* on caterpillars and for hosing off aphids and mites with water.
4. *Apply a pesticide as a last resort or if it is the only available control.* Choose the chemical carefully so that it is as nontoxic, nonpersistent, and specific to the pest as possible. For example, miticide should be used for mites, not an all-purpose combination. The chemical should be applied to the damaged plant. Spraying unaffected neighboring plants is not preventative, it is overkill.

APPLYING PESTICIDES TO OUTDOOR PLANTS

Most gardeners use chemical pesticides, and applying them safely and correctly will maximize their effectiveness.

Application Equipment

DUSTERS

Dusting is less common than spraying with liquid, but some gardeners prefer it simply because a duster does not need to be emptied and cleaned after every use (Figure 13-18).

ATOMIZER SPRAYER

This applicator is for small jobs and holds only about a pint (0.5 liter) of spray. The plunger pulls in air with each draw and forces it out with pesticide in fine droplet form at each stroke (Figure 13-19).

Figure 13-18 A duster for applying pesticides.

Figure 13-19 An atomizer sprayer.

Figure 13-20 A gardener using a compressed-air tank sprayer.

COMPRESSED-AIR TANK SPRAYER

The tank of the sprayer is partially filled with pesticide, and the remainder contains compressed air. Pulling the trigger on the spray wand releases the pressure and incorporates pesticide in a fine spray. A tank sprayer is probably the best all-around sprayer for the home gardener. The wand makes it possible to reach both high into trees and beneath the foliage of low-growing plants (Figure 13-20).

HOSE-END SPRAYER

This sprayer uses concentrated pesticide in the container and dilutes it with water under pressure from the hose. However, because it proportions pesticide into water relatively imprecisely, it is not recommended for general pesticide applications.

Pesticide Safety

The importance of safety precautions in storing, mixing, applying, and disposing of pesticides cannot be overemphasized. As potentially deadly chemicals, they merit respect and careful use.

STORING

1. Store pesticides out of reach of children and animals and at a moderate temperature. A locked storage cabinet is suggested.
2. Store them only in the original containers. Although it may seem practical to share a bottle of pesticide with a neighbor, don't do it. Storing pesticides in bottles other than their original containers is one of the main causes of poisoning.

MIXING

1. *Read the label* before opening the container. It is your best source of information on the pesticide, its toxicity, its uses, and so forth.
2. Do not use kitchen measuring utensils for pesticides. Keep a separate set expressly for pesticide use.
3. Do not mix pesticides. Although many are compatible, some are not and can seriously damage plants.
4. Never use a pesticide at stronger than the recommended rate. It is unnecessary pol-

lution of the environment, will not make the chemical any more effective, and could cause damage to the plant.

APPLICATION

1. Think about whether the pesticide will be a danger to pets. Cover pet food and water bowls and fish ponds when applying pesticide nearby.
2. Avoid breathing the pesticide. The lungs take up pesticide quickly. Stand back from the plant being sprayed or use a respirator.
3. If pesticide is spilled on the skin, wash it away immediately with soap and water. For spills on clothes, change, and launder the garments separately.
4. For ground spills, flood the area with water to dilute the pesticide to a minimally toxic level.

DISPOSAL

When some pesticide remains after spraying, there are several alternatives for disposing of this material.

1. Seal the excess in an unbreakable container, and put it in the trash.
2. Pour the excess onto a gravel drive or beside the road to be absorbed and broken down in the soil.

Do not pour excess pesticide down drains, and do not burn the containers.

Selected References for Additional Reading

Agrios, G. N. *Plant Pathology.* New York: Academic Press, 1969.

Forsberg, J. L. *Diseases of Ornamental Plants.* 3d ed. Urbana: University of Illinois Press, 1975.

Pirone, P. P. *Diseases and Pests of Ornamental Plants.* 4th ed. New York: Rodale Press, 1970.

Ware, G. W. *Pesticides, An Auto-Tutorial Approach.* San Francisco: W. H. Freeman, 1975.

Westcott, C. *The Gardeners Bug Book.* New York: Doubleday, 1973.

Yepsen, R. G., and the Editors of *Organic Gardening and Farming Magazine. Organic Plant Protection.* Emmaus, PA: Rodale Press, 1976.

PART III

GROWING PLANTS INDOORS

Introduction

No other type of horticulture has undergone the rise in popularity that indoor plant growing has during the past few years. Almost every home now displays at least a few indoor plants, whether they are in hanging baskets, large freestanding specimens, or in windowsill pots. Indoor plant enthusiasm continues to grow as more people discover the pleasure of living with plants.

The decorative and mood-creating qualities of indoor plants are probably the factors which have contributed most to their popularity. The changing appearance of plants passing through their growth and flowering periods attracts attention and makes them a source of constant enjoyment. Lush green foliage creates a cool, tranquil feeling and mixes with any decor from modern to traditional. Indoor plants link the outdoor world of nature and the indoor domain of people in a mutual living space.

Of course, plants are at a definite disadvantage indoors because human comfort will always be the primary consideration. Temperature will be controlled to the liking of people, as well as light intensity, air circulation, and other factors. The conditions which contribute to the health of plants sharing that living space are considered secondarily, if at all.

THE INDOOR ENVIRONMENT

There are many environmental conditions to which a plant must adapt if it is to live indoors. These include light, temperature, humidity, water quality, containerization, and air circulation.

Light

Probably the most important factor is reduced light. Light intensities comfortable for people to live and work in are too low for most plants. If plants do not receive sufficient light to manufacture the carbohydrate required for respiration, they will eventually die. Accordingly, indoor plants are usually placed near windows for maximum light or kept close to an artificial light source.

Temperature

Outdoors there is usually a 5–15°F (2.8–8.4°C) difference between day and night temperatures, with the temperature dropping slowly after sunset. These cooler night temperatures reduce the rate of respiration and prevent the rapid depletion of stored carbohydrate reserves necessary for plant growth. However, plants grown indoors seldom are exposed to cool night temperatures. As a result, their carbohydrate reserves are used more quickly, and the plants have little to use for new growth.

Low Humidity

Low humidity is a third factor contributing to the less than ideal conditions indoor plants experience. Home heating creates dry air conditions indoors in most parts of the country.

This causes plants to lose water rapidly by transpiration through the leaves. The plants absorb water to replenish that which has been lost, but often the water is transpired more rapidly than it can be replaced. The tips then turn brown, and the plant develops symptoms indicative of low humidity.

Water Quality

The quality of the water used on indoor plants is also unlike that which they would receive outdoors. Under natural conditions rainwater supplies moisture. It has relatively few impurities and, generally, a neutral pH (neither acid nor alkaline reaction). But tap water is used indoors. Depending on its source and prior treatment, it may contain numerous chemicals such as chlorine, sodium, or fluoride. In addition, it may be acidic or basic, which will affect the availability of nutrients.

Containerization

Plant roots restricted to a pot live under unnatural growing conditions. The nutrients and water available to the plant are limited by the small amount of medium in the pot and must be more carefully regulated than for plants outdoors.

Some characteristics of the potting medium also change by being restricted. Soil that is loose and drains quickly outdoors may pack down and drain slowly in a pot. Consequently, garden soil is not often suitable for houseplant use unless it is mixed with other materials (Chapter 15).

Air Circulation

In the outdoor environment air movement is fairly constant due to changing temperatures and natural winds. This circulation moves oxygen and carbon dioxide over the leaves of the plants, aiding in photosynthesis and respiration and drying foliage after rainfall. But indoor air can be relatively stagnant during the winter months when windows remain closed. Any fresh air is likely to come through the opening of outside doors and may be 40°F (22.4°C) colder than room temperature. Accordingly, plants in the direct path of doors often suffer undiagnosed illnesses from rapid air temperature changes.

With so many modified environmental factors to cope with, it would seem that growing plants indoors would be difficult. But the size and superb condition of many tropical plants growing in homes disprove this. Their owners obviously understand the requirements of plants and meet those requirements. Their skill can be acquired by anyone through reading, observing, and, most important, experience in growing many plants.

Chapter 14

Indoor Plant Maintenance

PURCHASING INDOOR PLANTS

With plants, as with any other purchase, it pays to shop around. There are bargains and lemons among indoor plants, and the way a plant looks in the store is not necessarily the way it will look after 2 or 3 months under home growing conditions.

One of the biggest disappointments occurs with purchases of flowering plants. Such plants are usually grown in greenhouses under high light intensities which trigger flower induction. Under indoor growing conditions the light intensity is seldom strong enough to cause new buds to set, so after its present buds are exhausted, the plant is not likely to bloom again. Table 14-1 lists some reliable houseplants which bloom indoors without the use of supplemental lights.

When you are considering buying a houseplant, take the time to examine the plant carefully. A well-grown houseplant will be compact and have richly colored foliage, strong new growth, and a full crop of leaves. The stem should be strong enough to easily support the weight of the top and have no weak or discolored areas, especially at the base. An insect-infested plant is never a good buy, so check the growing tips, leaf undersides, and potting medium for hiding pests or signs of their damage.

The price charged for the plant will vary according to where it is purchased, species, and, sometimes, proximity to the large indoor plant-growing areas of Florida and California. Mass merchandising outlets such as supermarkets and discount stores normally have the lowest prices, florists the highest, and local plant shops and greenhouses are in between. However, mass merchandising outlets usually have the least selection. Also, if the plants have been displayed in the store for long, they may have been improperly cared for and be in poor condition.

The price charged for a plant is often a reflection of the ease with which it is propagated and the time required to grow it to selling size. Wandering Jew and coleus, for example, propagate very easily from cuttings and grow rapidly. They are among the least expensive indoor plants. Palms, on the other hand, are grown from seed. A kentia palm seed takes an average of 8 to 12 months to germinate.

Before transporting a newly purchased plant, take the time to wrap it in paper or enclose it in a paper bag before taking it outdoors. In below-freezing weather the paper will act as insulation against the sudden temperature drop. In summer this procedure will protect the plant from excess sun exposure through the car windows.

Then find out the plant's growing requirements for light, watering, humidity, potting medium, and the like. Check a reliable reference, and then try to match the ideal growing conditions as best you can in your home.

ACCLIMATIZATION AND CONDITIONING

Plant "shock" is a common occurrence in newly acquired houseplants. It usually results from moving plants directly from the production area to home growing conditions without a conditioning period. The sudden change to lower light and humidity disturbs the metabolism of the plant and temperarily halts growth. Leaf drop may occur, and some plants may even die.

Properly grown houseplants are "conditioned" or "acclimated" prior to sale. The light and humidity are gradually lowered to the levels commonly found in indoor conditions, and watering and fertilization are less frequent. Plants which have not been acclimated often betray the fact by their appearance: they are pale green and extremely lush appearing, with much new growth and dense foliage. Basically, they are extremely healthy and vigorous plants and would continue to grow well in a greenhouse. But they are not able to adjust well to indoor conditions.

It is possible to acclimate a tropical plant which was not conditioned before sale. First, give the plant the maximum amount of light and humidity available and keep the soil moist. This will approximate the former growing environment and minimize shock symptoms. Then gradually move it further from the light and decrease the humidity and water to the recommended range over 1 to 2 months. This long period of adjustment will allow the plant to acclimate to an indoor growing environment, but even then some foliage drop should be expected. After acclimatization the plant should grow slowly but satisfactorily provided the correct light level and other requirements are met.

MAINTENANCE ACTIVITIES

Routine maintenance is an essential part of houseplant care if the original health and beauty of plants are to be preserved. Two major aspects of maintenance, watering and fertilizing, are significant enough to require discussion in individual chapters, but the less commonly practiced maintenance activities are no less important.

Inspection

Regular inspection for pests and diseases is a part of maintenance which is commonly overlooked, to the decline and ruin of many beautiful plants. Its purpose is much the same as annual checkups for people: to detect and cure minor health problems before they become bigger and more difficult to control. Inspection of houseplants should take place every 1 to 2 weeks and should consist of a general search of the plant for any signs of poor health. New growth, in particular, should be carefully scrutinized under bright light for signs of pests such as spider mites (see Chapter 18); a small magnifying glass is helpful for this

Table 14-1 Selected Houseplants Which Flower without Artificial Lighting

Common name	Botanical name
Bromeliad	*Aechmea, Billbergia, Nidularium,* and so on
Begonia	*Begonia* spp.
Spider plant	*Chlorophytum* spp.
Coleus	*Coleus × hybridus*
Crocus	*Crocus* spp.
Orchid	*Cypripedium, Cattleya,* and so on
Crown of thorns	*Euphorbia milii varieties*
Rose-plaid cactus	*Gymnocalycium* spp.
Amaryllis	*Hippeastrum* hybrids
Hyacinth	*Hyacinthus orientalis* varieties
Pincushion cactus	*Mammillaria* spp.
Grape hyacinth	*Muscari* spp.
Daffodil	*Narcissus* spp.
Geranium	*Pelargonium* spp.
Swedish ivy	*Pletranthus* spp.
Crown cactus	*Rebutia* spp.
Easter cactus	*Rhipsalidopsis* spp.
African violet	*Saintpaulia ionantha* varieties
Strawberry begonia	*Saxifraga sarmentosa*
Christmas cactus	*Schlumbergera bridgesii* varieties
Thanksgiving cactus	*Schlumbergera truncata* varieties
Christmas cherry	*Solanum pseudocapsicum*
Spathe flower	*Spathiphyllum* spp.
Cape primrose	*Streptocarpus saxorum* varieties
Tulip	*Tulipa* spp.

purpose. Likewise, more mature leaves should be examined both above and underneath, where insects frequently hide. The leaf axils are another favored spot of concealment for pests.

At the same time, the general condition of the plant should be evaluated. If young growth is spindly or lower leaves are beginning to yellow, the growing conditions should be altered to remedy the problem before the plant deteriorates any further.

The soil surface and roots should be checked for signs of soil-dwelling insects. Possibly toxic fertilizer accumulations (with their characteristic white coating) will be noticeable.

Plant inspection may appear to be time-consuming, but it actually will take only a few minutes. The time will be well spent if an insect infestation can be detected before the pests spread to other plants.

Washing

Leaf washing is the second aspect of maintenance. Outdoors, rainfall regularly washes the leaves of plants. It removes the dust, dirt, and smoke film which block out light and decrease photosynthesis. But indoors most water is applied directly to the soil, and so the leaves are never cleansed. Periodic washing is a substitute for natural cleansing by rainfall and should be performed whenever the leaves show an appreciable amount of dust. In addition, it will help remove insects and their eggs, lessening pest problems.

Several methods may be used to clean the leaves, and the one chosen should be determined by personal preference and the size of the plant. Small plants are perhaps most easily cleaned by the spray nozzle found on most kitchen sinks. Excess water will conveniently drain in the sink, and leaves should be sprayed thoroughly on both the tops and the undersides.

A shower is very useful for large-specimen plants. Place the plant in the tub, and adjust the water to approximately room temperature. Then leave the plant under a low-pressure spray for about 5 minutes, turning the plant occasionally to make sure all the leaf surfaces are adequately washed. If there is a particularly stubborn residue on the plant, the leaves can be sponge cleaned with dilute detergent, but the plant should be thoroughly rinsed afterward.

Many indoor gardeners find that carrying large plants to the shower every month is inconvenient, if not completely impossible, so they use a damp cloth or sponge for cleaning instead. This method is satisfactory provided the leaves are smooth and not easily damaged. It is especially recommended for areas in which the water has a high soluble salt content. High-salt-content water will leave unsightly residue on sprayed or showered plants, so wiping should be substituted.

Misting can be another form of leaf cleaning, although it is generally used in hopes of raising the humidity around a plant. For cleansing purposes, misting should continue until water runs off the leaves. The dust will then be flushed from the leaf surface instead of wet down.

Leaf Shines

Many commercial "plant shines" or "leaf cleansers" are now available in aerosol sprays, fiber-tipped applicator form, and foil-encased towelettes. All claim to improve the appearance of foliage and enhance the looks of plants. Most garden books, though, contend that they are bad for the health of the plants because they cause "pore" clogging. Neither side is completely right.

Most plant shines and leaf cleaners will not harm houseplants. However, a few will, especially on young foliage, and this is most often attributable to an aerosol propellant. The pore-clogging claim is not valid: first, because the "pores" (technically stomata; see Chapter 2) are not like human pores and do not respond in the same manner to clogging. In fact commercial nurseries often use chemical sprays called *antidesiccants* during transplanting. These antidesiccants are solely for the purpose of sealing the stomata and preventing moisture loss. Second, most stomata are found on the undersurfaces of leaves, where leaf shine is not applied. Consequently, the majority will not be affected.

Nor are leaf shines or cleaners necessarily the best way to cleanse plant leaves. Their chemical makeup varies greatly among the different brands. Some leave an artifical-looking gloss and an oily residue which attracts dust, while others leave no residue and only a slight shine. People who prefer to use a leaf shine or cleaner should compare several brands before deciding.

Homemade leaf shines such as milk or salad oil were commonly used in the past. Salad oil cannot be recommended because of the resulting greasy film. Milk does not have this objectionable characteristic and is safe and inexpensive.

The hair-covered leaves of tropicals such as purple velvet plant (*Gynura aurantiaca*) and panda plant (*Kalanchoe tomentosa*) can be either cleaned with a spray of water or dusted with a soft paintbrush. Leaf shines should not be used, and neither should hand wiping, which can flatten or break off the hairs. When rinsing the hairy-leaved members of the African violet family, always use water at room temperature or slightly warmer. Gesneriads are extremely sensitive to temperature changes. Water more than 10°F (5°C) cooler than room air temperature will cause yellow splotching on the leaves wherever it touches. The discoloration that results is caused from the destruction of leaf chlorophyll and is permanent.

Whereas some indoor tropicals have hair-covered leaves, others possess a natural filmlike coating that should not be rubbed or covered over with leaf shine. Bromeliads are one example. The foliage of many bromeliad species is covered with flaky silverish scales (Figure 14-1). The scales are decorative, but their function is highly practical. They are cells for absorbing water directly through the leaves.

Cacti and succulents frequently have a waxy cuticle that rubs off easily with the fingers. The coating serves as an insulation layer against moisture loss and is an adaptation factor that helps them survive in arid environments. As a general rule, leaf polishes or cleansers should not be used on cacti and succulents.

Figure 14-1 White bands of water-absorbing scales on a *Billbergia* bromeliad.

Pruning

Pruning is commonly regarded as an outdoor plant maintenance chore. Consequently indoor plant pruning is often neglected, resulting in many poor-looking houseplants.

Pruning has several functions. First, it causes branching and new growth, producing a fuller, more attractive plant. Second, it removes dead, insect-infested, or diseased plant parts to halt the spread of the organisms.

PINCHING

The most commonly practiced form of pruning for indoor plants is a technique called *pinching* (Figure 14-2). Pinching consists of breaking off up to about 3 inches (7.6 centimeters) of the growing tips.

The meristems of plants are known to produce a number of chemicals, including a plant hormone called *auxin* (see Chapter 2). This hormone constantly moves down the stem of the plant from its source of manufacture. Its chemical makeup inhibits growth of buds lower on the stem and prevents branching. This hormonal effect on growth is termed *apical dominance*, indicating that the top bud (the apical meristem) controls the growth of the buds below it. Buds directly below the growing point are kept fully dormant, since the auxin is in strongest concentration there. As auxin continues moving down the stem, the concentration decreases. The lowest buds on the stem receive auxin in least concentration and sometimes overcome dormancy and grow into shoots.

Auxin manufacturing rates and movement vary between species, accounting for why some plants branch naturally when others do not. Pinching removes the source of auxin manufacture on plants which will not branch of their own accord.

When a growing tip is removed, it is nearly always the buds nearest the tip which begin growing (Figure 14-2). Although sometimes only one will grow, usually two to three

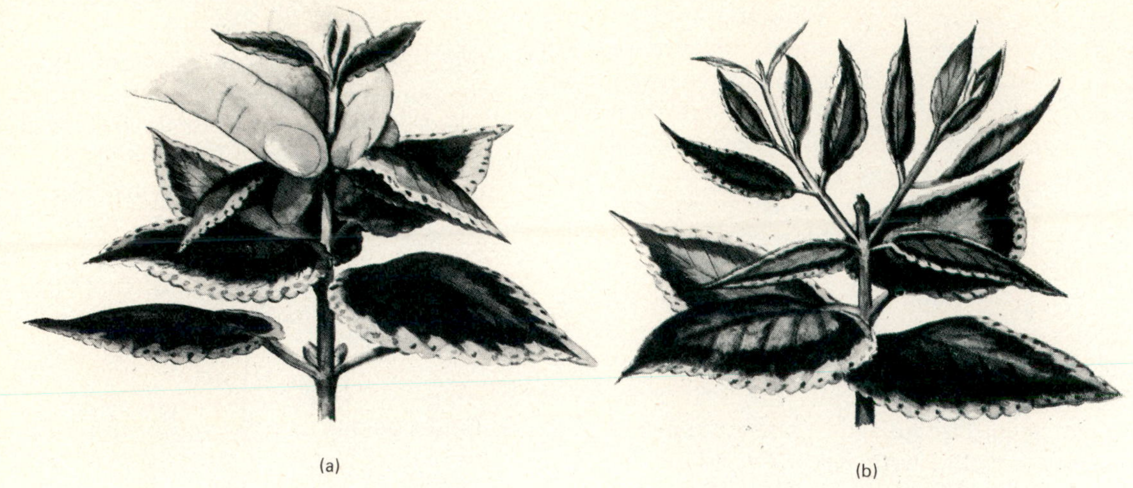

Figure 14-2 (a) Pinching a coleus. (b) One month later.

buds will surge into growth at the same time, producing several new leafy shoots. Since they are also growing tips, they soon begin auxin manufacture, keeping the buds below dormant, and the cycle continues.

Pinching must be done regularly to promote constant branching and to develop a full plant. As new shoots lengthen, the tips should be removed continually to make the plant full and dense.

Leafy, vining houseplants most often require regular pinching to prevent a stringy appearance. Ideally they should be pinched frequently after potting to develop a large number of cascading stems. The most common vining and upright-growing houseplants requiring frequent pinching are listed in Table 14-2.

RENEWAL PRUNING

Renewal pruning is used to rejuvenate a plant that has become unattractive and bare-stemmed through neglect, insufficient light, or some other cause. It involves removing all or nearly all of the top growth, leaving only 1- to 2-inch (2.5- to 5-centimeter) stems (Figure 14-3). In several weeks new growth will begin from the leafless stubs. With regular pinching and improved growing conditions, the plant will become attractive again in a few months.

Renewal pruning is a drastic measure and should be employed only if the plant is basically healthy, with a strong root system and stored carbohydrate reserves. For plants in a severely weakened condition, a different treatment should be used. First, the source of the problem shoud be corrected. This will encourage healthy growth and will often be sufficient to break the dormancy of lower buds. After the plant is partially refoliated, the remaining unattractive portions can be cut back to healthy growth. By allowing some foliage to remain, carbohydrate depletion will be prevented.

PRUNING BY THINNING

Not all indoor plants are pruned by pinching. Many houseplants such as ferns and spider plants grow in a form called a rosette (see Chapter 2) and produce new foliage from a

Table 14-2 Most Common Houseplants Requiring Frequent Pruning

Common name	Botanical name
Coleus	*Coleus* × *hybridus*
Goldfish vine	*Columnea* spp.
Tahitian bridal veil	*Gibasis geniculata*
Purple passion vine	*Gynura aurantiaca* 'Purple Passion'
Pocketbook plant	*Hypocyrta* spp.
Polka-dot plant	*Hypoestes phyllostachya*
Bloodleaves	*Iresine* spp.
Aluminum plant	*Pilea cardierei*
Swedish ivy	*Plectranthus* spp.
Christmas cherry	*Solanum pseudocapsicum*
Wandering Jew	*Tradescantia* or *Zebrina* spp.

point at soil level called the crown. Each new shoot that develops is a single leaf, and the dormant buds are found at the crown. Pinching a leaf back will never cause braching. Instead, it should be cut off at the base to promote new growth from the crown.

In pruning rosette houseplants, usually only a few leaves are removed at a time, as needed to improve the appearance of the plant. However, in cases in which the entire plant is unsightly, renewal pruning may be advisable. As with renewal pruning of stemmed houseplants, all the foliage is cut off at ground level, and the plant is allowed to refoliate

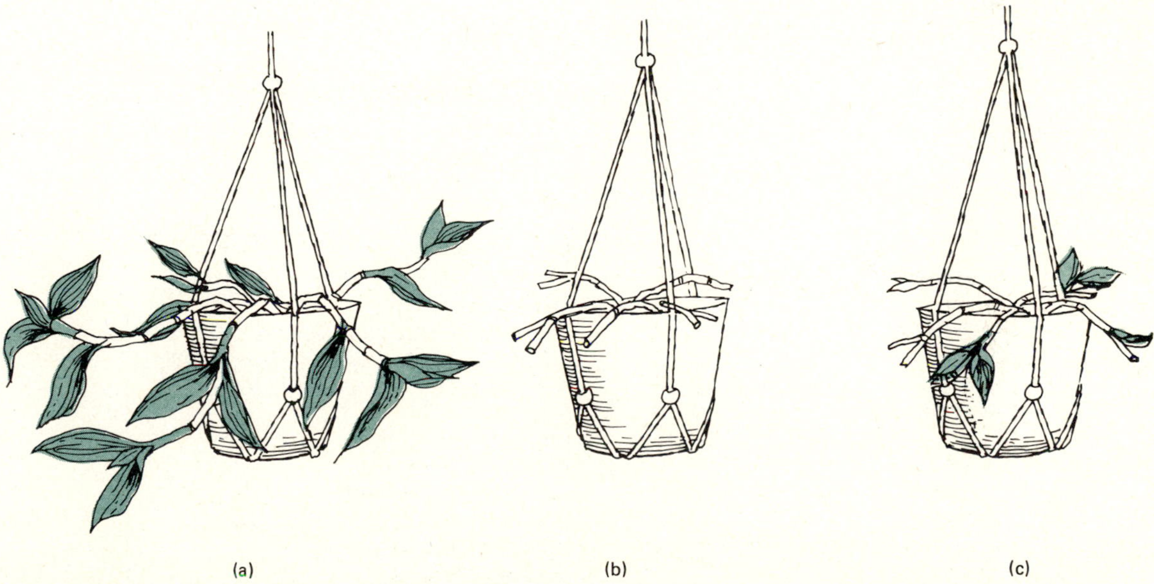

Figure 14-3 Renewal pruning. (a) Before pruning. (b) Immediately after pruning. (c) Six weeks after pruning.

itself under improved growing conditions and care.

PLANTS WHICH SELDOM REQUIRE PRUNING

A moderate percentage of houseplants seldom requires any pruning except removal of dead leaves. Pinching or renewal pruning of these plants will often destroy their form and create an unnatural appearance. Selected houseplants which require minimal pruning maintenance are listed in Table 14-3.

PRUNING FLOWERING HOUSEPLANTS

Flowering houseplants follow the basic rules for pruning which apply to foliage plants. For many plants, however, the flowering period can be lengthened by removing the flower heads and seed pods as soon as the blooms fade. If the seed pods are not removed, energy will be channeled into seed production rather than into the formation of additional flower buds.

Exceptions to this practice are plants in which the fruit is an ornamental feature, such as Christmas cherry. Blooms should also be left, if you are trying to raise seeds to grow additional plants. Table 14-4 lists plants from which blooms should not be removed.

PRUNING FOR DISEASE AND INSECT CONTROL

Pruning for the control of diseases and insects is not a common practice because in most cases the entire plant is infected before the problem is discovered. However, these problems often do start on one or two leaves or a shoot tip. Quick removal of the unhealthy parts can be preferable to a series of chemical sprays. Pruning is particularly valuable in dealing with houseplant foliage diseases because chemical cures are rarely available. When pruning for this purpose, all diseased portions should be cut back to healthy areas.

The clipping of browned leaf tips is commonly used to improve the appearance of houseplants such as dracaenas and palms. Although temporarily effective, the leaf tips usually brown again within weeks.

Transplanting and Repotting

Transplanting (or repotting, as it is sometimes called) is a routine part of indoor plant maintenance. It is performed repeatedly from the seedling or cutting stage through maturity.

Primarily, it is to provide crowded roots with additional potting medium from which to absorb water and nutrients. Less often,

Table 14-3 Selected Plants Requiring Minimal or No Pruning

Common name	Botanical name
African violet	*Saintpaulia*
Bromeliad	*Aechmea, Neoregelia, Vriesia*
Cactus	*Cereus, Gymnocalcycium, Mammillaria*
Chinese evergreen	*Aglaonema commutatum* varieties
Dracaena	*Dracaena*
Dumbcane	*Dieffenbachia* spp.
Norfolk Island pine	*Araucaria heterophylla*
Orchid	*Cattleya, Paphiopedilum*
Palm	*Chamadorea, Howiea, Phoenix*
Rosette-type succulent	*Agave, Aloe, Haworthia*
Snake plant	*Sansevieria* spp.
Spathe flower	*Spathiphyllum*
Umbrella plant	*Brassaia*

Table 14-4 Plants from Which Faded Flowers and Fruits Should Not Be Removed

Common name	Botanical name	Remarks
Bromeliad	*Aechmea, Nidularium,* and other genera	Colorful fruit often follows blossoming
Asparagus fern	*Asparagus* spp.	Ornamental red berries follow bloom
Spider plant	*Chlorophytum* spp.	New plants form on flower stalks
Cactus	*Mammillaria, Rebutia,* and other genera	Red or orange fruits frequently follow blossoming
Christmas cherry	*Solanum pseudocapsicum*	Red fruits form from white blooms
Chinese evergreen	*Aglaonema commutatum* varieties	Flowers followed by red berries; seldom flowers indoors
Wax plant	*Hoya carnosa* varieties	Next season's blooms derived from older growth

transplanting is used as a general tonic or cure-all for an undiagnosed plant ailment suspected of being caused by the roots or the potting medium.

Most people repot more often than necessary, switching to bigger pots when frequent fertilization would have been sufficient. Although unnecessary repotting is seldom harmful, it can cause a considerable amount of extra work.

TIMING OF REPOTTING

Several factors must be evaluated to determine if a plant should be moved to a larger container. First, does the plant look top-heavy in its pot and tip over easily? A plant with a large foliage mass will grow best if given an adequate volume of potting medium to support that top growth. Keeping it in an undersized pot will stunt its growth.

Next, does the plant require frequent watering to keep the soil moist, and does it wilt within a few days of watering? Roots which are overcrowded extract available water and nutrients very quickly, necessitating more frequent watering and fertilization.

Third, is the health of the plant declining with no change in care or growing location? Some houseplant media pack and crust on the surface after a period of months or years. Repotting in fresh media may be advisable.

In extreme cases of underpotting, roots will be found growing on the surface of the medium. Roots may also grow vigorously at the base of the pot, pushing the plant out of the container. This phenomenon is occasionally encountered in dracaenas and spider plants.

Some indoor gardeners use the presence of roots growing from the drainage hole or circling the root ball as an indicator for repotting. This method is not always reliable. Roots grow most vigorously in areas which hold both moisture and air. The soil near drainage holes and around the inside surface of unglazed clay pots provides these conditions most ideally. Accordingly, roots will grow there in abundance, while the bulk of the potting medium is undisturbed. However, if large numbers of roots have grown in circles, repotting is justified.

The season chosen for repotting indoor plants is not important. Most plants go through seasonal dormancy and growth spurts like outdoor plants, and transplanting during either of these periods is acceptable. Most transplanting is performed in spring or summer because the increased vigor makes plants outgrow their pots quickly during these periods.

Transplanting of blooming plants should be restricted to a period when they are not flowering, since it is common for a plant to drop buds and flowers prematurely when the roots are disturbed. Repotting blooming plants immediately after flowering is a reliable practice. It provides room for new root growth and, consequently, better foliage growth before the next flowering period.

UNPOTTING

Difficulty in removing a plant from its pot depends on the container and plant involved. Plants in 8-inch (20-centimeter)-diameter or smaller-sized containers can usually be unpotted by a technique known as "knocking out." To do this the medium should be moist (but not freshly watered) to keep it from falling away from the roots. Supporting the top of the soil with one hand, the pot is turned upside down, and the edge lightly tapped on a counter or table (Figure 14-4). The soil and roots should slide out intact. If they do not, harder tapping, shaking, or running a knife around the rootball may be necessary. Only in very difficult cases should a plant be loosened from its pot by pulling on the stems. Root or stem damage is likely to result.

Unpotting a plant that has been growing in a narrow-mouthed pot is always difficult. A portion of the roots will nearly always be lost when the root ball is pulled through the narrow opening. As with knocking out, the potting medium should be moist and the plant dug and eased through the narrow opening. Laying the pot on its side is often an aid in easing the roots out. The medium will fall to the bottom side, and a tool can be inserted in the space above to pull the roots free.

Figure 14-4 Knocking a spider plant out of its pot.

Flooding is another way to unpot a plant from a narrow-mouthed container. The pot should be submerged in water and a stream of water directed at the mouth to loosen the medium and flood it from the roots. Eventually enough will flood away to allow the roots through the opening with minimal damage. Because unpotting by flooding causes less root damage, it is valuable for delicate or highly prized specimens.

Unpotting a large plant is often a two-person job. Since it is too large to be held upside down and knocked out, the usual procedure is to lay the plant on its side and ease the roots out horizontally (Figure 14-5). Foliage breakage can be minimized by wrapping

Figure 14-5 Unpotting a large palm.

the top of the plant snugly with newspaper tied in place with string.

A good time to inspect the roots is after removing a plant from its container. Healthy young roots are firm and usually white. (A few are other colors, such as the orange of *Dracaena* roots.) But they generally should not be brown or mushy. Such roots are dead and decaying and should be trimmed off before the plant is repotted. Soil insects may also be detected in a repotting operation, since they frequently live around the roots.

Roots which have circled the root ball should have special treatment at repotting. If repotted in that condition, they may continue to wrap around the soil ball, never branching into the fresh soil and its available nutrients. Wrapped roots may be loosened by pulling them away with the fingers (Figure 14-6).

SELECTION OF THE NEW POT

When selecting a new pot for transplanting, both size and drainage should be considered. The new pot should preferably be only 1 to 2 inches (2.5–5 centimeters) deeper and wider than the previous pot. This additional area will hold an adequate volume of fresh potting medium for new root growth. If a larger pot is chosen, it is likely the new potting medium will stay waterlogged because the roots have not grown through it.

Drainage is a characteristic of the potting medium, but where drainage water is deposited is determined by the pot. A container with a drainage hole or holes allows excess water to pass out of the pot. In a pot without holes, water filters through the potting medium and excess eventually collects in the bottom. This excess water keeps the medium

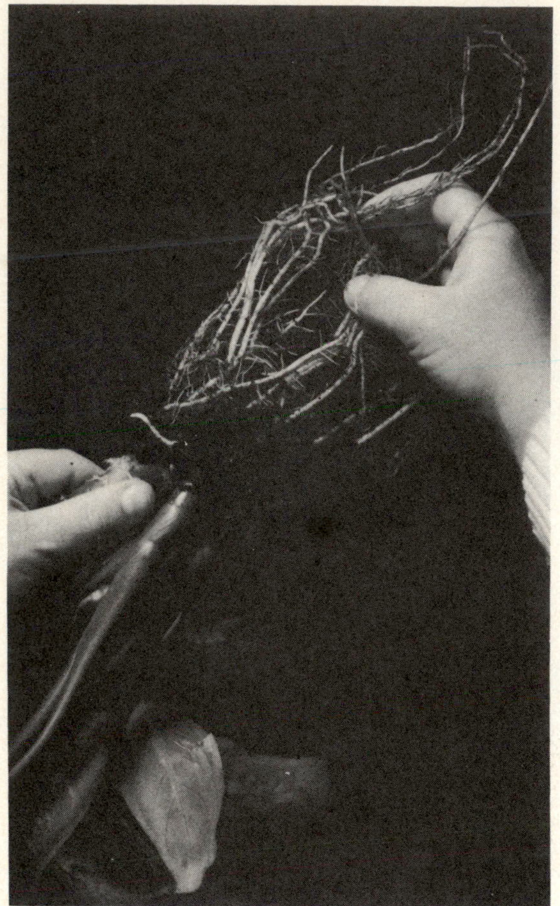

Figure 14-6 Loosening the roots of a pot-bound plant.

saturated, excluding oxygen necessary for root respiration. For this reason, pots with holes are preferable to those without unless watering is carefully regulated.

A water reservoir can be created in holeless pots to keep drainage water away from the roots. Stones or gravel at least a half-inch (1 centimeter) in diameter should be poured in a 1-inch (2.5-centimeter) or deeper layer in the bottom of the pot. The layer is then covered with nylon net, a sheet of sphagnum moss, or screen to keep out potting medium and leave air spaces open for water. In repotting large specimens, consideration should be given to the weight which gravel adds. A large volume of gravel can make moving a plant difficult.

It is a common misunderstanding that gravel in the bottom of any pot, having holes or not, will improve the drainage. The gravel layer does not improve drainage (see Chapter 15); it only provides a reservoir in pots from which the water has no way of escaping. Gravel is unnecessary for pots with holes. The only improvement these pots need is a single piece of broken clay pot (round side up) or screen over the drain hole to keep in the potting medium.

REPOTTING

Repotting is a simple process. The container is filled with potting medium to a level that will permit the plant to be growing as it was previously. The plant is then added and additional potting medium added around the sides and firmed with the fingers. Watering afterward settles out air spaces and lessens shock.

Watering may cause the soil to settle somewhat, and soil can then be refilled around the top. Enough room should be left to make watering convenient. A space of ½ to 1 inch (2–3 centimeters) is sufficient to avoid water runoff.

A plant should never be set deeper than it was originally growing to improve its appearance. This exposes the stems to soil fungi and bacteria, which can invade the tissues and kill the plant.

DOUBLE POTTING

Double potting (Figure 14-7) involves inserting a clay or plastic pot with drainage holes inside a holeless decorative pot. The spaces between the two pots are filled with sphagnum moss, gravel, or other material. The decorative pot then serves as the water reservoir.

Potting by this method provides the advantages of a free-draining pot with the beauty of a decorative one.

POSTTRANSPLANTING CARE

Sometimes plants shock after transplanting. They wilt and may remain wilted for days or weeks. To lessen shock, keep newly transplanted plants out of direct sunlight for a few days. Delicate plants can also be enclosed in a plastic bag for a week or two following transplanting. This increases the relative humidity while the roots are reestablishing and regaining their normal absorption potential.

In cases where bagging is ineffective and the plant remains wilted, pruning may be necessary. One-quarter of the foliage should be cut off and the plant should be observed for several days. If it remains wilted, more foliage will need to be removed.

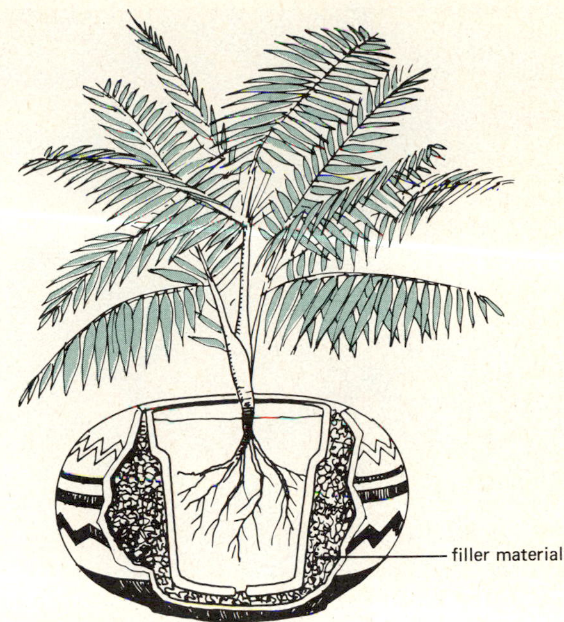

Figure 14-7 Double potting.

PLANTS WITH SPECIAL REQUIREMENTS

Bonsai

Bonsai is the art of growing and training trees to be miniature versions of the form they achieve in nature (Figure 14-8). It originated before the eleventh century in China, and some bonsai are hundreds of years old. However, a bonsai need not be old to have the appearance of age. Instead, aging is accomplished by pruning, training, and restricting the soil volume and fertilizer available to the roots.

Bonsai are not difficult to create and care for, but they do require a larger investment of time than many other houseplants. The first step in creating a bonsai is to select the plant. Visit a local nursery and choose an easy-to-grow plant such as pyracantha, pine, or juniper in a 1-gallon can. A large, bushy plant is not necessarily the best choice. Instead, purchase a plant with an irregular form, less foliage, and even some dead wood, since this will help create an aged appearance. The container selected for the bonsai should be smaller than its present pot. Shallow ceramic dishes with drainage holes designed for bonsai culture are traditional.

Branch pruning (Figure 14-8) is the next step in bonsai assembly. There are no inflexible rules regarding pruning, but enough of the limbs and foliage should be removed to create an open, aged feeling. If possible, prune to an irregular, windswept appearance, leaving only tufts of foliage on the tips of the branches.

Root pruning (Figure 14-8) follows. The tree should be unpotted and the root ball cut across horizontally. The layer of soil left should be slightly taller than the depth of the bonsai pot. It is then cut to conform to the shape of the pot.

The drainage hole should be covered with screen and the plant fitted into the container. The soil is pulled back from the stem slightly so that the plant is elevated above the

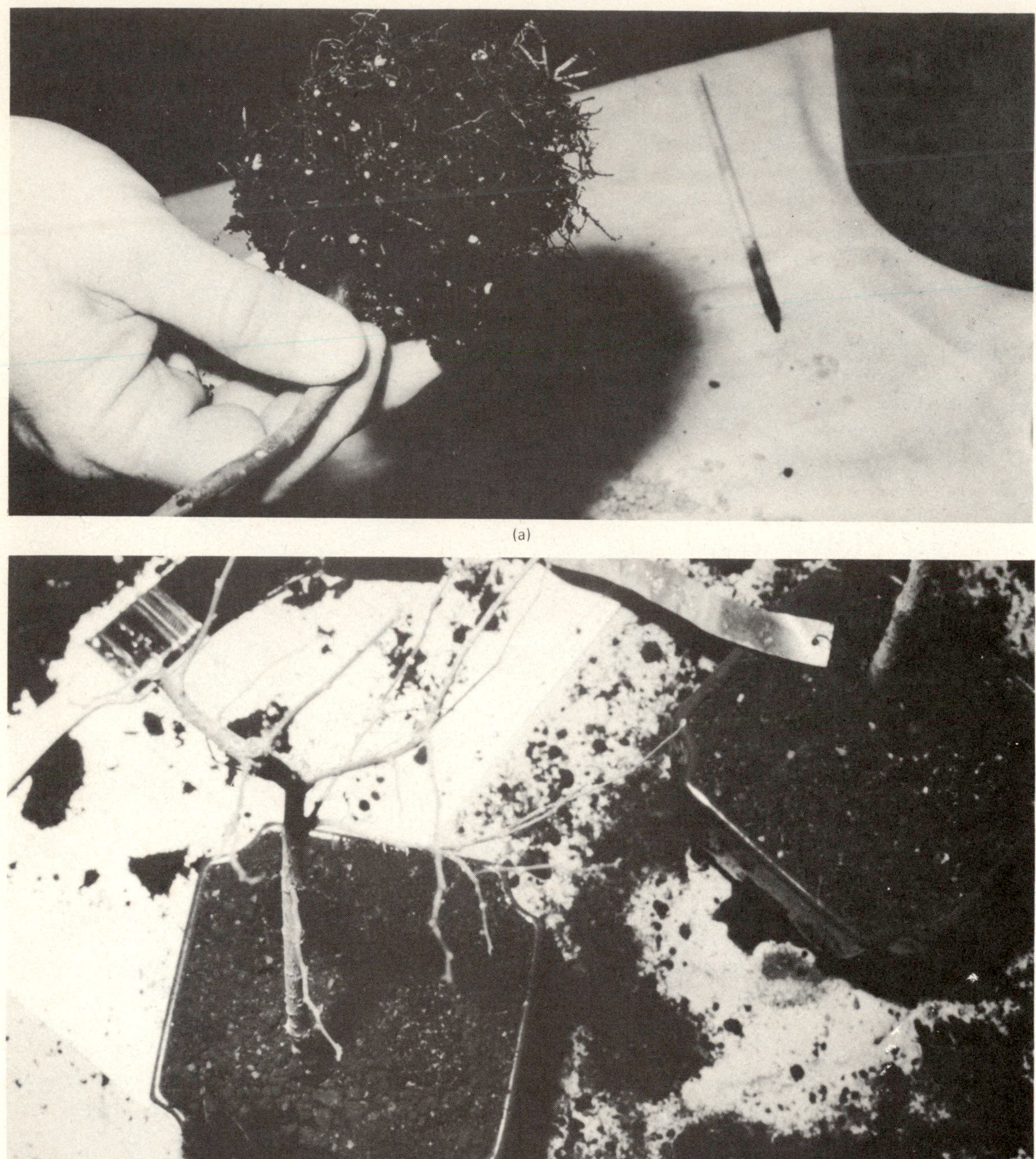

(a)

(b)

Figure 14-8 Steps in bonsai assembly. (a) Removal of the majority of the roots and foliage. (b) Potting in the bonsai container. (c) Training the tree by use of wrapped wire.

(c)

soil level, as it might be in nature. Remaining soil around the edge is removed to lower the level ¼ inch (0.5 centimeter) below the pot lip. Finally rocks, moss, or other ornaments are added, and the plant is watered.

For a posttransplanting period of at least a week, the bonsai should be kept out of direct sunlight. After this initial adjustment period, light should be increased gradually to the normal range for the species.

Bonsai are trained by wiring the branches to change their growth directions and form (Figure 14-8). Copper wire of a 10 to 20 gauge is wrapped in spiral fashion around the trunk and branches, which can then be bent to the desired shape. The wiring can be removed in several months, and the tree will retain the new form.

Watering of bonsai will depend on the species. Because of the limited soil volume, bonsai dry out relatively quickly. They must be watered frequently to prevent death from drought but should not be kept constantly moist since this would encourage undesired, vigorous growth.

Fertilizer for bonsai should be in liquid or water-dissolving form (see Chapter 15) applied from spring through fall. A dilute solution applied once per month is usually adequate, since the objective is to maintain the plant at its present size, not to encourage growth.

Bromeliads

Members of the family Bromeliaceae (Figure 14-9) are termed *bromeliads.* In nature they are tree-dwelling (epiphytic) plants. As houseplants they are dependable and easy to care for. All are valued for their foliage, and many have long-lasting flowers and colorful fruits.

Figure 14-9 ***Neoregelia carolinae*** **'Tricolor,' a popular bromeliad for indoor growing. The leaves are banded green and white. At flowering, the center cup becomes red.**

Bromeliads survive under a wide range of light levels from bright to dim, but the foliage will show the best coloration under the proper light level for the species. As a general rule, plants in the genera *Ananas, Dyckia, Orthophytum,* and *Tillandsia* are high-light types. *Aechmea, Neoregelia,* and *Billbergia* are moderate- to high-light-requiring, and *Crypanthus, Guzmania, Nidularium,* and *Vriesea* are moderate- to low-light-requiring.

The watering requirements of bromeliads differ from those of other houseplants. Potted types use soil moisture but also frequently have a center cup formed by the leaves which should be kept full of water. Both potted and plaque-mounted species absorb water efficiently through their foliage, and misting is beneficial. Since misting is the only water source for plaque-mounted bromeliads, daily spraying is advisable.

Bromeliads should be fertilized with a water-soluble fertilizer mixed at the strength recommended for foliar application. The fertilizer should be applied to the medium of potted specimens and also to the cup (Figure 14-9). Misting of both potted and mounted species with the solution is recommended.

Potted bromeliads generally have small root systems and require repotting infrequently. Use the Cornell epiphytic mix discussed in Chapter 15 or another similar lightweight medium. Table 14-5 is a list of bromeliads particularly suitable for growing as houseplants.

Table 14-5 Selected Bromeliads for Indoor Growing

Common name	Botanical name
No common name	*Aechmea* 'Burgundy'
Burgundy vase plant	*Aechmea maculata*
Urn plant	*Aechmea fasciata*
No common name	*Aechmea gamosepala*
No common name	*Aechmea maculata*
Patriotic plant	*Aechmea mertensii*
Pineapple	*Ananas comosus*
Queen's tears	*Billbergia nutans*
Muriel Waterman	*Billbergia* 'Muriel Waterman'
Earth star	*Cryptanthus* varieties
Cryptbergia	*Cryptbergia rubra*
Orange star	*Guzmania lingulata*
Flasklike neoregelia	*Neoregelia ampullacea*
Blushing bromeliad	*Neoregelia carolinae*
Fireball	*Neoregelia* 'Fireball'
Black Amazon bird nest	*Nidularium innocentii* 'Innocentii'
No common name	*Nidularium regelioides*
Pink quill	*Tillandsia cyanea*
Blushing bride	*Tillandsia ionantha*
No common name	*Tillandsia xerographica*
Flaming sword	*Vriesea splendens*
Sword plant	*Vriesea* hybrids

Cacti and Succulents

Cacti and succulents (Figure 14-10) are classified as either xerophytic or epiphytic. The former includes all desert cacti and all commonly grown succulents. The epiphytic cacti include Christmas cactus, Easter cactus, orchid cactus, and related jungle-dwelling species. The care for the two types differs greatly. Xerophytic cacti require high light, a coarse, fast-draining potting medium, and infrequent watering for most of the year. The epiphytes grow best in moderate light and a highly organic and constantly moist potting medium.

Improper watering is a major cause of death in cacti and succulents. Xerophytic species should be watered infrequently from fall through spring, their dormant period. The medium should be allowed to dry out thoroughly between waterings, and if possible, they should be kept at a cool 45–65°F (7.2–18.3°C) temperature. As soon as new growth begins, the watering frequency should be increased so that the medium is rewet as soon as it approaches dryness.

Figure 14-10 A group of succulents.

Because of their sensitivity to overwatering, xerophytic cacti should be transplanted only when actively growing in spring or summer. They should be repotted in a dampened media in lieu of postpotting watering.

Transplanting spiny cacti and succulents can be difficult; however, they require repotting only every 2 to 4 years. The process can be made easier by handling the plants with a band of folded newspaper, as shown in Figure 14-11.

Table 14-6 lists cacti and succulents suitable for growing as houseplants. When purchasing xerophytic cacti, beware of large cacti at bargain prices. They are often dug from the desert and their roots severed. Such plants are difficult to reroot and usually die over several months.

Ferns

Most ferns (Figure 14-12) are moderate- to low-light-requiring houseplants. During winter, when light intensities are low, they can be placed directly in an east, north, or west window. But in summer they should be either moved back from east and west windows or given protection by a sheer curtain.

Figure 14-11 A newspaper aids in transplanting a cactus.

With the exception of a few genera such as *Pellaea* and *Cyrtomium,* ferns will not survive complete drying of the potting medium. The medium should contain some moisture at all times. A quick-draining but moisture-retentive potting medium is best for root growth. Most prepackaged houseplant media are suitable. Humidity level is more critical for ferns than other houseplants, and a 30–60 percent relative humidity range is necessary for the healthy growth of most species. Table 14-7 lists selected ferns recommended for indoor growing.

Florist Plants

Florist plants (Figure 14-13) such as chrysanthemum, poinsettia, and forced bulbs are a handsome, long-lasting addition to the indoor environment. A potted plant will usually remain in bloom for at least a week and often up to a month. A few species can be adapted to growing as houseplants, and others can be transplanted outdoors and used as garden perennials.

The life of the flowers can be extended considerably with correct care in the home. The plant should be placed in bright light but preferably out of direct sunlight and drafts. The potting medium should be kept moist.

Fresh Cut Flowers

Fresh flowers will live longer if they are placed out of direct sunlight, away from heating vents, and preferably in a cool area.

Boxed flowers should be placed in water as soon as they arrive. A vase should be selected and filled with warm water [approximately 100°F (38°C)]. Preservative chemi cals will usually be included with the flowers and should be added to the water according to the package directions; otherwise use 1 tablespoon (15 milliliters) each of sugar and vinegar per quart (1 liter) of water. All leaves

Table 14-6 Selected Cacti and Succulents for Indoor Growing

Common name	Botanical name
Cacti	
Sea urchin cactus	*Astrophyton asterias*
Bishop's cap	*Astrophyton ornatum*
Silver torch	*Chamaecereus silvestri*
Peanut cactus	*Cleistocactus strausii*
Golden barrel cactus	*Enchinocactus grusonii*
Rainbow cactus	*Echinocereus pectinatus* var. *neomexicanus*
Orchid cactus	*Epiphyllum* hybrids
Fire barrel	*Ferocactus acanthodes*
Rose-plaid cactus	*Gymnocalycium mihanovichii* var. *friedrichii*
Agave cactus	*Leuchtenbergia principis*
Pincushion cactus	*Mammillaria* spp.
Ball cactus	*Notocactus* spp.
Bunny ears	*Opuntia microdasys*
Crown cactus	*Rebutia* spp.
Easter cactus and others	*Rhipsalidopsis* hybrids
Christmas cactus and others	*Schlunbergera* hybrids
Succulents	
Agave	*Agave angustifolia*
Aloe	*Aloe* spp.
Ponytail	*Beaucarnea recurvata*
Jade plant	*Crassula argentea* varieties
Watch chain	*Crassula lycopodioides*
String of buttons	*Crassula perforata*
Hen and chickens	*Echeveria* hybrids
Crown of thorns, pencil plant, and others	*Euphorbia* spp.
Ox tongue	*Gasteria* spp.
Wart plant	*Haworthia* spp.
Wax plant	*Hoya carnosa* varieties
Panda plant, air plant, and others	*Kalanchoe* spp.
Mother-in-law's tongue	*Sansevieria trifasciata* varieties
Burro tail and others	*Sedum* spp.
Wax ivy, string of beads, and others	*Senecio* spp.

which will be below the waterline should be picked off, and 1 to 2 inches (3–5 centimeters) cut from the flower stems. Water should be added daily as it is absorbed. After 3 or 4 days the flowers should be removed, their stems recut, and the flowers replaced in fresh water.

Arranged flowers will arrive in preservative solution. The only care they usually require is replenishing the water in the vase. However, if the flowers arrive wilted, recutting the stems and putting the flowers in warm water will help revive them.

Figure 14-12 The popular Boston fern.

Gesneriads

The family Gesneriaceae includes many outstanding houseplants such as episcias, African violets, and gloxinias. Many popular flowering houseplants are found in this family (Table 14-8).

Most gesneriads are warm-temperature plants and should be grown at temperatures from 55 to 80°F (12.8–26.7°C). Nighttime temperatures in the lower half of this range and daytime temperatures in the upper half will provide the best conditions for growth. Light should be moderate to high to achieve the bloom, but long periods of strong sunlight should be avoided. Less-intense winter sunlight is acceptable, and gesneriads are known to grow well with artificial lighting as the only light source.

The potting medium of gesneriads should remain slightly moist at all times. All water should be applied at room temperature since cold water can cause permanent leaf spotting.

The humidity range most acceptable for gesneriad growing is 50–80 percent. They can be successfully grown at lower humidities provided other growing conditions are met satisfactorily. Because of the high humidity desirable for raising gesneriads, they are particularly suited to terrarium growing.

Table 14-7 Selected Ferns for Indoor Growing

Common name	Botanical name
Maidenhair fern	*Adiantum raddianum*
Maidenhair fern	*Adiantum tenerum*
Bird's nest fern	*Asplenium nidus*
Mother fern	*Asplenium bulbiferum*
Holly fern	*Cyrtomium falcatum* varieties
Boston fern	*Nephrolepis exaltata* varieties
Boston fern	*Nephrolepis duffii*
Button fern	*Pellaea rotundifolia*
Green cliff brake	*Pellaea viridis*
Hart's tongue	*Phyllitis scolopendrium*
Common staghorn fern	*Platycenium bifurcatum*
Rabbit's foot fern	*Polypodium aureum* var. *areolatum*
Rabbit's foot fern	*Polypodium aureum* 'Undulatum'
Climbing bird's nest fern	*Polypodium punctatum*
Holly fern	*Polystichum tsus-simense*
Cretan bracken	*Pteris cretica* varieties
Australian bracken	*Pteris tremula*
Ladder bracken	*Pteris vittata*
Tongue fern	*Pyrrosia lingua*
Leatherleaf fern	*Rumohra adiantiforms*

Table 14-8 Selected Gesneriads for Indoor Growing

Common name	Botanical name
Columnea	*Columnea* 'Chanticleer'
	Columnea 'Early Bird'
	Columnea 'Mary Ann'
Flame violet	*Episcia* 'Moss Agate'
	Episcia 'Acajou'
	Episcia 'Cygnet'
	Episcia 'Jinny Elbert'
Nautilocalyx	*Nautilocalyx forgettii*
Candy corn plant	*Nematanthus wettsteinii*
African violet	*Saintpaulia ionantha* varieties
Slipper plant	*Sinningia pusilla* varieties
Cape primrose	*Streptocarpus saxorum*
	Streptocarpus 'Wiesmoor hybrids'
	Streptocarpus 'Constant Nymphs'
	Streptocarpus 'Diana'
	Streptocarpus 'Fiona'
	Streptocarpus 'Karen'
	Streptocarpus 'Marie'
	Streptocarpus 'Louise'
	Streptocarpus 'Paula'
	Streptocarpus 'Tina'
	Streptocarpus 'Margaret'

Figure 14-13 Typical florist pot plants: (back left) tulip, cineraria (middle), and Easter lily (back right).

Hanging Plants

The care of hanging plants differs from the care of those set on tables, windowsills, or the floor for several reasons. First, the air temperature is usually higher around hanging plants. Heat will rise to the top of a room, and consequently the air surrounding a plant hung at eye level may be 5°F (2.8°C) higher than the air at floor level. There is also greater air movement around a suspended plant since there is no supporting surface to block air flow.

Because of these environmental differences, hanging plants lose moisture from both the medium and the leaves at an accelerated rate. More frequent watering will thus be required. Since many hanging plants are drought-sensitive ferns, it is important to check these plants frequently for signs of moisture stress.

Orchids

Although many orchids are epiphytes, a number are terrestrials. Selected species of both types grow well as houseplants (Table 14-9). The anatomy and growth pattern of many orchids are unusual. Monopodial orchids (Figure 14-14) have a rosette form, or leaves on opposite sides of the stem, giving them a two-dimensional appearance. Sympodial orchids (Figure 14-14) have storage organs called *pseudobulbs*, with each pseudobulb bearing one leaf. These orchids grow by extension of rhizomes and generally gain one new pseudobulb and one new leaf per year.

Table 14-9 Selected Orchids for Indoor Growing

Common name	Botanical name
Pink pine	*Bletia purpurea*
No common name	*Brassavola cucullata*
Lady of the night	*Brassavola nodosa*
No common name	*Brassia maculata*
No common name	*Calanthe vestita*
Tulip cattleya	*Cattleya citrina*
Summer cattleya	*Cattleya gaskelliana*
Autumn cattleya	*Cattleya labiata*
No common name	*Coelogyne cristata*
No common name	*Coelogyne massangeana*
No common name	*Cycnoches ventricosum* v. chlorochilon
Lady slipper	*Cypripedium insigne*
No common name	*Dendrobium aggregatum*
Spice orchid	*Epidendrum atropur pureum*
No common name	*Gongora armeniaca*
No common name	*Hexisea bidentata*
No common name	*Laelia anaceps*
No common name	*Laelia autumnalis*
No common name	*Laelia rubescens*
No common name	*Lycaste aromatica*
Tiger orchid	*Odontoglossum grande*
Columbia buttercup	*Odontoglossum cheirophorum*
Dancing-lady orchid	*Odontoglossum ornithorhynchum*
Butterfly orchid	*Oncidium papilio*
Dancing-lady orchid	*Oncidium sarcodes*
Moth orchid	*Phalenopsis equestris*
No common name	*pPholidota chinensis*
Coral orchid	*Rodriguezia secunda*
No common name	*Stanhopea oculata*

Orchids (Figure 14-15) fall into the following three growing-temperature categories:

Cool	(45–58°F)	(7.2–14.4°C)
Intermediate	(55–70°F)	(12.8–21.1°C)
Warm	(62–80°F)	(16.7–26.7°C)

However, as an all-round temperature range, 56 to 62°F (13.3–16.7°C) night temperature and 62–80°F (16.7–26.7°C) day temperature are acceptable.

Epiphytic orchids are potted in osmunda fiber, bark, or similar lightweight potting medium. Watering should remoisten the potting medium just before it completely dries. As water runs through these loose media freely, care must be taken to ensure that the entire medium is wet. Watering thoroughly with a sink sprayer will accomplish this, as will submersion watering (Chapter 17).

Terrestrial orchids may be grown in an all-purpose potting medium suitable for foliage plants. They should be watered whenever the medium becomes nearly dry.

When repotting epiphytic orchids, a moist medium is more convenient. The bark or osmunda should be soaked in water overnight prior to the day intended for repotting. The orchid should be gently eased from its previous pot and the old medium cleaned from its roots. Next, any dead roots should be cut

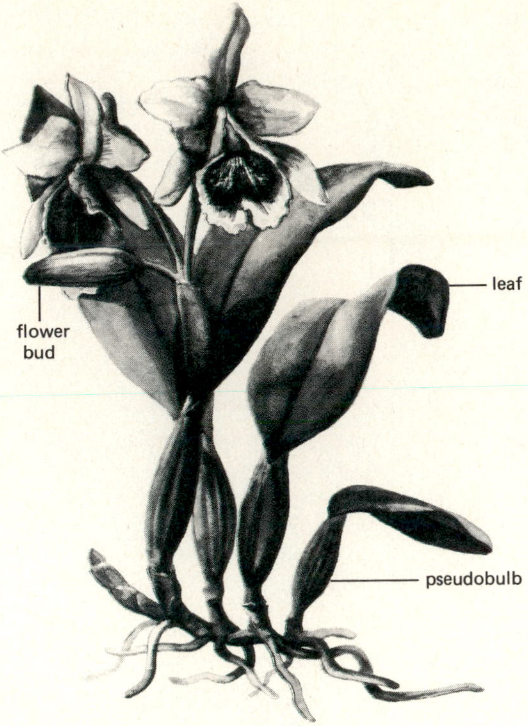

Figure 14-14 The monopodial (a) and sympodial (b) growth forms of orchids.

away and the plant repotted in the fresh medium. Finally, staking should be done, if necessary. The orchid should not be watered until the medium dries.

The light requirements of orchids can be met by placing the plants in a south-facing window during the fall and winter months, and in an east or west window during the spring and summer when light intensities become brighter. Orchids receiving insufficient light will be surprisingly attractive, with sleek, bright green leaves, but they are unlikely to flower.

The humidity range for orchids should be high (this is the most difficult growing condition to meet indoors). Misting will improve the humidity for a few minutes; it is more

Figure 14-15 Cattleya orchids.

important as a source of moisture that is absorbed by the foliage. Other methods for increasing household humidity are discussed in Chapter 17.

Fertilizer should be applied to orchids every other week while they are in active growth. The fertilizer should be of a liquid or dissolvable powder type and applied to the medium, leaves, and pseudobulbs at dilute strength.

Selected References for Additional Reading

Behme, R. L. *Bonsai, Saikei, and Bonkei.* New York: William Morrow, 1969.

Davenport, E. *Ferns for Modern Living.* Kalamazoo, MI: Merchants, 1977.

Elbert, V. F., and G. A. Elbert. *The Miracle Houseplants.* New York: Crown, 1976.

Graf, A. B. *Exotica III.* 9th ed. East Rutherford, N J: Roehrs, 1976.

Hoshizaki, B. J. *The Fern Growers Manual.* New York: Alfred Knopf, 1975.

Kramer, J. *Growing Orchids at Your Windows.* New York: Hawthorn Books, 1976.

Lamb, E., and B. Lamb. *Popular Exotic Cacti in Color.* New York: Macmillan, 1976.

Logan, H. B. *Orchids You Can Grow.* New York: Hawthorn Books, 1971.

Rice, L. W. *Cacti and Succulents for Modern Living.* Kalamazoo, MI: Merchants, 1976.

Wilson, L. *Bromeliads for Modern Living.* Kalamazoo, MI: Merchants, 1977.

Chapter 15

Potting Media and Fertilizers

INDOOR PLANT POTTING MEDIA

The word "medium" (plural, media) is not familiar to most people when used in relation to indoor plant growing. A medium is any material used for rooting or potting plants. It is soillike in that it performs the same functions as soil does for plants outdoors: supporting the roots and acting as a reservoir for moisture, nutrients, and air. But a medium can be a mixture of many ingredients, some manufactured and some naturally occurring. It need not be a natural product like soil. In fact most commercially available media used for houseplants contain no soil at all.

Specially mixed potting media have several advantages over most pure soils for growing houseplants. First, they are made up of materials which create large pore spaces, and they are therefore less likely to pack around the roots of plants. Danger of root death due to air exclusion is largely eliminated. Second, they have a loose structure that drains well. Water passes through quickly, yet some moisture is retained in the organic components of the medium included for that purpose. Third, they are usually free of disease organisms, weed seeds, and insects which could damage indoor plants. Finally, they are lighter in weight, which makes moving large plants easier and decreases shipping costs.

Overall, potting media are superior to almost all soils for growing plants in pots and are the best choice for rooting and growing indoor plants.

Premixed Potting Media and Their Components

Many brands of premixed potting media are available. They are sold under the name "houseplant soil" or sometimes "cacti soil" or "African violet soil," although most contain no soil at all. They are completely satisfactory for indoor plant growing, although they can be relatively expensive when used in quantity. A few, particularly the types sold for cacti, could be improved by the addition of more sand. When purchasing a premixed potting medium, read the label for assurance that the product has been sterilized. If it has not, it will need to be home sterilized.

Premixed potting media do not have one set formula or list of ingredients. They vary widely between manufacturers, depending on the available materials.

Any potting medium will generally consist of two to three types of components. The first is organic, plant-derived material. Its water-absorbing and -holding capacity will be high, and it will frequently have a spongy structure resistant to packing. In addition, its nutrient-holding ability or "cation-exchange capacity" will be high, enabling it to function as a reservoir for nutrients.

The second component is called the *coarse aggregate.* The aggregates improve the drainage of the medium because the angular shape of each particle wedges them apart, creating large or macro pore spaces (Figure 15-1). These pore spaces facilitate the free flow of air and water through the medium and increase the drainage rate. In general, coarse aggregates have a hard structure and a low water and cation-exchange capacity.

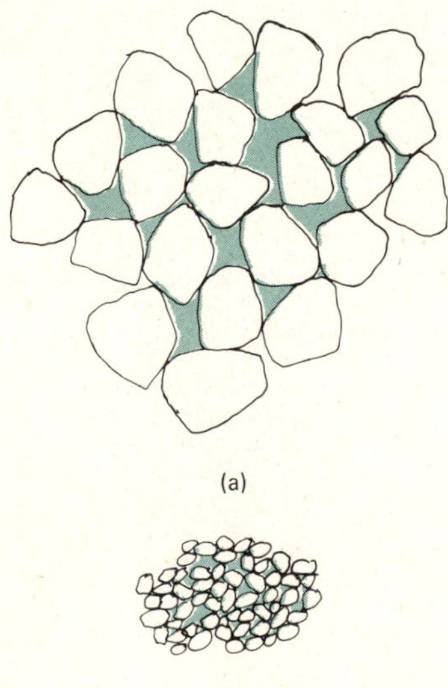

Figure 15-1 Macro (a) and micro (b) pore spaces.

A third optional component is soil. It is used to hold and supply nutrients and acts as a reservoir of water in addition to the organic material. Soil particles are usually smaller than the coarse aggregate or organic material particles. Consequently, they fit close to each other and other materials, forming micro pore spaces (Figure 15-1) which restrict the flow of air and water. However, these micropores are efficient at retaining water due to a phenomenon known as *capillary action.* The concepts of pore space, capillary action, and cation exchange are discussed in more detail in Chapter 6.

The following are the most common ingredients used in making prepackaged mixes. They are divided into three categories.

ORGANIC, PLANT-DERIVED MATERIALS (TABLE 15-1)

Peat Moss. Sold under the names Canadian peat moss, sphagnum peat moss, and just peat moss, this bog plant is harvested, dried, and used in the United States in enormous quantities each year. In its raw state, it can be bought in mats or bags, is light brown, and has a stringy texture (Figure 15-2). It is also commonly available shredded or "milled," with fibers broken down so they mix easily with other ingredients.

Peat moss is the main ingredient in many prepackaged houseplant mixes, often making up 50 percent of the volume of the medium. It is even used alone for some purposes such as air layering (Chapter 3) and rooting cuttings.

When dry, peat moss is a very fluffy, lightweight material. When wet, it is capable of holding a large volume of water without packing down, therefore providing plant roots with the ideal combination of moisture and air.

Peat moss is usually very acidic, with a pH range of 4.0 to 5.0. This range is too low for most houseplants, so pH-increasing materials are normally added when peat moss is part of the potting medium. One of the few

Table 15-1 Common Ingredients of Houseplant Media

Ingredient	Type	Remarks
Compost	Organic, plant-derived	A combination of decayed leaves, twigs, grass, and so on made at home; must be sterilized before use
Activated charcoal	Organic, plant-derived from partially burned wood but functions as a coarse aggregate	Absorbs odors, chemicals from the soil for several months before becoming saturated; can be reactivated by heating in an oven at 350°F (177°C) for about 30 minutes
Coarse sawdust	Organic, plant-derived material	Available only in certain areas of the country; should be coarse (⅛″; 3 mm) if possible
Bark	Organic, plant-derived material	By-product of lumber mills like coarse sawdust; holds water, nutrients
Bark chunks	Organic, plant-derived material	By-product of lumber mills; holds moisture, nutrients
Sphagnum peat moss	Organic, plant-derived material	Holds water and nutrients well, acid pH, breaks down very slowly
Leaf mold	Organic, plant-derived material	Flaky texture; decomposes in the medium, releasing nutrients; must be sterilized before use
Humus or muck soil	Organic, plant-derived material	Highly decomposed with a powdery feel; can be difficult to rewet if allowed to dry out
Sand	Coarse aggregate	Inert (not chemically reactive); use only coarse grades and not ocean sand
Aquarium gravel	Coarse aggregate	Heavy, inert, and very coarse; excellent for use with cacti and succulents
Perlite	Coarse aggregate	Lightweight and coarse; contains flourides which are toxic to some plants
Vermiculite	Coarse aggregate; water- and nutrient-holding component	Lightweight, capable of holding water and nutrition; also useful as a coarse aggregate, although it packs down over time
Scoria or lava rock	Coarse aggregate	Found in natural deposits near volcanos; relatively lightweight with a pleasing dark color

Figure 15-2 Milled (left) and natural (right) sphagnum moss.

disadvantages of peat moss is that if it is ever allowed to dry out thoroughly, it will be very difficult to rewet. Water will run over the fibers rather than being absorbed. The moss or medium must be gradually rewet by submersion or bottom watering or by the use of a chemical "wetting agent."

Leaf Mold. Leaf mold (Figure 15-3) is a second possible component of prepackaged houseplant media. Like peat moss, it is derived from plants and is made up of partially decayed leaves and twigs. It is dark brown with a flaky texture. Leaf mold is capable of holding water and nutrients, although not as much as peat moss. Leaf mold may be used in premixed potting media in addition to or as a complete substitute for peat moss.

Leaf mold is a natural product in the decomposition of leaves. It is seldom sold as a pure product but can be made at home by composting fallen leaves (Chapter 6). It can also be gathered from the ground in deciduous woods.

Humus Peat or Muck. This ingredient is actually a soil type. It is found in low-lying areas and is composed largely of organic material in the form of highly decayed vegetable matter such as leaves, stems, and roots. It is black and fluffy in appearance and has a fine, powdery feel. It is commonly available under the name of garden peat or Michigan peat, although it is not peat at all. Muck is a highly decomposed material that continues decomposing rapidly when used in a potting medium. Its fineness may also create aeration problems.

Do not confuse humus with sphagnum peat moss. The latter is a lighter-brown fibrous material and is the more expensive of the two.

Bark. Shredded or ground bark is occasionally used as an organic potting medium component. Fir, redwood, pine, hemlock, and oak are all satisfactory sources of bark. Bark of incense cedar and walnut should be avoided since it may contain toxic substances.

Compared to peat moss, bark has less water-holding capacity and a higher pH. However, pH-altering chemicals are still needed to raise the pH to an acceptable level. In addition, nitrogen fertilizer is generally included in bark-based potting media to compensate for fertilizer lost as the bark decomposes.

Figure 15-3 Leaf mold.

One type of bark derived from pines has been shown to be mildly fungicidal, suppressing the growth of fungi in the potting medium. However, this property does not render it significantly superior to bark from other species.

COARSE AGGREGATES (TABLE 15-1)

Sand. Sand is a naturally occurring aggregate used occasionally in premixed potting media. Its heavy shipping weight discourages more frequent use. It is primarily quartz which is not chemically reactive in the medium. Not all grades of sand are satisfactory for use in potting media. Only the coarser grades known as builder's or mason sand will improve medium drainage and aeration when used in sufficient proportions.

Perlite (Figure 15-4). This aggregate is commonly used in commercial media because of its extremely light weight. It is made by heating volcanic rock until it expands and then grinding the resulting material into irregular chunks 1/16 to 1/8 inch (1.59–3.18 millimeters) in diameter. Like sand, perlite does not hold appreciable amounts of water or nutrients and does not react in the potting medium. In appearance and size it is very similar to white aquarium gravel.

One of the disadvantages of perlite is its tendency to float to the top of a potting medium when a plant is watered. Another is that, due to its light weight, it may not provide enough anchorage for the roots of larger plants, and they may fall out of their pots.

Figure 15-4 Perlite.

This is a particular problem with cacti potted in a largely perlite medium, since large-bodied cacti frequently have sparse root systems which provide little anchorage.

Perlite has also been proved to contain levels of fluoride which are toxic to some houseplants. As a precaution, it should be rinsed in water three times before it is used. Perlite-containing potting media should be avoided for members of the Liliaceae family since they are particularly susceptible to fluoride injury.

Vermiculite (Figure 15-5). Vermiculite is made by a heat expansion process similarly to perlite, but a micalike material is used in lieu of volcanic rock. The resulting material is silvery-brown in color, lightweight, and with a spongy feel. The pH is neutral to slightly alkaline, in the range of 7.0 to 7.5. Vermiculite naturally contains appreaciable amounts of potassium and magnesium, two nutrients essential to plant growth.

Vermiculite is unusual among the coarse aggregates in that it is an active part of a medium. It has a high capacity for holding water and nutrients, although the size of the individual particles still facilitates good drainage.

Several size grades of vermiculite are manufactured, although not all of them may be available in garden centers. The fine grade

Figure 15-5 Vermiculite, an occasional component of potting media.

is useful for seed germination, and the medium for rooting cuttings or adding to media. The coarse grade is occasionally used for packing plants for shipping. The main disadvantage of vermiculite is its tendency to compress and lose its drainage-enhancing quality with time or when it is watered frequently. Nonetheless, vermiculite is still a popular material for potting media.

Charcoal (Figure 15-6). Charcoal in small chunk form (such as used for aquariums) is sometimes used as a potting media additive

Figure 15-6 Horticultural charcoal.

in place of other aggregates. It can be beneficial in some instances in that it absorbs and holds odors, toxic pesticides, and other undesirable chemicals. But its use is not practical or recommended because of its relatively high cost and short-term effectiveness.

Scoria or Lava Rock. Scoria (Figure 15-7) is a natural volcanic rock ground to a size of 1/16 to 1/8 inch (1.59–3.18 millimeters). It is gray or black in color and intermediate in weight between perlite and sand. Where readily available, it is an inexpensive and satisfactory coarse aggregate.

SOILS

Loamy Soil. Loamy soil is a well-balanced combination of the basic soil particles sand, silt, and clay in conjunction with a moderate amount of humus or other organic matter. It is the rich topsoil found in many successful gardens. Loamy soil holds water well, does not pack or crust when dry, and can be easily crumbled in the hands when it is moderately damp. It is an excellent potting medium additive even in relatively large quantities but should not be used alone.

Clay Soil. Clay soils are composed largely of tiny platelike particles which pack together closely, forming many micro pore spaces. Clay soils are highly water- and nutrient-retentive but poorly aerated. They are seldom added to premixed potting media but can be used in restricted amounts when balanced carefully with organic and coarse aggregate components.

Figure 15-7 Scoria, a coarse aggregate used in potting media.

Sandy Soil. If the sand it contains is coarse enough, a sandy soil may substitute in part for the coarse aggregate portion of a potting medium. However, fine sandy soils are less widely recommended since the particles create relatively small pore spaces.

Silty Soil. This soil has a floury feel when dry and has poor aeration like clay. It has neither the nutrient-holding capacity of clay nor the aggregate function of sandy soil. It should not be used in houseplant media.

Homemade Potting Media

Many people make their own potting media for various reasons: first, buying premixed potting media in bags becomes expensive if there are more than a few plants in the household. Second, premixed media are often formulated with more thought to eye appeal and shipping costs than plant growth. They can be heavily composed of dark, rich-looking humus but lacking essential aggregates like sand.

There is no one formula or list of ingredients that must be adhered to when making a potting media at home. The mix composition is largely determined by the availability of different ingredients in the area and their relative cost. The organic part of the mix need not necessarily be made up of sphagnum moss, for example. Leaf mold gathered from the floor of a woods, compost, or bagged humus can be substituted in whole or in part. Likewise, leftover aquarium gravel, crushed charcoal, cinders, or sand from a gravel pit or lake can be used in place of more expensive perlite and vermiculite. Ocean sand contains salt even after thorough washing and should not be used.

In addition to the bulk ingredients used in homemade media, other ingredients will sometimes be needed to alter pH or improve fertility. Although in quantity they constitute a very minor component, they are essential parts of a good potting medium.

pH ALTERATION

Many potting media formulas call for additions of small amounts of powdered dolomitic limestone to a mix with a peat moss, leaf mold, or bark base. The limestone counteracts the acidity of the organic component to give the media the slightly acid pH at which most tropicals grow best. It also contains quantities of calcium and magnesium needed for growth.

FERTILITY IMPROVEMENT

Many fertilizer materials, both natural and synthetically made, are recommended for incorporating into homemade potting media. The natural materials such as bone meal and dried manure decompose slowly and release their nutrients slowly into the potting medium. Plants are supplied constantly with fertilizer over several months. Other dry chemical fertilizers such as ammonium nitrate, super phosphate, or a typical three-way combination of nitrogen, phosphorus, and potassium are less expensive but effective for a shorter period of time.

Incorporated fertilizers are largely optional except when bark is used as a potting medium component. Occasional fertilization should also supply all the nutrients required by plants.

POTTING MEDIA FORMULAS

The following media formulas are useful for the majority of houseplants. They show the variation that is acceptable when mixing a growing medium. They can be taken either as bases from which to experiment or as standard formulas to be followed exactly. For the experienced gardener, the feel of the medium may indicate its suitability, with the desired product being loose, with a rich supply of organic matter to produce a spongy feel.

Soilless or Artificial Media. Mix 1. The following is an adaptation of the foliage plant mix developed by Cornell University.

4 qt (3.8 liters) shredded sphagnum moss
2 qt (1.9 liters) vermiculite
2 qt (1.9 liters) perlite
1 tbsp (15 ml) ground dolomitic lime
1 tsp (5 ml) 10-10-10-granular fertilizer.

Mix 2. University of California Mix "C," using fine sand:

1 part (by volume) fine sand
1 part (by volume) peat moss.

Mix 3. Bark-based mix:

3 parts milled pine bark
1 part sphagnum peat moss
1 part sand.

Add for each cubic foot (0.03 m^3) of mix:

6 oz (200 g) dolomitic lime
2 oz (65 g) hydrated lime
1/4 tsp (1.2 ml) fritted trace elements.

Soil-containing Media. Mix 1. For use with heavy soils which are clay-based:

1 part (by volume) soil
2 parts (by volume) organic matter
2 parts (by volume) coarse aggregate.
Plus for each bushel (35 liters), add
6 to 8 oz (200–250 g) ground or dolomitic limestone.

Mix 2. For use with medium-weight soils such as loams:

1 part (by volume) soil
1 part (by volume) organic matter
1 part (by volume) coarse aggregate.
Plus for each bushel (35 liters), add
6 to 8 oz (200–250 g) ground or dolomitic limestone.

Mix 3. For use with light, sandy soils:

1 part (by volume) soil
1 part (by volume) organic matter.
Plus for each bushel (35 liters), add

6 to 8 oz (200–250 g) ground or dolomitic limestone.

Mix 4. For stretching commercially prepared mixes:

2 parts (by volume) premixed potting medium
1 part (by volume) loamy garden soil of good quality
1 part (by volume) coarse aggregate.

Specialized Potting Media. The formulas just listed are sufficient for most houseplants. However, few plants require special media, whereas others will do better with slightly different media.

High organic medium—for ferns and African violets:

3 parts shredded sphagnum moss
1 part sand.

Xerophytic cacti and succulent medium—fast draining and coarse:

1. 1 part coarse sand or aquarium gravel
   1 part topsoil
   1 part peat moss.
2. 1 part premixed commercial medium
   1 part coarse sand.

Acidic medium—for plants requiring a low pH (5.0–5.5) (see Table 15-2):

Table 15-2 Acid-loving Houseplants Which Commonly Show pH-caused Chlorosis

Common name	Botanical name
Azalea	*Rhododendron hybrid*
Camellia	*Camellia japonica*
Citrus	*Citrus* spp.
Climbing fern	*Lygodium palmatum*
Flame of the woods	*Ixora coccinea*
Gardenia	*Gardenia jasminoides*
Heath, heather	*Erica* spp.
Hydrangea (blue)	*Hydrangea macrophylla*
Sasanqua	*Camellia sasanqua*
Sweet potato	*Ipomaea batatas*
Venus fly trap	*Dionaea muscipula*

1. 3 parts sphagnum peat moss
   1 part vermiculite
   1 part sand.
2. Pure sphagnum peat moss.

Rooting medium—fast draining with a moderate moisture-holding capacity (not for xerophytic cacti):

1. Pure vermiculite (medium grade).
2. 1 part perlite
   1 part shredded sphagnum moss.

Seedling media—fine texture and a moderate moisture-holding capacity:

1. 1 part vermiculite (fine grade)
   1 part finely milled sphagnum moss.
2. 1 part sand
   1 part premixed commercial medium
   1 part finely milled sphagnum moss.

Epiphytic orchid media—extremely fast draining and well aerated:

1. Pure bark chunks (½ inch; 1–2 centimeters) in diameter.
2. Pure osmunda fiber (Figure 15-8).

Epiphytic bromeliad medium—fast draining:

1. 1 part sphagnum peat moss
   1 part bark.
2. University of California "C" Mix.

MIXING POTTING MEDIA

For the most thorough mixing, the ingredients for a houseplant medium should be damp but not wet. Mixing ingredients which are completely dry will cause dust to fly as well as the finer particles to settle to the bottom of the medium.

Dampening sphagnum peat moss can be difficult when it is fully dry. It can be allowed to soak in a bucket overnight and then squeezed out to the correct dampness the next day. As an alternative, hot water will wet it quickly.

Figure 15-8 Osmunda fiber derived from the osmunda fern. The pole at the top is used to support climbing vines. The loose fiber at the bottom is used for potting orchids and other epiphytes.

STERILIZING POTTING MEDIA

Sterilizing houseplant media is done to prevent the introduction of unwanted disease organisms, insects, and weed seeds to indoor plants. It is particularly important for seedling and rooting media, where soil-borne diseases are common.

Not all media require sterilization to be considered safe to use. Vermiculite and perlite are disease-free because they are heat-treated in the manufacturing process. However, all other ingredients including peat moss, leaf mold, soil, and sand require sterilization before use.

Heat sterilization is the most common soil sterilization method. The damp mix is placed in a large baking pan, covered with foil or a lid, and baked at about 140°F (82°C) for 30 minutes or more until heated to the center. It is then allowed to cool and sealed in a plastic bag to prevent recontamination.

Soil-containing and soilless media give off an odor during baking which some people find objectionable. The unpleasantness can be avoided if the medium is baked in a sealed roasting bag instead of a pan.

Any fertilizer that is to be included in the potting medium should be added after it is sterilized. Some fertilizers are affected by high temperatures and will break down and release all their nutrients in the baking process.

INDOOR PLANT FERTILIZERS

The growing medium in which a plant lives is a reservoir for the nutrients it needs for growth. As these nutrients are absorbed or leached from the medium, they must be replaced with fertilizers.

Major Nutrients for Houseplant Growth

There are many nutrients needed by plants in varying amounts, but the ones used in the greatest amounts are nitrogen, phosphorus and potassium, calcium, magnesium, and sulfur. They are termed *plant macronutrients.*

NITROGEN

Nitrogen (abbreviated N) is the nutrient used in greatest quantity by houseplants. Plants absorb nitrogen constantly and use it for manufacturing many chemical compounds within the plant, including chlorophyll and amino acids. Without a constant supply of nitrogen, new leaves are unable to synthesize chlorophyll and become pale yellow in appearance.

PHOSPHORUS

Phosphorus (abbreviated P) is the second essential nutrient for plant growth. It plays a part in many plant activities including energy transfer. Phosphorus is particularly noted to be essential for proper flower and fruit formation.

POTASSIUM

Postassium (abbreviated K) is the third element nutrient for plant growth. It is involved in carbohydrate translocation and protein synthesis. Potassium also enhances coloration, strengthens stems, and aids in strong root development.

CALCIUM

Calcium is needed for cell division and expansion. An insufficient supply of this element also slows root growth and can cause death of the meristem.

MAGNESIUM

Magnesium is a constituent of chlorophyll and, like potassium, is necessary for protein synthesis.

SULFUR

Sulfur is a component of many amino acids and two of the B vitamins found in plants.

Of the six elements considered as macronutrients, nitrogen, phosphorus, and potassium are most generally included in houseplant fertilizers. Each element is included as part of a chemical compound which can be easily shipped and stored. Nitrogen, for example, can be included as calcium nitrate ($Ca(NO_3)_2$) or ammonium nitrate (NH_4NO_3). Phosphorus can be formulated as phosphoric acid (P_2O_5) or monocalcium phosphate (CaP_2O_5). Potassium can be included as potassium nitrate (KNO_3), potassium chloride (KCl), or potassium sulfate (K_2SO_4).

The salts of N, P, and K are used to make houseplant fertilizers dissolve readily in water, and in fact some are packaged predissolved as concentrated liquids. The dissolving ability makes it possible for them to be carried through the medium in water to reach the roots of plants.

Interpreting Fertilizer Labels

Because of the importance of the three elements just mentioned for healthy plant growth, they are contained in almost all houseplant fertilizers. By law, any product sold as a fertilizer of "plant food" must list on the front label the percentage of these elements it contains in bold type and separated by dashes. These numbers are known as the fertilizer analysis and are of more help in determining the value of the fertilizer than any claims made by the manufacturer about his product.

For example, a fertilizer whose analysis reads 5-10-5 contains 5 percent of its weight in nitrogen, 10 percent in a phosphorus-containing compound, and 5 percent in a potassium-containing compound, for a total of 20 percent nutrients by weight. The 5-10-5 formula is a relatively common analysis for houseplant fertilizers, but they range from 1.6 to 60 percent total nutrient content, depending on the manufacturer. Table 15-3 shows the analyses of a number of houseplant fertilizers which were being sold in a garden center in 1977.

When purchasing a fertilizer, you do not necessarily get what you pay for. Price appears to be governed more by packaging and volume than by the total amount of nutrients the fertilizer contains. A low-analysis fertilizer may be nearly the same price as a high-analysis one. When selecting a fertilizer, look for the highest analysis at the lowest cost to get the best buy.

The comparison of three fertilizers in Table 15-4 illustrates this point. Brand A is the best buy, brand B the second, and brand C the least economical. Comparison shopping can save considerable money.

Although it is not crucial, a "balanced" fertilizer with equal portions of N, P, and K compounds is widely used by commercial houseplant growers with excellent results. A 5-5-5 fertilizer would be an example of one balanced formula. Barring this, a formula should be chosen which has a nitrogen content equal to or greater than its phosphorus and potassium portions, such as 18-6-12.

Fertilizer Forms

Houseplant fertilizers are available in many forms, from tablets to granules and ready-to-use liquids. Each has advantages with regard to price and ease of application, but all are equally acceptable fertilization methods.

DISSOLVING POWDER CONCENTRATES

Powder fertilizers designed to be mixed with water are one fertilizer form. Economically they are usually the best buy since they are the most concentrated. The nutrients dissolve and are applied to plants in place of plain water, are carried down to the root zone in the water, and are immediately available for uptake by plant roots. A disadvantage with older brands was a poor dissolving ability due to filler materials which settled out. But this is uncommon now. Some brands contain bright-colored dyes which can stain skin or clothes when they are used in concentrated form.

TABLETS AND SPIKES

Dry fertilizers can be purchased as tablets designed to be laid on the medium or as small spikes to be pushed down into it (Figure 15-9). As a plant is watered, these solids break apart, dissolving their nutrients into the water as it flows down to the roots of the plant.

Solid fertilizers avoid the bother of mixing and are usually moderate in price.

Figure 15-9 Houseplant fertilizer in tablet and spike form.

Table 15-3 Examples of Houseplant Fertilizers

Brand name	Analysis	Formulation	Weight	Retail price
Greenaid All-Purpose Houseplant Food	3-1-3	Concentrated liquid	18 oz	$1.69
New Plant Life	2-1-2	Concentrated liquid	16 oz	1.49
New Plant Life Fish Emulsion	5-5-1	Concentrated liquid	9 oz	.99
Marsh's VF-11 Plant Food	0.15-0.85-0.55	Concentrated liquid	8 oz	.69
Kelly Green No-Mix Fertilizer	1-2-1	No-mix liquid	32 oz	1.59
Plantabbs Terrarium Plant Food	18-6-12	Granular slow release	4 oz	.99
Stern's Miracid Soil Acidifier and Plant Food	30-10-10	Dissolvable powder	8 oz	1.29
Stern's Miracle-Gro Water-Soluble Plant Food	15-30-15 and micro-nutrients	Dissolvable powder	8 oz	1.29
Precise Timed-Release Plant Food	12-6-6	Encapsulated slow release	3.7 oz	1.29
Plantabbs	11-15-20	Tablet	2.9 oz	.89
Kelly Tabs	10-17-11 and micro-nutrients	Tablet	1.5 oz	.59
Sole Tabs	8-3-1	Tablet	Not given	.69
Ortho Houseplant	5-10-5 and micro-nutrients	Concentrated liquid	0.78 lb	1.89
Frank's Plant Food For All Houseplants	5-10-5	Concentrated liquid	10 oz	.89
Fish Emulsion Fertilizer	5-1-1	Concentrated liquid	9 oz	.99
Plantabbs Orchid Soluble Plant Food	18-18-18	Dissolvable powder	2 lb	3.75
Osmocote Time-Release Plant Food	14-14-14	Encapsulated slow release	1 lb	2.65
Osmocote Long-Feeding Fertilizer	18-6-12	Encapsulated slow release	3 lb	5.25
Rapidgro	23-19-17	Dissolvable powder	8 oz	1.65
Peter's	20-20-20	Dissolvable powder	16 oz	2.50

Table 15-4 Cost Comparison of Three Houseplant Fertilizers

	Analysis	Nutrients	Weight	Total nutrition	Package price	Price per ounce of nutrient
Brand A	5-10-5	20%	10 oz	2 oz	$.89	$ .45
Brand B	11-15-20	46	3 oz	1.3 oz	.89	.68
Brand C	5-5-1	11	9 oz	0.99 oz	.99	1.00

LIQUID CONCENTRATES

Fertilizers in liquid concentrate form are actually dissolvable powders to which water has been added. They are further diluted before use and applied in place of plain water at watering time.

Liquid concentrates mix in water instantly but are more expensive than almost all other fertilizer forms based on the amount of nutrients.

READY-TO-APPLY LIQUIDS

Ready-to-apply liquids are poured directly on a plant in place of water and, accordingly, always have a very low nutrient concentration. They avoid mixing inconvenience but are the worst buy because of the high percentage of water they contain.

ENCAPSULATED SLOW RELEASE

Encapsulated fertilizer (Figure 15-10) can be worked into the medium surface or mixed into the medium at planting with equal effectiveness. The material consists of pinhead-sized beads containing concentrated liquid or solid fertilizer. Covering each bead is a water-permeable plastic coating. The slow-release feature is made possible by moisture present in the soil seeping in and out of the beads. It carries in it dissolved nutrients, which are moved to the roots of the plant at each watering. In the end only the plastic coating is left. The covering does not break down readily and may last for several years in the medium, but it is not harmful.

Figure 15-10 A microencapsulated fertilizer for houseplants.

Slow-release encapsulated fertilizers require the least work and mess of any fertilizer type. The beads do not stain, and one application is effective for 3 to 4 months. The expense is moderate when compared to that of other forms.

GRANULAR SLOW RELEASE

Granular slow-release fertilizers are enclosed in small beds and look very much like garden fertilizers. However, they do not possess the plastic coating of encapsulated slow-release types. Instead they release their nutrients slowly because they are a blend of several chemical compounds. One compound releases nutrients immediately, whereas the other must be chemically broken down before it can be absorbed by the plant. Thus a plant receives nutrition over a long span of time from one application.

Granular slow-release fertilizers are moderately expensive. They have no particular advantage over other fertilizer types applied at proper rates and intervals.

Fertilizer Frequency and Rate

Because of the wide variety of concentrations of fertilizers available, it is impossible to prescribe a standard dilution rate of application frequency. The container mixing directions are the main source of mixing information.

However, many houseplant fertilizers advise weekly or monthly applications of fertilizer on a year-round basis, and it is seriously questionable whether this is advisable. In some cases adherence to container directions could result in serious overfertilization because the frequency recommendation is based on a rapidly growing plant.

The fertilizer needs of plants growing indoors vary greatly, depending on the plant species and its growing conditions. A naturally slow-growing species like yew pine (*Podocarpus* spp.) would require much less fertilizer than a fast-growing Wandering Jew (*Zebrina* spp.), for example. Fertilizing both at the same rate is therefore not desirable.

Plant nutrient requirements are also greatly affected by the light intensity they receive. In the north, winter light intensities are very low, and indoor plants experience a slowdown in growth. At this time fertilizer should be applied sparingly or not at all. However, increased light intensity in the spring causes an increased growth rate, boosting water and fertilizer requirements. Fertilizer frequency may need to be increased to keep pace with the rapidly growing plant and to replace those nutrients leached by frequent watering.

As a general rule, fertilization should be done only when a plant is actively growing. The dilution rate should be equal to or weaker than that the package directions stipulate. Application should be infrequent, with the fertilizer applied only during periods of growth and at intervals equal to or longer than those recommended. Few houseplants are harmed seriously by underfertilization, but all are susceptible to damage from overfertilization.

General Guidelines for Fertilizing Houseplants

One of the most important facts to remember when fertilizing a plant is not to increase the fertilization rate or frequency unless you are sure that the plant is suffering from nutrient deficiency. Even if nutrient deficiency is suspected, fertilizer should never be mixed stronger than package directions indicate. Instead, increase the fertilizer frequency slightly, and wait for signs of improvement.

A second precaution is never to apply fertilizer to a dry medium, with the exception of extremely weak ones designed to be used in place of water. A concentrated fertilizer solution can draw water from the roots, and fertilizer "burn" will result.

Third, fertilizing should not be considered a cure for plants which are declining from an unknown cause. It can do actual harm by creating stress on the plant because of excess fertilizer buildup around the roots. Instead, stop fertilizing entirely, resuming it only when the problem is corrected and the plant is growing well again.

Special Fertilizers

A number of special fertilizers for houseplants are sold today. Some are strictly for African violets, some are derived from fish, and others claim that one drop in water satisfies all plant requirements.

ORGANIC FERTILIZERS

These fertilizers are extracted from plant or animal sources instead of being obtained by mining or chemical reactions. Fish emulsions, seaweed extracts, and manure formulations are several types of organic fertilizers.

Organic fertilizers contain the same nitrogen, phosphorus, and potassium as other fertilizers, but only in naturally occurring chemical forms. In the process of nutrient uptake, plants absorb fertilizer only as simple inorganic molecules. It is therefore necessary that microorganisms break down organic fertilizer components before absorption can take place. The organisms work slowly, so organic fertilizers are available over a long period of time.

By analysis, organic fertilizers usually contain a lower percentage of nitrogen, phosphorus, and potassium than inorganic ones, but they sell for about the same price per package. They also contain small quantities of "micronutrients," but these are seldom lacking in houseplants. Although organic fertilizers are no less beneficial for plants than inorganic types, they have no particular advantages over them. It should be noted that the microorganisms needed to break down organic fertilizers into usable chemical forms may be scarce in sterilized media. This can cause nutrients to become available extremely slowly and create nutrient deficiency problems in spite of regular fertilization.

AFRICAN VIOLET AND FLOWERING PLANT FERTILIZERS

Fertilizers sold as specially formulated for blooming plants contain the same N, P, and K compounds as all-purpose houseplant fertilizers but usually in a different ratio. Low nitrogen, high phosphorus, and moderate to high potassium percentages typify a flowering plant formulation.

Although a deficiency of phosphorus can be a cause of poor flowering, using a high-phosphorus fertilizer will not guarantee a blooming plant. Other requirements such as light intensity or duration, maturity of the plant, and temperature must be satisfied. Nor can a high-phosphorus fertilizer guarantee a longer bloom life or be an insurance against bud and flower drop, for other factors are also involved in these processes.

In price, flowering plant fertilizers are about the same as all-purpose fertilizers. There is no research to support claims that these high-phosphorus fertilizers are better than general-purpose ones for flowering houseplants. An evenly balanced (N, P, and K equal) fertilizer will generally supply all the nutrients needed for flowering.

Micronutrient Fertilizers

Whereas macronutrients are used in moderate amounts by plants, the micronutrients are needed only in minute quantities. The micronutrients include such elements as iron, copper, zinc, manganese, boron, and molybdenum. Micronutrient fertilizers can be purchased either with micronutrients alone or as a part of a regular all-purpose fertilizer.

Potting media usually contain enough micronutrients to supply a plant indefinitely due to the included soil and the decomposition of organic materials. But occasionally plants will develop nutrient deficiency symptoms which are not corrected by all-purpose fertilizers, and a micronutrient deficiency is the likely cause. In these cases a micronutrient fertilizer is advisable, and it can be applied either to the roots or as a spray on the affected foliage.

Many times, micronutrient fertilizers will state that their nutrients are in "chelated" form. Chelated nutrients are used when the pH of the potting medium is causing micronutrients to be "tied up." Even though the micronutrients are present, at an abnormally high or low pH, they are held to the medium and are unavailable to a plant. However, chelated nutrients are less affected by pH and will remain available even at unfavorable pH's.

ACID-FORMING FERTILIZERS

Acid-forming fertilizers react in the medium to make it more acidic or lower the pH. They are also sold under the name of azalea fertilizer, since azaleas are common plants which require a highly acid potting medium. But other plants listed in Table 15-2 are also "acid-loving" and will show deficiency symptoms for iron or manganese if the medium pH rises out of the acid range. For these plants, acid-forming fertilizers are useful to keep the pH in the 5.0–5.5 range.

FOLIAR FERTILIZERS

Some fertilizers are sold exclusively as foliar fertilizers, and others give instructions on mixing a solution for use as a foliar fertilizer. These fertilizers usually contain N, P, and K plus micronutrients or just the micronutrients alone.

Foliar fertilizers are successful because plants absorb small amounts of nutrients directly through their leaves. The relative absorption ability is dependent on both the species and the thickness of the waxy cuticle that covers the leaf surface.

Fertilizers applied to the foliage are used primarily for two purposes: to apply micronutrients and to fertilize epiphytic plants adapted to receiving nourishment chiefly in that manner. Foliar fertilizers are very weak in concentration and are applied with a mister, spraying until the solution runs off the foliage. Repeat applications are usually necessary before enough nutrient is absorbed to eliminate deficiency symptoms.

VITAMINS AND ONE-DROP-DOES-ALL CHEMICALS

Such chemicals cannot properly be termed fertilizers because they do not contain the elements which have been proven essential for plant growth. Instead they contain vitamins and plant hormones (see Chapter 2). Although these chemicals are necessary for healthy plants, they are manufactured in the plant itself and generally do not need to be additionally supplied. Vitamins and plant hormones should not be considered as fertilizer replacements or substitutes.

MEDIUM- AND FERTILIZER-RELATED INDOOR PLANT PROBLEMS

Medium Problems

Any problem present in the potting medium will affect the roots of the plant, and this in turn will produce symptoms on the top growth. Careful observation of the medium for any problems which might be developing is essential when growing plants.

COMPACTION

Compaction is the excessive packing or settling of potting medium. This packing can exclude air from the roots and results in rotting of roots or a decline in plant vigor.

Compaction is a characteristic of a medium rather than a result of any improper plant care, and is common in soil-containing media. The symptoms are yellowing foliage, weak new growth, and a decrease in growth rate. The only cure is to remove the plant from its pot, gently break or soak off the majority of the medium from the roots, and repot in fresh medium. To make the new medium less prone to compaction, add a greater portion of peat moss and coarse aggregates and less or no soil.

POOR DRAINAGE

Poor drainage (assuming the pot has a drainage hole) is again a symptom of an incorrectly mixed medium. Media which are prone to compaction are also prone to poor drainage.

A plant growing in a medium with poor drainage will exhibit the same symptoms as one growing with compaction, and for the same reason: air is excluded from the roots, causing them to decay and die. The cure is also the same; remove the old medium from the roots and repot.

WATER REPELLENCY

Water repellency is not common, but it can be a problem in media composed of high percentages of sphagnum moss. Although the moss is capable of holding up to 20 times its weight in water, when completely dried it will shrink and become water-repellent. Water applied to the medium will run through or down the sides of the shrunken soil mass with little being absorbed. Rewetting can be achieved by submersion or bottom watering, and the medium should not be allowed to become as dry between waterings again. When the problem is recurring, apply a wetting agent mixed in the regular water.

UNSUITABLY HIGH OR LOW pH

Although most media start at a slightly acid pH suitable for growing most plants, changes in pH can occur due to the pH of water used on plants.

Tap water in most parts of the country is neutral to basic, and in time, acidic potting media become alkaline to the point of creating nutrient deficiencies. Use of distilled water will eliminate changes of media pH due to water, and acid-forming fertilizers which include micronutrients will also help.

Fertilizer Problems

Both over- and underfertilization can be injurious to the health of houseplants. The conditions are relatively easy to correct provided the problem can be identified.

OVERFERTILIZATION (TOXICITY)

Overfertilization results from an excess concentration around plant roots of compounds called *fertilizer salts.* This high concentration pulls water out of the roots and creates, in effect, a drought condition. Wilting, browned leaf tips, and stunted growth are symtoms of both maladies. But unlike a drought-stricken plant, an overfertilized plant will not immediately recover when watered, since the cause will still be present. The only cure is to remove the excess salts as quickly as possible. This is done by leaching. Large amounts of water should be poured through the medium to flush the fertilizer salts out. See Chapter 17 for a more complete description of the leaching process.

NITROGEN DEFICIENCY

Lack of nitrogen shows up first as yellowing of the older leaves of a plant, but it can eventually cause both older and younger leaves to discolor (Figure 15-11). Yellowing is also symptomatic of low light and overwatering, but in these instances the leaves soon drop, whereas in the case of nitrogen deficiency they do not. Additional deficiency symptoms are small new leaves and thin stems.

PHOSPHORUS DEFICIENCY

The first symptoms of phosphorus deficiency are a stunting of growth and the production of undersized new leaves.

POTASSIUM DEFICIENCY

Potassium deficiency appears on older leaves first, frequently as a browning of the leaf margins and tips. The areas between the veins may also be affected.

IRON DEFICIENCY

A lack of iron is the most common micronutrient deficiency found in houseplants. It occurs frequently in acid-loving houseplants growing in media of an unsuitably high pH. The first symptom is yellowing between the veins of the young leaves. The yellowing will

Figure 15-11 Typical nitrogen-deficiency symptoms of a Wandering Jew. The shoot on the left is bleached looking, whereas the one on the right has a healthy dark green color.

eventually include the entire leaf, and the symptoms are known as *iron chlorosis.*

Iron deficiency (Figure 15-12) can be temporarily corrected by foliar sprays of micronutrients and permanently corrected by the suggestions listed for media pH problems in Chapter 18.

FERTILIZER SALT DEPOSITS

Buildups of whitish fertilizer deposits on the medium surface or the outside of clay pots (Figure 15-13) are a common occurrence. They are formed when fertilizer salts dissolved in water move to the outside of the soil ball in response to evaporation of water. The water leaves as vapor, but the salts remain.

Salts on the pot or medium surface do not automatically indicate that there is excess fertilizer in the medium; however, a leaching of the medium is a worthwhile precaution to take.

Figure 15-12 Iron deficiency in a coffee plant (*Coffea arabica*).

Figure 15-13 Fertilizer salt deposits on the outside of a clay pot.

Selected References for Additional Reading

Bunt, A. C. *Modern Potting Composts, A Manual on the Preparation and Use of Growing Media for Pot Plants.* University Park: Pennsylvania State University Press, 1976.

Furuta, T. *Environmental Plant Production and Marketing.* Arcadia, CA: Cox, 1976.

Lawrence, W. J. C. *Seed and Potting Composts.* London: G. Allen and Unwin, 1962.

Mastalerz, J. W. *The Greenhouse Environment.* New York: Wiley, 1977.

Chapter 16

Light and Indoor Plant Growth

The sunlight that reaches Earth each day is the key to the survival of all life. Without it, plants could not photosynthesize and would die. Without this plant food source, humans and other animals would soon expire.

Outdoors, sunlight is not frequently considered when planning gardens. It is simply assumed that unless an area is tree-shaded, sufficient light will be available to grow whatever is planted. When plants are placed indoors, ample light for growth is no longer automatically available. In separating themselves from nature with walls and ceilings, people restrict the available light and must properly manage and supplement the remaining light to grow plants successfully.

EFFECTS OF LIGHT ON PLANT GROWTH AND DEVELOPMENT

Light is essential for photosynthesis, and without it, chlorophyll, carbohydrates, hormones, and many other plant-manufactured chemicals could not be made. Nor would energy be available for respiration and growth.

A logical conclusion would be that plants cannot grow without light. But this is not strictly true. Plants grown in insufficient light do not stop growing, but they do not produce healthy growth. Instead, they develop abnormally long stems with few leaves (Figure 16-1) and develop a stringy or lanky look. During this process, called *etiolation*, the plant respires its stored carbohydrates. When most of the carbohydrate has been respired, it will die.

Etiolation is the result of action of the plant hormone auxin. Just as auxin keeps buds dormant, it also keeps stems from lengthening. It is manufactured in the presence of light; a plant kept in insufficient light will have a low auxin content and long stems.

Phototropism and Similar Responses

Phototropism is the leaning of a plant toward light (Figure 16-2). It is similar to etiolation in some ways, except that only the side of the stem away from the light gains auxin and

Figure 16-1 Etiolation of a plant due to insufficient light. Both plants have the same number of leaves and were propagated at the same time, but the plant on the right received less light.

lengthens. Because the sides of the stem are growing at different rates, the stem curves in the direction of light. Thus the common belief that plants "seek" light is actually hormonally caused.

In other plants, differences in light level change the leaf form. A weeping fig *(Ficus benjamina)* grown in high light will have large numbers of thick, leathery leaves. When moved to a shadier location, it will drop many of these and refoliate with "shade leaves." The shade leaves will be larger, thinner, and fewer in number than the previous "sun leaves."

Cutleaf philodendron *(Monstera deliciosa)* shows a different reaction to varying light levels (Figure 16-3). The number of splits which new leaves contain decreases at lower light levels. A plant growing in low light may have no splits at all, whereas one growing in moderate light may have several in each leaf.

Light and Flowering

In addition to affecting photosynthesis and hormone manufacture, light also plays a key role in the flowering of many plants.

INTENSITY

Insufficient light can inhibit the flowering of many plants grown indoors. Relatively high light intensities are necessary to trigger flowering in most cultivated plants, and those grown indoors are no exception. Unfortunately, high light intensity is not available through most windows.

PHOTOPERIOD

Photoperiod combined with sufficient light intensity causes or hastens flowering in some indoor plants. The word "photoperiod" (discussed in Chapter 3) refers to the day/night ratio in each 24-hour period and can affect not only flowering but also tuber and bulb formation.

Plants can be classified with regard to photoperiod as short day, long day, or day neutral. Many indoor plants are from tropical regions near the equator where day length varies little throughout the year. Accordingly, most are day neutral and flower due to combinations of age, light intensity, and temperature. But a few indoor plants such as chrysanthemum, poinsettia, and Christmas cactus require specific photoperiods to bloom (see Table 16-1 for additional listings).

Figure 16-3 Splitting of philodendron. (*Monstera deliciosa*) leaves. The unsplit leaf developed in low light, the split leaves in high light.

Figure 16-2 Phototropism caused by light reaching the plant from only the left side.

The number of weeks a photoperiodic plant must be kept under correct day and night lengths to trigger flowering varies between species. If the light/dark cycle is not maintained long enough, the plant may revert to its vegetative state, ceasing bud development in the middle of the flowering process. If kept too long under light/dark control, no harm will usually be done. However, a short-day plant on a long-night schedule too long will lose photosynthetic time and may lose vigor. As a general rule, a photoperiodic schedule

Table 16-1 Photoperiodic Houseplants

Common name	Botanical name	Photoperiodic classification	Remarks
Hardy begonia	*Begonia grandis*	Short day	Forms aerial stem tubers
Rex begonia	*Begonia rex*	Long day	Increases leaf area and stem elongation only
Wax begonia	*Begonia semperflorens*	Long day	Flowers under long days at high temperatures but day neutral at low temperatures
	Begonia socatrana	Long day/ short day	Flowers under long days; forms aerial stem tubers under short days
Tuberous begonia	*Begonia* × *tuberhybrida*	Long day/ short day	Forms blooms under long days and underground stem tubers under short days
Pocketbook plant	*Calceolaria crenatiflora*	Long day	Accelerates flowering but is not essential
Cattelya orchid	*Cattleya* spp.	Short day	Causes flowering
Chrysanthemum	*Chrysanthemum* × *morifolium*	Short day	Flowering may also be hastened by cool temperatures
Poinsettia	*Euphorbia pulcherrima*	Short day	Causes flowering
Fuchsia	*Fushcia* × *hybrida*	Long day	Causes flowering in some varieties
Kalanchoe	*Kalanchoe blossfeldiana*	Short day	Causes flowering and increased leaf succulence
Sensitive plant	*Mimosa pudica*	Not applicable	Leaflets open in light but close in dark
Evening primrose	*Oenothera* spp.	Long day	Flowering accelerated by low temperatures also
Shamrock	*Oxalis* spp.	Short day	Forms underground tubers
Sedum	*Sedum spectabile*	Long day	Causes flowering
Cineraria	*Senecio* × *hybridus*	Short day	Short days hasten flowering but are not essential
Christmas cactus	*Schlumbergera bridgesii*	Short day	Needs dry period also to hasten flowering

should be maintained until the flower buds are completely developed.

Techniques for Controlling Photoperiod. Most photoperiodic houseplants are short-day plants. They require daily periods of uninterrupted darkness before flowering will take place. Plant species vary in their reaction to light. Poinsettias are very sensitive, and bloom will be delayed if the dark period is ever broken by light or if the darkness is not absolute.

One common home procedure for supplying a short-day plant with a long dark period is placing the plant in a closet each day at 5 P.M. and taking it out the next morning at 8 A.M. The schedule can be varied to accommodate personal schedules provided at least 14½ continuous hours per day are spent in darkness. The darkness should be checked by standing inside with the door closed. If light can be seen around the door, the leak should be covered or another closet should be used.

Flowering long-day plants is less exacting. The plant can simply be left in a room that is used in the evenings. The artificial lights will supplement the natural day length with enough light to total 14 or more hours, and blooms will be initiated.

Light and Vegetative Growth

The minimum amount of light a plant must receive to remain in a healthy condition is dependent on its species. However, the following generalizations can be made regarding light requirements.

FOLIAGE COLOR

In many instances the darker the foliage of a plant, the less light it requires. The most common low-light plants—snake plant, rubber plant, cast iron plant, Chinese evergreen, and philodendron—all have deep green foliage. When grown in low light, these plants develop a dense chlorophyll layer near the leaf surfaces. This makes the plants efficient users of light.

In the opposite fashion, many variegated plants such as variegated spider plant, tricolor dracaena, and variegated Wandering Jew require more light than their nonvariegated relatives. They lack chlorophyll in the white areas of the leaves and require greater light to compensate.

Many plants which have red coloring pigments in addition to their chlorophyll are also high light requirers. Coleus, ti plant, and croton have their chlorophyll masked by red pigment and require strong light to penetrate through to the chlorophyll below.

Variegated plants frequently show their most attractive coloration only in moderate to high light. Under lower-light conditions they often revert to solid green for more efficient photosynthesis.

FLESHINESS AND SUCCULENCE

Plants with thickened leaves and/or stems such as jade plant, kalanchoe, and cacti require more light than those with thin leaves.[4] Their fleshy structure minimizes the surface area vulnerable to transpiration water loss and enables them to survive in dry climates. However, at the same time the photosynthesizing area is reduced. Consequently, these plants can be grown indoors more successfully in bright-light locations.

NATIVE GROWING CONDITIONS

A plant that grows naturally in the shade of other plants is usually better adapted to indoor growing than a plant accustomed to unobstructed bright lights. Ferns are low-light-requiring plants which illustrate this well. Because they grow naturally in the shade of trees, they also grow well in a low-light area indoors. This generalization holds for many tropical plants including devil's ivy, peperomia, and arrowhead.

CONCLUSIONS

Ultimately, the amount of light required to maintain a plant species or variety indoors is best determined from a reliable reference book. The generalizations stated would hold true for the majority of indoor plants, but exceptions are not uncommon. A plant might decline due to an improper light environment if these guidelines were followed in lieu of a specific reference.

LIGHT INTENSITY

Light intensity is one factor that contributes to what can be called *total light*. The other is light duration. A lack of total light is frequently the cause of poor growth in houseplants, particularly in northern areas and in winter.

Natural light intensity is affected by many factors. Some are the result of geograph-

[4]Exceptions include Christmas, Easter, and orchid cacti, which grow best with less light.

ical or astrological variation, and some are climatic. Understanding these factors and how they interrelate to affect light intensity will help the indoor gardener understand his growing handicaps or assets with respect to light.

Latitude

Because of the way the Earth is oriented in space, the sun will rise exactly in the east and set exactly in the west only at the equator. For people living in the Northern Hemisphere, the sun always appears slightly in the southern sky (Figure 16-4). How far off the southern horizon it appears depends on the distance from the equator (latitude), with the sun dropping lower toward the southern horizon the further north you go.

This orientation of the sun in the southern sky affects its intensity substantially. By hitting the Earth's atmosphere at an angle instead of directly above, the distance the light travels through the atmosphere to reach the Earth is increased. Light is lost in the atmosphere, and its intensity when it ultimately reaches Earth is much less. For this reason, the further north you live, the less light intensity reaches the Earth.

Time of Year

Time of year also affects light intensity because of the sun's angle. Although the sun stays in the southern sky throughout the year, it drops lowest toward the southern horizon in winter and approaches closest to directly overhead at midday in summer (Figure 16-4). How much of an angle it changes over again depends on the distance from the equator. In the southern United States it will change only slightly from summer to winter, and intensity will vary only moderately. But in the far North the change will be very pronounced. The sun will appear low on the horizon and be low in intensity in winter, while rising to nearly

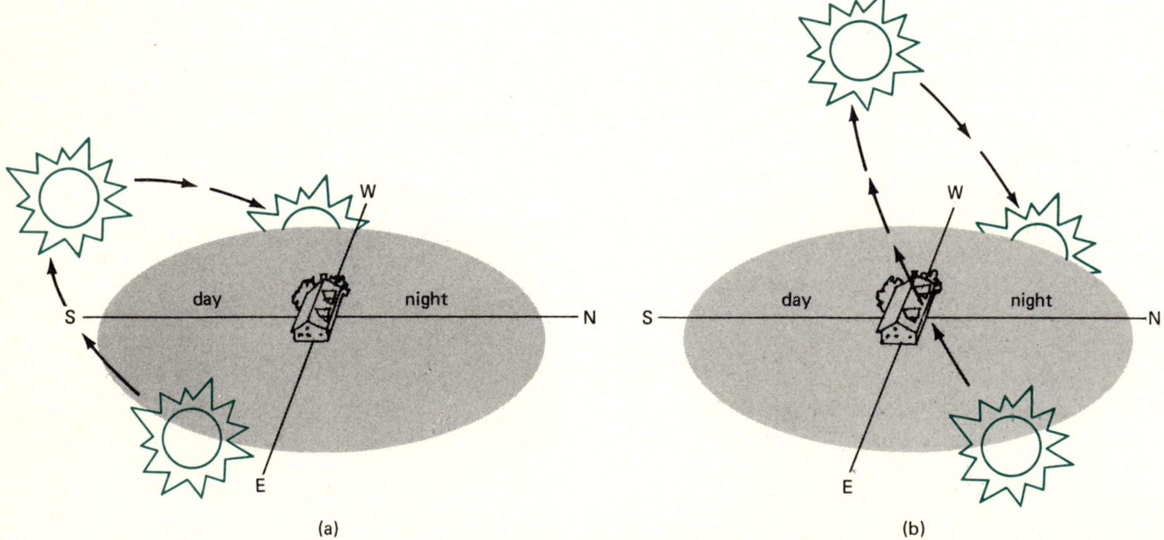

Figure 16-4 Typical positions of the sun at noon in winter (a) and summer (b). Note that the sun remains in the Southern Hemisphere and that the positions of rising and setting differ with the season.

overhead with a high intensity in summer. In spring and fall it will be midway between and moderate in intensity.

Time of Day

Light intensity is also affected by time of day. Midday, when the sun is at its highest point, provides the greatest intensity of light. In either morning or evening the sun is lower toward the horizon, and its intensity will decrease accordingly.

Although latitude, time of year, and time of day are the main influences on light intensity, other factors can raise or lower the intensity drastically from the expected.

Altitude

People living at high altitudes [greater than 1000 feet (304.8 meters) above sea level] frequently receive greater light intensity than would be expected at their latitude. The effect of the atmosphere on lowering sunlight intensity is the reason. Air is less dense and the atmospheric layer is thinner at high altitudes, and less intensity is lost passing through it. Denver, Colorado, illustrates this effect of altitude perfectly. Located at over 5000 feet (1524 meters) in altitude, its light intensity year-round is substantially higher than that of Boston, Massachusetts [elevation 21 feet (6.4 meters)], even though the two are less than 4° apart in latitude.

Clouds, Fog, and Smog

In cloud-, fog-, or smog-covered areas, light intensity is habitually lower than in clear areas at the same latitude. The water vapor or pollution haze acts as a reflector, sending much of the sunlight back toward space. In addition, clouds, fog, and smog diffuse sun rays, giving the earth shadowless illumination rather than direct sunlight.

Snow Cover

Persistent snow cover affects light intensity by reflecting sunlight that strikes it. Cold areas which have snow cover most of the winter receive a slightly increased amount of light through all windows because of this reflection.

This principle also applies to houses surrounded by light-colored sand. While dark soil will absorb light, light sand reflects it and will increase the light entering a house.

LIGHT DURATION

Light duration (day length) is a second component of total light. To some extent an increased duration can compensate for a low intensity in providing plants with adequate light for photosynthesis. However, below a certain minimum light intensity, an increased duration will not compensate.

Days are longer in summer than in winter, and again this is a result of latitude. Days and nights at the equator are equal in length all year round. But in the North it is not uncommon for winter days to be as short as 9 hours and summer days as long as 15 hours. The briefness of winter days and the increased length of summer days become more pronounced the further north you go. In Alaska, for example, days are 20 hours or more long in the summer.

Unfortunately, the shortest winter days correspond with the dimmest light intensity, reaching a climax on December 21, the winter solstice. At this point the sun again begins moving toward the northern sky.

Summary

Latitude, day length, altitude, cloud and snow cover, and the other topics discussed all combine to give total light. Because of these many variables it is almost impossible for a person to determine whether his area is generally bright, moderate, or dim except by observation. However, the Environmental Sciences Service Administration has recorded light intensity and duration throughout the United States for the months of July and December. Their findings give a rating of total light for the United States in midwinter and midsummer.

Figure 16-5 codes the continental United States into high-, medium-, and low-light areas for winter and summer. Bear in mind that for most areas, summer light is three to four times as great as winter light.

DETERMINING THE AMOUNT OF TOTAL SOLAR RADIATION ENTERING WINDOWS

The factors affecting the intensity and duration of light that reaches the Earth are important to understand. But it is the total light entering windows which is of the greatest importance to an indoor gardener.

Several devices are used to measure light intensity including plant light meters and photographic light meters. Both are expensive, but they can give reasonably accurate readings if used properly.

Readings will need to be taken several times a day, and under cloudy and sunny conditions, to determine the average light a plant is receiving. After determining this, the light requirement for the plant must be found. Although this has been determined for many plants and is published in Cooperative Extensive Service bulletins, the list is far from complete.

In some cases the light requirement of a species will be listed in a range of footcandles. The footcandle is an old measurement[5] denoting the illumination of a surface 1 foot from the light of a standard candle. For reference purposes, a comfortable light for reading is 50 footcandles, and a bright summer day yields about 10,000 footcandles. The United States Department of Agriculture classified over 100 houseplants by their light requirements, with the range of 75–200 footcandles (767–2153 luxes) deemed low, 200–500 footcandles (2153–5382 luxes) moderate, and 1000+ footcandles (10,764 luxes) high. Compensation-point intensities were evaluated as 25 footcandles (229 luxes) for low-light plants, 75–100 footcandles (767–1076 luxes) for medium species, and 1000 footcandles (10,764 luxes) for high-light types.

For most gardeners, simple high, medium, or low designations of light based on the direction a window faces and distance from the light are more workable than footcandles. Exact light measurements are not necessary to grow houseplants successfully. The use of a few reference books, the ability to quickly spot the signs of light deficiency, and trial and error are the tools of most successful indoor gardeners.

Window Direction

Several factors make south-facing windows the brightest and best for growing many indoor plants. First, since the sun shines from the southern sky all year, unobstructed south-facing windows receive the longest duration of daily direct sunlight all year. Only south-facing windows receive sun from midmorning through late afternoon, and an indoor gardener with southern windows is fortunate. He has enough light for almost any plant and can modify the light by moving plants further from the window.

East- and west-facing windows provide an intermediate amount of light, less than

[5] The footcandle measurement of light intensity is outdated and has been replaced by the metric lux. However, most publications dealing with the light requirements of plants still use this measurement.

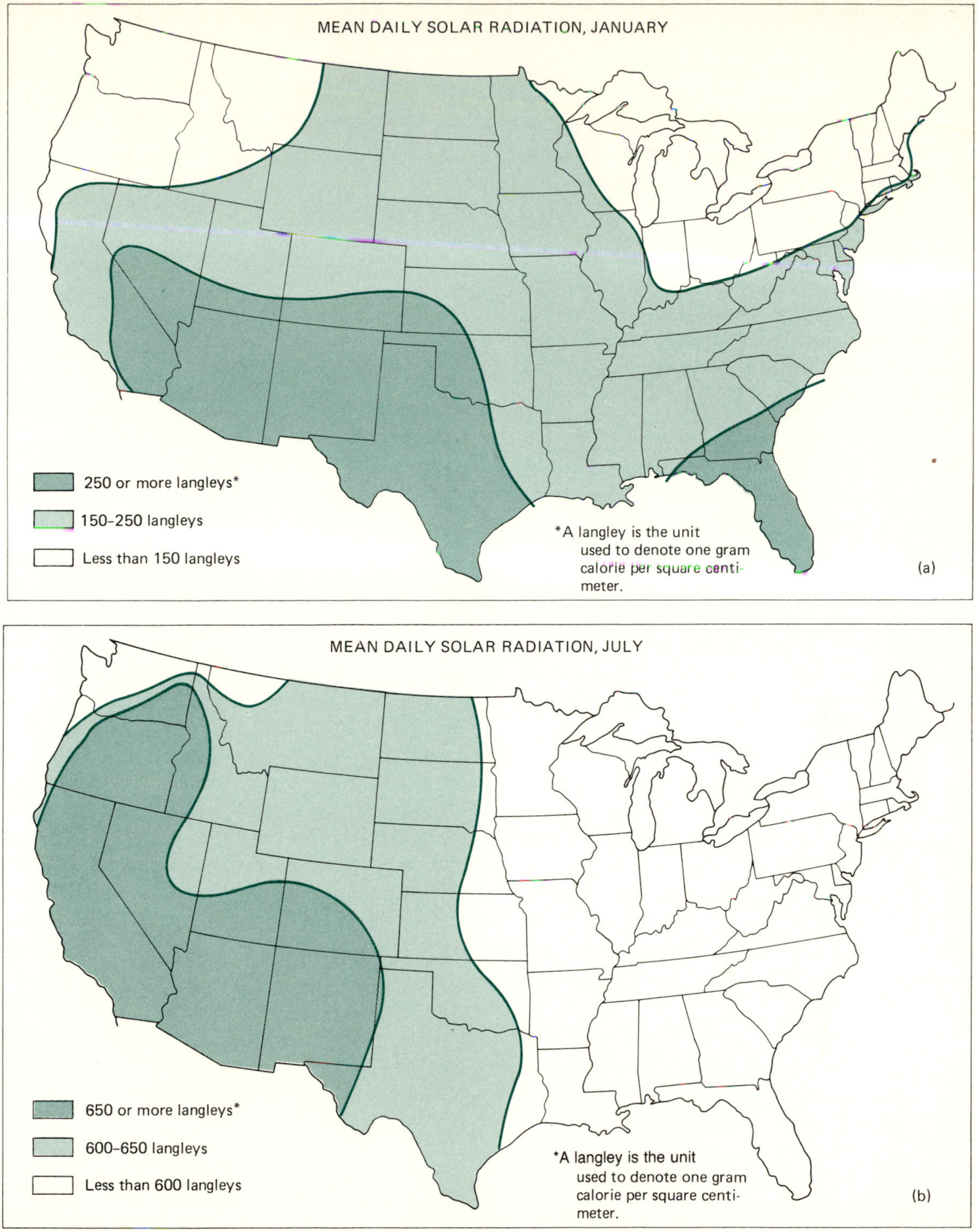

MEAN DAILY SOLAR RADIATION, JANUARY

250 or more langleys*

150–250 langleys

Less than 150 langleys

*A langley is the unit used to denote one gram calorie per square centimeter.

(a)

MEAN DAILY SOLAR RADIATION, JULY

650 or more langleys*

600–650 langleys

Less than 600 langleys

*A langley is the unit used to denote one gram calorie per square centimeter.

(b)

Figure 16-5 Mean daily solar radiation throughout the United States in January (a) and July (b).

southern but more than northern. Each receives direct sunlight for about half of the day and only reflected light for the remainder. The total light entering from east- and west-facing windows is sufficient for growing the majority of houseplants, provided they are not placed too far from the glass.

North-facing windows provide the least light but are still acceptable for growing a number of species. Their dimness results from the fact that they receive no direct sun, only light reflected from the ground or nearby buildings.

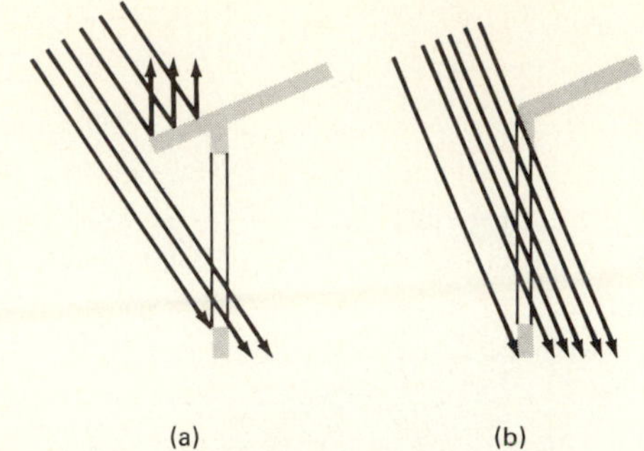

Figure 16-6 The effect of an eave on the amount of sunlight entering through a window. Note the depth of penetration of sunlight into a window with (a) and without (b) an eave.

Factors Decreasing Sunlight Entry through Windows

Several factors, such as the presence of overhanging eaves or tall buildings, can decrease the sunlight entering through windows. These factors should be taken into account in determining the light level of windows. They are discussed in the following sections.

OVERHANGING EAVES

Wide eaves (Figure 16-6) can decrease the time or distance direct sunlight enters a room appreciably. They shade the window when the sun is high in the sky, at midday and several hours before and after. This effect is more pronounced in the summer than in the winter because the sun is higher off the southern horizon.

BUILDINGS OR FENCES

Tall buildings or fences close to windows can also decrease direct sunlight. As with overhangs, the problem is one of shading from the south, east, or west, but unlike blockage due to eaves, the problem will be more pronounced in winter.

[6]In the western states west-facing windows may provide slightly more total solar radiation than eastern.

SHRUB OR TREE PLANTINGS

Shrubs or trees planted close to a window can seriously limit light entry, sometimes reducing it to only reflected light. Trees and shrubs which are deciduous will defoliate in fall and cause little blockage of light, but evergreens exclude light continuously. Removal or pruning of the plants is the only solution. Limbing up or thinning will work on trees. Shrubs should be thinned and headed back to minimize blockage of sunlight. Chapter 11 describes all these pruning styles.

CURTAINS AND SHADES

Most homeowners open their curtains or shades in the morning and close them after dark, but valuable light can be lost this way during the early morning hours. East-facing windows, in particular, can lose a substantial portion of their direct sunlight when curtains remain closed after sunrise.

Even when curtains are open, they can decrease the light area around the outside edge of the window. To maximize light, hang curtains so that they expose the entire window area when opened. Avoid overhanging valances, which block overhead light.

TINTED OR STAINED-GLASS WINDOWS

These windows can affect light entry only slightly or block it almost completely, depending on the shade of glass used. Although little can be done short of removing them, their effects should be taken into account when positioning plants for proper lighting.

PLACEMENT OF PLANTS FOR OPTIMUM LIGHT

Just as windows are classified as providing high, medium, or low light, plants can be classified as high-, medium-, or low-light-requiring and positioned within a room to satisfy their requirements. Generally the dilemma revolves around how far from a window a plant can be grown successfully. This minimum light level at which a plant can maintain itself is called the *compensation point.* The light it receives is sufficient to photosynthesize carbohydrate for its respiration but leaves no surplus carbohydrate for growth. At lower light levels the respiration use of carbohydrate will exceed its rate of manufacture, and the plant will slowly die. At higher light levels excess carbohydrate will be used for growth.

The concept of compensation point is very important in indoor plant growing. Plants being used for decoration can be bought at the desired size and maintained in good condition with the correct amount of light. Because the plant is growing very slowly, the time devoted to maintenance can be kept at a minimum. Also, plants grown at compensation-point light levels can be placed further back from windows than is optimum for growth, creating space for additional plants.

The following descriptions and figures of bright, moderate, and low light are intended only as rough estimates of available window light. However, they include the range from a compensation-point amount of light through the highest light advisable or available for plants indoors.

Bright Light

The designation "bright"- or "high"-light-requiring is given to a number of plants including most flowering species, cacti, and succulents (Table 16-2). Plants which require bright light are, as a whole, the hardest to grow successfully indoors. Difficulty in providing them with sufficient sunlight during the darker winter months is the main problem.

Light conditions vary enormously, depending on the area. As a general rule, bright-light plants should be placed no further than 3 feet (1 meter) from a southern window or 2 feet (0.6 meter) from an east or west window (see Figure 16-7). It is nearly impossible to give bright-light-requiring plants too much light. The more light they receive, the healthier they will be. Most adjust readily to full sunshine and will grow vigorously outdoors during warm weather.

Moderate Light

Most foliage houseplants fall into the medium light requirement category (Table 16-3). They can be provided with adequate light if placed up to 6 feet (2 meters) from a southern window, up to 4 feet (1.2 meters) from east or west windows, or up to 2 feet (0.6 meter) from a north window. As with bright-light-requiring plants, moderate-light users grow better when placed close to the glass, especially in winter. However, as summer approaches, those in south or west windows will be subject to increasing amounts of light. They should be watched for symptoms of excess light and moved back from the window if symptoms appear.

Table 16-2 High-Light Plants

Common name	Botanical name
Flowering maple, Chinese lantern	*Abutilon* spp. *[a]
Chenille plant	*Acalypha hispida**
Copperleaf	*Acalypha wilkesiana* varieties*
Lipstick vine	*Aeschynanthus* spp.
Agave	*Agave* spp.*
Bamboo	*Bambusa* spp.*
Ponytail palm	*Beaucarnea recurvata**
Wax begonia	*Begonia* × *semperflorens-cultorum**
Cacti	*Xerophytic* spp.*
Camellia	*Camellia* spp.
Natal plum	*Carissa grandiflora**
Cattleya orchid	*Cattleya* spp.
Rosary vine	*Ceropegia woodii*
Chrysanthemum	*Chrysanthemum* × *moriflorium**
Citrus	*Citrus* spp.*
Croton	*Codiaeum variegatum**
Coleus	*Coleus* × *hybridus**
Goldfish plant	*Columnea, Nematanthus*
Cigar plant	*Cuphea* spp.*
Cymbidium orchid	*Cymbidium* spp.
Tricolor dracaena	*Dracaena marginata* 'Tricolor'
Poinsettia	*Euphorbia pulcherrima**
Fatshedera	*Fatshedera lizei**
Purple velvet plant	*Gynura aurantiaca* varieties
Variegated ivy	*Hedera helix* varieties*
Amaryllis	*Hippeastrum* spp.
Wax plant	*Hoya carnosa* varieties
Polka-dot plant	*Hypoestes phyllostachya*
Bloodleaf	*Iresine herbstii*
Shrimp plant	*Justicia brandegeana*
Gold hop	*Justicia brandegeana* 'Yellow Queen'
Kalanchoe	*Kalanchoe blossfeldiana**
Ice plant	*Lampranthus, Mesembranthemum,* and other genera*
Lantana	*Lantana* spp.*
Wax leaf privet	*Ligustrum japonicum, Ligustrum lucidum**
Oxalis	*Oxalis* spp.*
Geranium	*Pelargonium* spp.*
Avocado	*Persea americana**
Yew pine	*Podocarpus* spp.*
Rose	*Rosa* spp.*
Stephanotis	*Stephanotis floribunda*
Succulents	Many species*

[a]An asterisk indicates no maximum light level.

Table 16-3 Medium-Light Plants

Common name	Botanical name
Maidenhair fern	*Adiantum* spp.
Anthurium	*Anthurium* spp.
Aucuba	*Aucuba japonica*
Begonia	*Begonia* spp.
Umbrella plant	*Brassaia* or *Tupidanthus*
Fishtail palm	*Caryota* spp.
Spider plant	*Chlorophytum comosum* varieties
Kangaroo and grape ivies	*Cissus* spp.
Coffee	*Coffea arabica*
Ti plant	*Cordyline terminalis*
Earth star	*Cryptanthus* spp.
Dumbcane	*Dieffenbachia* spp.
False aralia	*Dizygotheca elegantissima*
Dracaena	*Dracaena* spp.
Euonymus	*Euonymus* spp.
Fatsia	*Fatsia japonica*
Fig, weeping	*Ficus benjamina*
Fig, fiddle-leaf	*Ficus lyrata*
Fig, creeping	*Ficus pumila*
Ox tongue	*Gasteria* spp.
Prayer plant	*Maranta leuconeura* varieties
Boston fern	*Nephrolepis* spp.
Lady slipper orchid	*Paphiopedilum* spp.
Peperomia	*Peperomia* spp.
Dwarf date palm	*Phoenix roebelenii*
Pileas	*Pilea* spp.
Staghorn fern	*Platycerium bifurcatum*
Swedish ivy	*Plectranthus* spp.
Hare's foot fern	*Polypodium aureum*
Aralia	*Polyscias* spp.
Moses in the cradle	*Rhoeo spathacea*
African violet	*Saintpaulia ionantha* varieties
Strawberry geranium	*Saxifraga solonifera*
Epiphytic cacti	*Schlumbergera, Rhipsalis, Epiphyllum,* and other genera
Jerusalem cherry	*Solanum pseudocapsicum*
Baby tears	*Soleirolia soleirolii*
Piggyback plant	*Tolmiea menziesii*
Wandering Jew	*Tradescantia, Zebrina*
Ginger	*Zingiber officinalis*

Low Light

Only a small percentage of indoor plants grows satisfactorily in low light (Table 16-4), and most of these are better in a medium-light location. However, a number of species classified as moderate-light requirers can survive low-light conditions for a period of up to several months without harm.

Low-light plants can be grown 4 to 6 feet (1.2–2 meters) from east and west windows and anywhere up to 4 feet (1.2 meters) from north windows (Figure 16-7). They will also grow 6 to 8 feet (1.8–2.4 meters) from a south window. Unlike bright-light plants, some low-light plants can be injured by excess light. They will live well in moderate light but generally should not be grown in bright light.

MAXIMIZING AVAILABLE LIGHT

In general, an indoor gardener must work with whatever natural light is available. He cannot alter his climate, and short of making structural modifications for larger windows or cutting skylights, not much can be done to increase light entry. However, there are several "light management" practices which can make the most efficient use of available light.

Rotation

Plant rotation has proven its workability as a commercial enterprise, and is also practical for the home gardener. Many of the large foliage plants found in lobbies and public shopping malls are rented from greenhouse owners specializing in interior landscaping. Every 2 to 8 weeks the plants in the low-light display area are replaced and taken back to the greenhouse for a recuperation period. During this rotation in the greenhouse they accumulate carbohydrate which will be used for respiration when they are again in the display area.

The home gardener without ample light for all his plants can practice a modified type of rotation, moving some plants closer to windows and others further away every 2 to 3 weeks. This technique works best on fairly hardy indoor plants and on moderate- or low-light-requiring species. The requirements of high-light plants are less flexible. They can be placed in moderate or dim areas to a limited extent but should not be kept there more than a few days to avoid leaf drop. In addition, plants which are in bud should not be moved because the sudden change in environment can cause the buds to drop. However, after the buds open, the plant can be moved to be more effectively displayed.

Reflective Colors

Light is reflected by white surfaces, completely absorbed by black, and absorbed to some degree by all colors between. Once light enters through a window, its contact with plants can be maximized by light-colored walls, carpets, and drapes, which bounce light back rather than absorbing it. White walls reflect up to 90 percent of the light which they receive. They can contribute significantly to the amount of direct light that plants receive. An increase in reflected light will make it possible to grow moderate- and especially low-light-requiring plants further back from windows.

ARTIFICIAL LIGHTING FOR PLANT GROWTH

Another solution to the problem of insufficient light is the use of artificial lighting. The indoor gardener who branches out into artificial lighting has a practically unlimited choice of plants. He or she can grow almost any flowering plant successfully, as well as herbs, vegetables, and transplants for outdoor

Table 16-4 Low-Light Plants

Common name	Botanical name
Miniature flag plant	*Acorus gramineus*
Chinese evergreen	*Aglaonema* spp.
Norfolk island pine	*Araucaria heterophylla*
Asparagus fern	*Asparagus* spp.
Cast iron plant	*Aspidistra elatior*
Bird's nest fern	*Asplenium nidus*
Parlor palm	*Chamaedorea elegans*
Holly fern	*Cyrotomium falcatum*
Corn plant	*Dracaena fragrans* varieties
Pleomele	*Dracaena reflexa*
Devil's ivy	*Epipremnum aureum*
Rubber plant	*Ficus elastica*
Fittonia	*Fittonia verschaffeltii* varieties
Kentia palm	*Howea fosteriana*
Philodendron	*Philodendron* spp.
Hart's tongue fern	*Phyllitis scolopendrium*
Snake plant	*Sansevieria* spp.
Spathe flower	*Spathiphyllum* spp.
Arrowhead vine	*Syngonium podophyllum*
Bromeliads	*Vriesea, Guzmania, Nidularium*

use. "Light gardening," as it is called, can be a fascinating hobby.

The Makeup of Light

Sunlight is naturally suited for plant growth, but some artificial light is not. An understanding of the makeup of light and the kinds of light necessary for plant growth is necessary for a gardener to wisely purchase lighting equipment.

Light is a form of energy that travels in waves. The length of these waves varies, and Figure 16-8 shows the relationship between wavelengths of visible light and other types of energy. Of the energy types listed, only three—ultraviolet, visible, and infrared rays (heat)—are contained in solar radiation and are significant to plant growth.

Ultraviolet light causes tanning, and plants growing outdoors are exposed to it all day. However, it is not necessary for plant growth and can actually be considered harmful. The fact that ultraviolet light does not pass into glass greenhouses further proves that plants can be grown very successfully without it.

Infrared waves transmit heat by traveling through air or space and warming whatever objects they contact. Consequently the infrared rays in sunlight warm plants outdoors during the day, but at night infrared will be given off from previously warmed surfaces like the soil. Because heat is needed for plant growth, infrared rays are essential. But they do not need to be supplied by a light fixture and do not need to be considered when selecting plant lights.

This leaves only visible light necessary for plants, and even all the colors which make up that range of light have not been proved necessary. Light in the orange-red and blue-violet portions of the spectrum have the most significant influence on plant growth. The

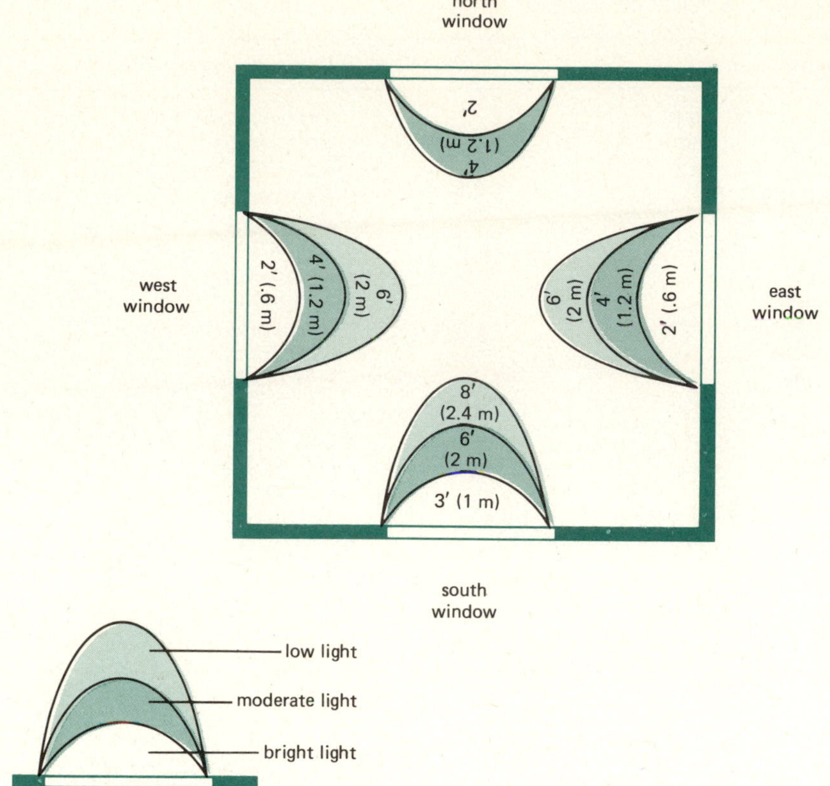

Figure 16-7 Suggested distances from windows for growing high-, medium-, and low-light-requiring plants.

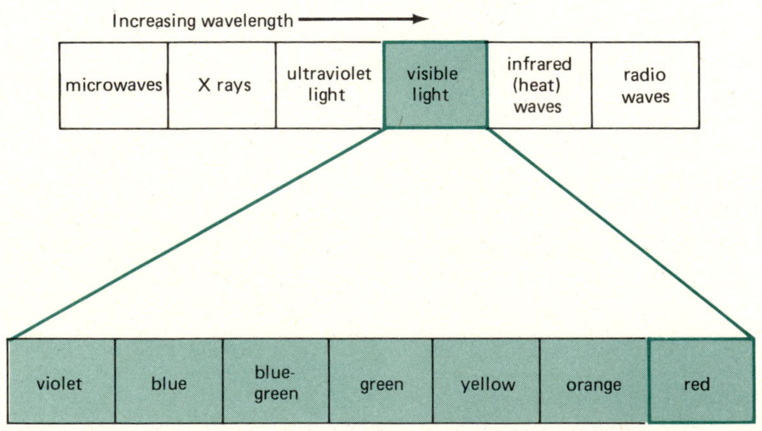

Figure 16-8 The radiant energy spectrum.

green and yellow wavelengths are commonly reflected back by plants, giving them their characteristic green appearance.

Both orange-red light and blue-violet light are used in photosynthesis. Red light has been proved to be the wavelength range that triggers flowering in photoperiodic plants, whereas blue-violet light is responsible for phototropic responses.

Types of Artificial Lights

Two basic types of lights are used for artificial lighting, incandescent and fluorescent.

INCANDESCENT LIGHTS

Incandescent lights were the first lights invented. They produce light by channeling electricity through a wire, which then heats up and glows. Compared to fluorescent lights, incandescent lights use more electricity to produce the same amount of light, the extra energy being given off as heat. In wavelengths of light emitted, they are high in orange-red but low in blue-violet, giving objects they light a warm appearance. Although the red-orange is useful for flowering plants and makes an attractive spotlight, the lack of blue-violet makes normal incandescent bulbs unsatisfactory for plants when used alone.

FLUORESCENT LIGHTS

Fluorescent lights are long tubes which can only be used in specially designed light fixtures. The inside of the tube is coated with phosphorescent material and has an electrode at each end. Electricity flows from each electrode and, by a chemical reaction with mercury, stimulates the phosphorescent coating to give off light.

Although initially more expensive, fluorescent lights are more economical to operate than incandescent lights, producing up to five times as much light from the same wattage of electricity. They give off almost no heat from the tubes, and only a small amount from the ballast, the part of the fixture that regulates the electricity flow to the electrodes. Finally, fluorescent tubes have a much longer life than incandescent bulbs.

The wavelengths of light emitted by fluorescent tubes vary, depending on the blend of phosphorescent materials used to coat the tube. Table 16-5 lists the most common types of fluorescent tubes and their emission in the orange-red and blue-violet wavelengths. Of the normal fluorescent tubes, cool white has been used most successfully for growing foliage plants. However, it is low in the orange-red area of the spectrum and should be supplemented by an incandescent light for flowering species.

PLANT LIGHTS

Plant lights in both incandescent and fluorescent styles are now commonly available. They are designed to provide light in both the blue-violet and the orange-red wavelengths in the best combination for plant growth. Because fluorescent lights are economical to operate and provide a higher intensity, most plant lights are of the fluorescent type.

Plant lights are sold in fewer numbers than normal fluorescent tubes, so they are more expensive. For a small light garden the cost may not be important. But for larger set-ups, combinations of household tubes from Table 16-5 can be substituted to provide both

Table 16-5 Wavelength Emissions of Normal Fluorescent Tubes

Tube type	Blue-violet	Orange-red
Cool white	Good	Poor
Cool white deluxe	Good	Fair
Warm white	Fair	Poor
Warm white deluxe	Good	Fair
Daylight	Excellent	Poor
Natural white	Good	Fair
Soft white	Fair	Fair

blue-violet and orange-red light. There will be little difference in plant growth.

Setting Up a Fluorescent Light Garden

Even with the efficiency of fluorescent lighting, more than one tube is generally needed to grow high-light-requiring plants such as flowering species and transplants. A minimum of two (and preferably four) tubes placed side by side is commonly used by experienced light gardeners. There is no set formula regarding the wattage of bulbs or how close to place them. In general, tubes of 40 watts or more are best, placed not more than 3 inches apart.

These minimum standards do not correspond with the lighting setups provided in most plant growth furniture. This can be explained by the fact that most common indoor plants are moderate-light-requiring foliage plants which need less light than flowering plants. Accordingly, a homemade light garden in which only foliage plants will be grown can contain fewer tubes.

A homemade light garden is not difficult to construct. One widely used setup is a shop light containing two tubes and a built-in reflector. The fixture is mounted on a metal stand or shelving unit, and a lamp timer regulates the hours of operation.

A more attractive light garden can be built using a reflectorless two-tube unit mounted in a bookcase, stereo cabinet, or room divider. The only stipulation is that the area to which the fixture is attached be painted white for maximum light reflection. Detailed diagrams for homemade light gardens are plentiful in light gardening and home building project books.

Growing Plants under Lights

The distance from the tubes at which plants are grown and the number of hours the lights are left on per day largely determine the success of a light garden.

DISTANCE FROM TUBES

For 40-watt household tubes, seedlings and high-light-requiring plants should be about 2 inches (5 centimeters) from the bulbs. With higher-wattage bulbs, the distance can be increased but should not be greater than 6 inches (12.7 centimeters). Since fluorescent lights do not give off heat, there is no problem of heat damage to plants placed too close to the bulbs. Some plants grow so rapidly that they touch the bulbs, but they are seldom injured.

Foliage plants can be grown further from the tubes since their light requirements are lower. Many foliage houseplants will grow well placed up to 2 feet (0.6 meter) from the light source.

HOURS PER DAY

The length of time lights should be on is not an absolute figure. In general, 12 to 16 hours per day is considered suitable if there is no additional source of light. Although foliage plants can grow with less, there is no advantage to running the lights a shorter time except for controlling the photoperiod. The amount of electricity used to operate the tubes is minimal.

MAINTENANCE OF FLUORESCENT LIGHTS

Although fluorescent lights bulbs may appear to burn out spontaneously, they actually lose intensity slowly over their entire lifetime of 12,000 or more hours. For this reason it is wise to replace the bulbs after about 75 percent of the stated hour-life has passed because the intensity will have decreased by 15 to 20 percent. New bulbs should be dated with indelible marker, and replacement time calculated by estimating the hours the bulb burns per day multiplied by the number of days of use. Replacing one bulb every 3 or 4 months will avoid excess light symptoms from the sudden intensity increase.

LIGHT-RELATED HOUSEPLANT DISORDERS

Insufficient Light

Lack of light is one of the most common causes of houseplant problems, and it is among the easiest to diagnose. Etiolation of new growth is one symptom. Yellowing and dropping of the older leaves usually also occur. Once the leaves have begun to yellow, they will fall regardless of any improvement then made in the light.

Plants in the advanced stages of light deficiency have thin weak stems and few leaves. The leaves remaining may be widely spaced, pale, and undersized. Such plants should be placed under increased light to recuperate.

Excess Light

Excess light is seldom a problem for indoor plants. However, a plant will occasionally develop symptoms when, for example, a low-light plant is grown in a south-facing window. The new leaves of a plant growing in excess light will be small and a pale yellowish-green. The internodes will be very short, giving the plant a compact but unhealthy appearance.

A more common case of excess light damage occurs when houseplants are placed outdoors after growing indoors all winter. The sudden increase in intensity destroys the top layers of cells on the leaf surfaces, and the leaves acquire a silvery cast (Figure 16-9). The

Figure 16-9 Bleached foliage caused by sunburn on a cast iron plant (*Aspidistra elatior*).

silvery cast turns brown or bleached yellow in a few days. Unfortunately, sunburned leaves do not heal, and the only way to improve the appearance of the plant is by removing these leaves.

Plants summered outdoors must have a lengthy acclimiazation period to avoid sunburn. Bright-light plants will benefit from summer sunlight, but medium- and low-light species should be placed in shaded or filtered sun areas. They will thrive under outdoor conditions without risking sunburn.

To acclimate houseplants to outdoor conditions, they should first be placed in the brightest indoor area for about a week. They can then be moved to a sheltered outdoor location that receives early morning or later afternoon sun but is shaded the rest of the day. After another week the high-light plants can be moved to a brighter outdoor area but should still receive protection from intense midday sun.

When bringing the plants indoors again, the acclimitization procedure should be reversed. Gradually adjusting the plants to decreased indoor light will prevent leaf drop and other signs of light deficiency. The plants should also be checked during this period for insects and diseases, since these problems can spread rapidly indoors.

Selected References for Additional Reading

Cherry, E. C. *Fluorescent Light Gardening.* Princeton, N J: D. Van Nostrand, 1965.

Elbert, G. A. *The Indoor Light Gardening Book.* New York: Crown, 1975.

Gaines, R. L. *Interior Plantscaping: Building Design for Interior Foliage Plants.* New York: Architectural Record Books, 1977.

Kramer, J. *Plants under Lights.* New York: Simon and Schuster, 1974.

Ortho Book Division. *The Facts of Light about Indoor Gardening.* San Francisco: Chevron Chemical Company, 1975.

Van der Veen, R., and G. Meijer. *Light and Plant Growth.* New York: Macmillan, 1959.

Vince-Prue, D. *Photoperiodism in Plants.* London: McGraw-Hill, 1975.

Chapter 17

Indoor Watering and Humidity

WATERING INDOOR PLANTS

Watering is the most frequent maintenance activity associated with growing houseplants. It is also (along with improper lighting) one of the main causes of houseplant failure, probably because absolute directions are not available. Proper watering depends on an understanding of the water requirements of the species, how water is held in and lost from the potting medium, and how it is used in plants.

Pathways of Water Loss in Houseplants

The water needs for the majority of indoor plants are met by water absorbed from the potting medium surrounding the roots. Water held in the medium can be thought of as a reservoir of moisture for the roots that needs to be refilled periodically to assure that adequate water is always available. When the reservoir is full, water can be lost in one of two ways: through transpiration or evaporation (Figure 17-1).

TRANSPIRATION

After water enters the roots, it moves upward through the stems and out to the leaf blades to be used in photosynthesis and the manufacture of other chemical compounds (hormones, pigments) in the plant. However, relatively little of the total absorbed water is used for these synthesis operations; the rest is lost as vapor through the leaves in a process called *transpiration.*

Transpiration occurs almost continuously in houseplants, but its rate is greatly affected by environmental conditions. First, transpiration speeds up as the temperature rises. This is true of almost all processes involved in plant growth. Second, it increases in rate as humidity decreases. A plant growing in a dry room would lose water faster than one growing in a humid greenhouse, for example.

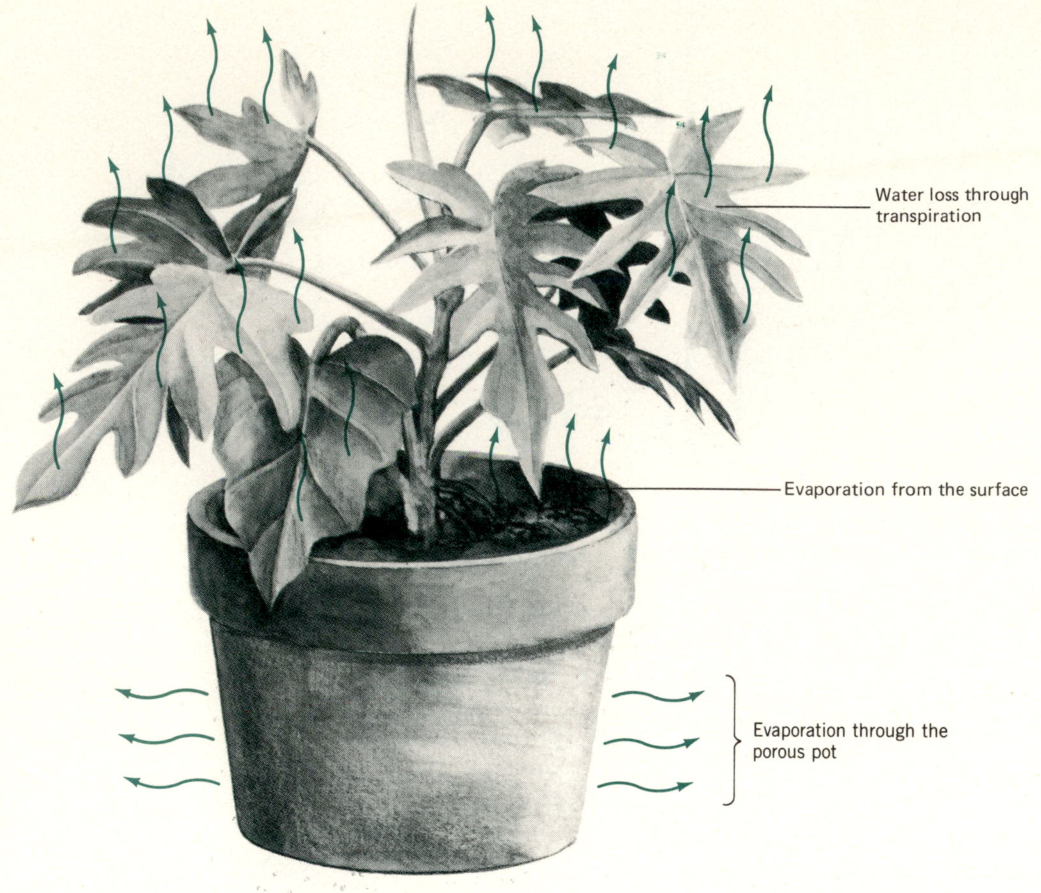

Figure 17-1 The pathways of water loss in a plant potted in a clay pot.

Transpiration rate is also affected by the plant species and form. Large-leaved or densely foliated tropicals transpire considerable water because of the large area of leaf surface exposed to the air. Cacti are more compact and have a limited amount of exposed surface. Consequently, they transpire less water.

EVAPORATION

Transpiration is one way in which water leaves the medium. Evaporation from the surface is another.

Like transpiration, evaporation occurs continually. Water leaves the area where it is concentrated (the wet potting medium) and moves as vapor into the drier adjacent air. Humidity and temperature cause changes in the evaporation rate of water as they do with transpiration, speeding it up as the temperature increases or the humidity drops.

With some types of pots, evaporation of water can occur through the sides of the container. Pots made of materials which are porous to water (unglazed clay or pressed wood fiber, for example) absorb water from the medium and release it into the air.

Watering Frequency

The ways in which temperature and humidity affect the rate of water loss demonstrate one reason why an exact watering schedule cannot be prescribed: the environmental conditions under which a plant is grown can vary radically. Water loss by a plant growing in a cool, humid bathroom will be much slower than that from an identical plant in a warm, dry living room, and the frequency at which each requires water will differ accordingly. Added to that is the change in rate of water loss depending on whether the plants are in porous or nonporous pots.

The environment and type of pot partially determine how frequently a plant must be watered, but there are other factors on which the watering frequency should be based.

POT SIZE

Almost without exception, the smaller the pot in which a plant is growing, the more frequently it will have to be watered. This is because a small amount of potting medium holds less moisture that a large volume. Even at the same evaporation and transpiration rates, the water will be depleted sooner from the small pot.

TYPE OF POTTING MEDIA

As discussed in Chapter 15, potting media differ in the amount of water they can hold because of differing ingredients. A potting medium largely composed of peat moss will be able to hold several times more water than one heavily fortified by perlite. Such a medium will take longer to dry out completely, and watering will not be required as frequently.

PLANT GROWTH RATE

Some plants naturally grow faster than others, and they require more water to photosynthesize and provide energy for that growth. These vigorous growers absorb water from the potting medium very rapidly, necessitating that it be replenished more frequently.

Even plants which are not vigorous growers may pass through changes in growth rate. A rapid spring and summer growth rate is common in houseplants. The watering frequency may have to be increased to keep up with the water demand brought on by the plant's size increase.

SIZE OF THE PLANT IN RELATION TO THE POT

The larger a plant is in relation to its pot, the more frequently it will require watering. As the top of the plant increases in size, more leaves require water for photosynthesis and transpiration. The root mass will also have increased, giving a much larger absorption area. However, the plant is still restricted to the same volume of medium from which to absorb water. It will deplete the available water quickly, and the water will need to be replenished sooner. Therefore when a plant is grown in the same pot for a long time and continues increasing in size, its watering frequency will also need to be increased.

PLANT SPECIES

The many plant species grown for indoor decoration differ widely in their needs for water. Cacti and succulents grow best when the potting medium is kept fairly dry through most of the year, whereas bog plants like umbrella sedge (*Cyperus alternifolius*) grow with their roots completely submerged in water. Most other plants tend to be somewhere in between. They grow best when there is a slight to moderate amount of water in the potting medium instead of when it is either fully dry or completely saturated.

With the exception of ferns and a few other species, most plants can survive up to a week with dry potting medium and not sustain permanent damage. Many experienced indoor gardeners grow plants very successfully on a "wet-dry" cycle, watering only when the medium is dry but just before the plant wilts. However, a wet-dry cycle may minimize new growth because of water stress. Watering more frequently is advisable when rapid growth is desired.

Soil Moisture Testing

Since differing environmental conditions rule out a prescribed watering schedule, the question of timing the watering frequency remains. The easiest answer is to evaluate the potting medium moisture by touch.

TOUCH METHOD

Feeling the potting medium (Figure 17-2) has proved to be a reliable method of testing the moisture and determining the watering frequency. The fingertips touch the surface of the potting medium or are pushed about ¼ inch (6 millimeters) deep to feel for moisture. A cool damp feeling indicates that moisture is still present near the surface and the medium surrounding the roots still contains adequate water. A dry feeling indicates that moisture surrounding the roots needs to be replenished and the plant should be watered.

COLOR METHOD

The color method is less reliable than the touch method, but it is used successfully by some gardeners. The method works on the principle that wet potting medium is darker in color than dry, and the moisture content is estimated visually. A drawback of this method is that the surface of the medium may appear dry several days after watering, but the medium underneath may still be damp.

WEIGHT METHOD

The weight method is reasonably reliable but requires more skill than the touch method. Immediately after watering, a potted plant will be heaviest because of the weight of water contained in the potting medium. As the water is transpired or evaporates, the medium loses water weight, and the potted plant will weight considerably less. With practice, a gardener can estimate water need by picking up the pot and comparing its present weight to the weight immediately after watering.

Figure 17-2 Testing moisture in the potting medium by feel.

COMMERCIAL MOISTURE SENSORS

Several types of chemical and electrical sensors are now available. The first type has a piece of blotter paper impregnated with a chemical that changes color in the presence of moisture. The paper is enclosed in plastic and mounted on a probe that is pushed into the medium. The changing color of the paper shows whether moisture is present in the root zone.

The electrical water sensor is an adaptation of a laboratory instrument called a *solubridge*, which is used to measure soil nutrient content. It measures the ease of conducting an electrical current through the medium, which in turn is affected by the ions present in the soil water. The more water present, the more ions will be dissolved, and the greater the ease of conductivity.

This type of water sensor is accurate at a normal nutrient range. However, it will read falsely (giving a wet reading in a dry medium) when the nutrient level is extremely high. Conversely, it will read "dry" when placed in distilled water because of the complete absence of ions in the water.

Applying the Correct Volume of Water

The goal when watering most houseplants is to wet all the potting medium thoroughly at each watering, allow it to dry to a point of slight dampness, and then rewet the medium again.

POTS WITH DRAINAGE HOLES

Assuming that the plant is watered from the top, the volume of water applied to a plant in a pot with a drainage hole is not critical. The makeup of the medium will allow it to hold only a limited amount of water, and the potting medium should be formulated so this capacity is not too high for healthy root growth. Water in excess of the capacity will then drain out the bottom holes.

With many commercial media it is practically impossible to overwater because the water-holding capacity of the media is so low. Damage from overwatering could result, however, if the pot remained standing in drainage water. With other media (particularly those containing soil) overwatering is possible. These media have a high water-holding capacity, and watering too frequently could cause rooting of the plant roots.

POTS WITHOUT DRAINAGE HOLES

The volume of water applied to plants in pots without drainage holes is critical. Because there is no way for excess water to leave the pot, any amount greater than that required to wet the potting medium will collect at the bottom. For this reason, it is advisable to provide a reservoir of pebbles or charcoal in pots without holes to serve as a collection area for excess water. Standing water in the pebbles is below the root growth level. The water there will not damage the plant and will slowly be absorbed back into the potting medium.

To decide how much water should be applied, the volume of the potting medium should be estimated. A volume of water one-fourth as great should then be applied. After about 5 minutes the water will have been absorbed, and the pot can be gently tipped sideways to allow the excess to run out. If more than a couple of teaspoons (10 milliliters) runs out, the water volume should be reduced next

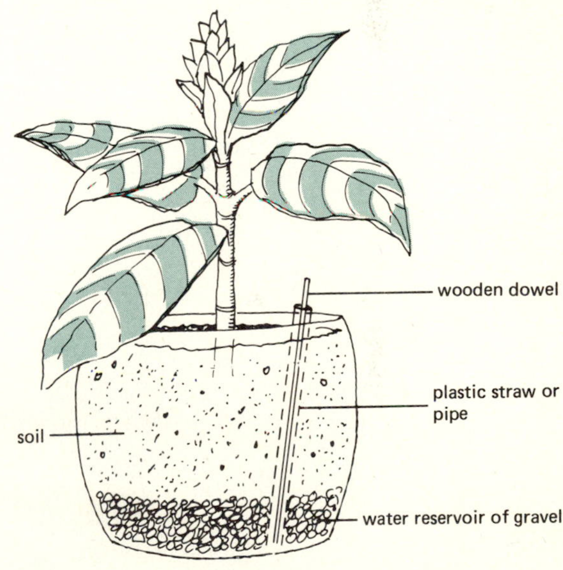

Figure 17-3 A dipstick system for checking the water level in the reservoir of a pot.

time. If none runs out, enough water may not have been applied to wet all the medium. More should be added and the pot tested again in a few minutes. Ideally just a few drops should flow out to indicate that the entire medium is wet.

Another way of determining the volume of water to apply is by using a dipstick. When the plant is potted, a length of plastic straw or pipe is buried with one end in the gravel reservoir (Figure 17-3). The other end remains above the surface. By inserting a wooden skewer or dowel into the straw, the presence of water in the reservoir can be determined. This technique is valuable for large specimens which are too heavy or cumbersome to be checked by tipping.

Watering Techniques

TOP WATERING

Most people water from the top. It is easy and completely satisfactory for almost all indoor plants. Top watering is so common that it requires little explanation, but two general rules should be followed. First, enough water should be applied to wet all the potting medium. For pots with holes this means applying enough so that some runs out the bottom. Second, a few minutes should be allowed for water to drain out, and then the drainage water should be discarded. Many plants develop root rot and die if they are left in drainage water.

Figure 17-4 Bottom watering of a dumbcane.

BOTTOM WATERING

Bottom watering (Figure 17-4) can be used only on plants in pots with drainage holes since the water is taken up through the bottom of the pot. It is very useful for plants whose leaves are susceptible to spotting from water and for rosette plants, in which the crown is likely to hold water and rot. Bottom watering is also good for watering large cacti whose bodies nearly fill the top of the pot. Water applied from the top frequently does not reach the roots of such cacti because it can be poured in only around the rim of the pot. The water flows down the outside of the root ball and out, never adequately wetting the roots. Bottom watering is also useful for rewetting a medium that has dried excessively and retracted from the pot.

For proper bottom watering, the pot must sit in a plant saucer or bowl. The saucer should be filled with water, and in 5 to 10 minutes the medium will have absorbed the water by capillary action. The saucer should be filled again and, if necessary, a third time until the medium stops absorbing. The excess should then be discarded.

One precaution must be taken with bottom watering. Because water does not flow through the medium and out the bottom, the fertilizer salts contained in the medium are never flushed out. There is a possibility that they will accumulate to toxic levels. As a precaution, a plant which is constantly bottom watered should be submersion-watered or leached every 1 to 2 months to rinse through excess fertilizer.

Figure 17-5 Submersion watering of a spathiphyllum in a kitchen sink.

SUBMERSION WATERING

Submersion watering (Figure 17-5) is also used only for pots with drainage holes and for the same purposes as bottom watering. It can also be useful for large, unwieldy hanging baskets.

Submersion watering is similar to bottom watering, in the sense that water enters through the drainage holes, but it is quicker. In submersion watering, the plant is set in water deep enough to come up to the rim of the pot. In 5 to 15 minutes the water will have penetrated to the surface of the medium. The pot is then taken out of the water, allowed to drain, and replaced in its growing area.

In submersion watering there is little likelihood of excess fertilizer buildup. During the wetting, the medium becomes saturated, and the fertilizer dissolves in the water. In the draining period it is carried out with the excess water.

WICK WATERING

Wick watering (Figure 17-6) is not the ideal solution for watering plants because it keeps the medium too moist for most species. However, it is successful with plants which must have constant moisture (African violets and most ferns, for example) and for those in which complete drying could cause death.

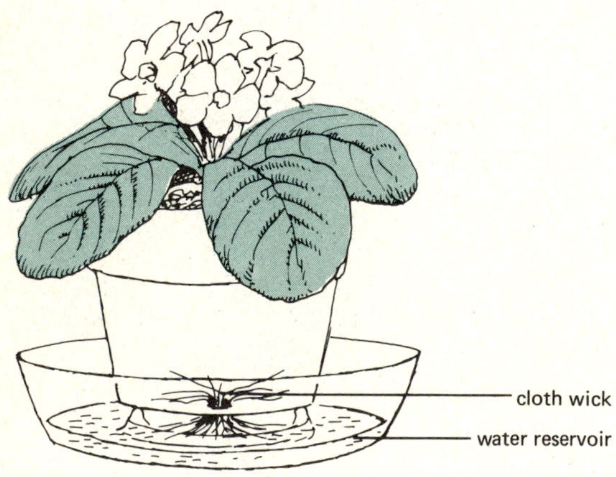

Figure 17-6 A wick watering system. Note that the pot is positioned above water level.

Wick watering follows the same principle of capillary action which pulls water up through the medium in bottom watering. In this case capillary action occurs in the wick. Water is absorbed into the wick from the water reservoir, an area of high water concentration. The water is pulled through the wick by capillary action to an area of low concentration (the potting medium). The wick action continues until the potting soil is saturated with water.

SHOWER WATERING

Showering, as discussed in Chapter 14, is principally a method of leaf cleansing which waters the plant at the same time. In some regions the mineral content of the water is very high, and watering by showering will leave unsightly mineral deposits. In these cases avoid showering or gently towel dry the leaves to prevent the deposits.

Special Watering Techniques

BULB- AND TUBER-BEARING HOUSEPLANTS

During their growth and flowering periods, bulb- and tuber-bearing houseplants such as amaryllis and gloxinia require no special watering techniques. They can be treated like most other houseplants. But as their blooming period ends, these plants may enter a yearly dormancy phase. The top growth dies, putting the bulb in a rest period for several months before growth begins again.

The dormant period for bulb- and tuber-bearing plants coincides in nature with climate changes such as decreased rainfall. This change should be simulated by the gardener to meet the plant's growth requirements. After the foliage begins yellowing, watering frequency and volume should be slowly decreased over about 2 months. During this period the foliage will slowly die back to the bulb. When all the foliage is dead, watering should completely stop for several months while the plant rests. It can then be divided or repotted while dormant and regular watering resumed to encourage fresh growth.

The key to correctly watering bulb- and tuber-bearing houseplants lies in careful observation of the plant during all its growth phases. Some bulbous plants never enter dormancy indoors and may be severely injured by underwatering. Others may be unintentionally overwatered during dormancy and rot. Bulbous plants will show signs of approaching dormancy such as yellowing and death of older leaves. These signs are signals to reduce watering. Plants emerging from dormancy often begin new growth before water is applied. As soon as this new growth begins, watering should be started. If watering is delayed, the bulb will respire excessive amounts of carbohydrate from the storage organ and may be stunted or die.

LEACHING

Leaching is a watering technique used to remedy the problem of excess fertilizer salts in potting media. It works on the principle that the salts are soluble in water and consists of watering a plant four or five times in succession at 5- or 10-minute intervals. Each watering dissolves some of the salts, and the following one flushes them out in the drainage water.

A succession of waterings is not the only form of leaching, although it is probably the most common. Plants can also be submerged and drained several times or left under a slow-running tap for 10 minutes for the same result.

Water Quality in Indoor Plant Growth

For most plants, the amount of water they receive is more important than the quality of that water. In most parts of the country, plants grow well given normal tap water, with only a few being adversely affected by its temperature, pH, or chemical content. One exception to this generalization can be found in the southwestern states, where high-pH and high-salt-content water can cause plant health problems.

TEMPERATURE

Correct water temperature is not critical for most houseplants provided it is within a 60 to 90°F (15–32°C) range. Letting water become room temperature is probably ideal for houseplants, but it is not practical if there are many houseplants to be watered.

African violets, though, have been proved sensitive to water temperature. Water more than 10°F (5°C) colder than air temperature will cause permanent yellow spotting on the leaves due to chlorophyll destruction (Figure 17-7). Bottom watering or warming the water to room temperature will avoid the problem.

pH

Acidity or alkalinity (pH) is a characteristic of water as well as potting media. It varies depending on the water source. Continual watering with highly acidic or alkaline water will eventually alter the original pH of the medium to an undersirable range.

Although not highly accurate, an inexpensive pH test kit can determine if water is exceptionally acidic or basic. The gardener will then be aware of media pH problems that could occur because of its use.

CHEMICAL CONTENT

Tap water, depending on its source, can contain any number of "impurities" including (but not limited to) sodium, fluoride, chlorine, calcium, and iron.

Sodium. Sodium often enters water through the softening process. It is used to

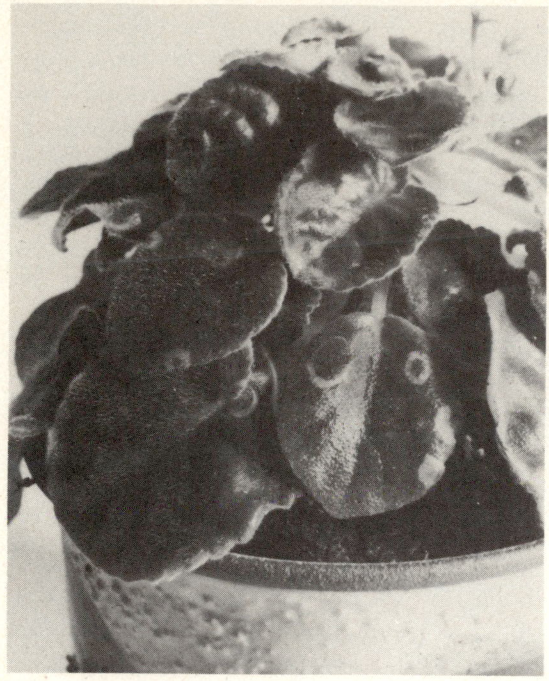

Figure 17-7 Spotting of African violet leaves caused by cold water.

replace the calcium in calcium carbonate, the mineral that often causes hard water. In areas of the southwest it may be naturally present in the groundwater.

The sodium found in tap water can occasionally build up to concentrations harmful to houseplants with normal watering. To avoid this problem, do not use water from the hot-water tap. In home water softeners, hot water alone is treated with sodium, and cold water remains sodium-free. If the sodium is present in the natural groundwater, or if all household water is softened, rainwater or distilled water should be used. Leaching every 2 to 3 months will also help flush the sodium from the potting medium.

Fluoride. Fluoride is put into most city water supplies as a preventative against tooth decay. The concentration used is very small—about one part per million. However, some untreated water supplies contain more than that concentration naturally.

Relatively small amounts of fluoride are capable of causing damage to plants, but only members of the lily family are known to be particularly sensitive. Common houseplants in this family include dracaenas, spider plants, and asparagus ferns. Excess fluoride can also be found as a result of amendments like perlite. The injury appears as browing of the leaf tips.

Chlorine. Chlorine is added to city water for sanitation purposes; it kills bacteria, fungi, algae, and other potentially harmful organisms. It is generally added as a salt that dissolves in the water or occasionally bubbled in as a gas under pressure.

Chlorine is essential to plant growth in small quantities and is normally supplied to plants outdoors in the soil and sometimes by rainfall. Tap water does not contain sufficient chlorine to damage houseplants.

Contrary to popular belief, allowing tap water to stand overnight will not eliminate the chlorine. Instead, it escapes extremely slowly as a gas, although the process can be speeded up by exposing the water to ultraviolet light.

Alternatives to Tap Water

RAINWATER

The belief that rainwater is pure is not true. It contains a surprising amount of chlorine and some nitrogen, sulfur, and other substances as well. It can also contain chemicals from air pollution and be affected by whatever roof surface it runs off before it is collected. Although rainwater is certainly no worse than tap water for plants, it has no special plant-growing properties because it came from the atmosphere instead of a well or reservoir. However, it is a valuable alternative in areas with highly salt-laden water.

DISTILLED WATER

Distilled water has been treated to remove all substances but the hydrogen and oxygen which form water. It is truly "pure" water that does not contain minerals or other foreign chemicals.

If the natural water supply is high in sodium, fluoride, or another chemical injurious to houseplants, distilled water is a good alternative water source. It is also useful for leaching plants because it can carry away more salts than the same volume of tap water.

SYMPTOMS OF IMPROPER WATERING

Underwatering. Plants which are receiving too little water or are not being watered frequently enough may incur root injury. They may show different symptoms, depending on the plant species. Usually they wilt but can be revived provided the leaves have not lost moisture to the point of becoming crisp. Even if the top portions die completely, many plants will regenerate from the roots if the medium is kept moist.

Other symptoms of underwatering are not as easy to diagnose. Slow growth, browned leaf tips and margins, and bud and flower drop can be due to insufficient water as well as several other causes. The symptoms of under- and overwatering can even be the same since both result in root injury.

Overwatering. Excess water excludes oxygen from the potting medium. Eventually root death will occur because the roots are unable to respire.

Overwatering can be caused by watering too frequently, by applying too much water to a pot without a drainage hole, or by a medium that holds too much moisture. Symptoms of overwatering include weak and slow growth, lower leaf drop, and discoloring of the roots. In the advanced stages plants wilt because the dead roots can no longer absorb water.

Vacation Watering of Indoor Plants

Keeping houseplants watered during vacation can be inconvenient. However, there are ways to keep them sufficiently moist without interim watering.

One appraoch is to slow down plant metabolism. This will decrease the water requirement and make the water in the medium last the longest possible time. Photosynthesis, transpiration, and respiration must all be slowed, and the following suggestions can be used separately or in combination to accomplish this.

LOWERING THE TEMPERATURE

As temperature decreases, biological processes slow down. A drop from 75°F (24°C) to 55°F (13°C) will decrease water use appreciably.

MOVING PLANTS OUT OF DIRECT SUNLIGHT

Direct sunlight warms the leaves, increasing transpiration. It also increases the photosynthesis rate and the water needed for that process. Most high-light plants can be maintained without full sunlight for up to 2 weeks without deteriorating, but they should be given bright diffused light to prevent leaf drop. A sheer curtain will provide this type of light.

GROUPING PLANTS TOGETHER IN THE BATHROOM

A small bathroom can be turned into a holding room for plants fairly easily. The windows should be closed and the tub and sink filled with water to maximize the humidity. All the plants should then be watered and set in the bathroom with the lights on (not sun or heat lamps) and the door closed. The high humidity will slow transpiration, and the lights will permit limited photosynthesis. Most plants can be left up to 2 weeks under these conditions. The main drawback is the production of weak growth on vigorous plants because of insufficient light. This undesirable growth should be pruned off when the plants are returned to their normal growing locations.

ENCLOSING PLANTS IN PLASTIC BAGS

Dry-cleaning bags work well for this purpose, creating a closed system from which no water is lost. Plastic drop cloths can be used for large numbers of plants. One is spread on the floor and the recently watered plants are grouped together on top of it. A second cloth is used to cover them, and the edges are sealed with tape or rolled and stapled to make an airtight envelope. One precaution must be followed: bagged plants should not be placed in strong sunlight because the sun can heat the air inside the plastic and cause heat damage.

For specimens too large for bagging, cover just the pot, sealing around the crown as snugly as possible with tape or string. The plastic will eliminate water evaporation through the medium and thus conserve part of the moisture.

Overall, bagging is the most reliable of all vacation watering ideas. Plants completely sealed in plastic can be left for a month or more without care (Figure 17-8).

Figure 17-8 Enclosing a plant in a plastic bag will enable it to survive for a long period without watering.

HUMIDITY AND INDOOR PLANT GROWTH

Humidity is the amount of water vapor in the air, and sufficient humidity is essential for healthy plant growth. Indoor plant species vary in their humidity needs. Cacti and succulents grow satisfactorily in relatively low humidities, but leafy tropicals grow best at higher humidities.

The effect of humidity on plant growth is due mainly to its influence on transpiration rate. Generally, the higher the humidity, the slower the loss of water through transpiration. Without constant and rapid water loss, plant roots do not expend as much energy absorbing water, and that energy can be used for growth and development.

Relative Humidity

The humidity content of the air is generally recorded in a percentage figure called the *relative humidity.* This figure takes into account the fact that the ability of air to hold water vapor varies with its temperature. Air can hold increasingly greater amounts of water as its temperature rises. The relative humidity is the percentage of water vapor the air is presently holding in relation to the total water it is capable of holding at that temperature. For example, 50°F (10°C) air at 80 percent relative humidity means the air contains 80 percent of the water vapor it is able to hold at that temperature.

The outdoor humidity of an area is influenced by the amount of precipitation (generally rain) which the area receives and by its proximity to large bodies of water. Accordingly, the East, Northeast, and Northwest have high natural humidites, and the dry inland portions of the West have less. But outdoor humidity is not necessarily equal to indoor humidity. It will be equal only if there is free air circulation between the two. Once windows are closed and heating or air conditioning is used, the humidity will change.

Low indoor humidity in winter is blamed for many houseplant problems. Actually indoor air contains the same amount of water as outside air, and the relative humidity of the outside air may be 60 to 70 percent . The temperature difference creates the problem. Cold outdoor air at a moderate humidity is brought indoors, warmed, and circulated to freshen the indoor air. But the water it contained at 20°F (−6°C) outside is much less than it is capable of holding at 65°F (18.3°C) indoors, and its relative humidity drops from a typical 60 to 20 percent.

Measuring Humidity

There are several instruments available for measuring relative humidity in the indoor environment. A human hair hygrometer is one. It shows the effect of rising or falling humidity on the length of a single human hair and indicates the humidity on a simple dial. It is a relatively expensive instrument, but it is the only dial-type device that is reasonably accurate.[7]

A sling psychrometer is the second instrument for measuring humidity. It consists of two identical thermometers mounted side by side on a small plaque. The mercury bulb of one measures ordinary temperature. The bulb of the second is covered with a small piece of wick (Figure 17-9). When measuring the relative humidity, the wick covering the one thermometer bulb is first wet. The two thermometers are then whirled in the air. When both thermometers show constant readings, the difference between the two temperatures is noted. The "wet-bulb thermometer" with the wick will always show a lower temperature because of the evaporation of the water around the bulb. Using the temperature difference and special tables, the relative humidity can be quickly calculated.

In general, a reading of 35 percent or less can be considered low humidity, 35 to 70

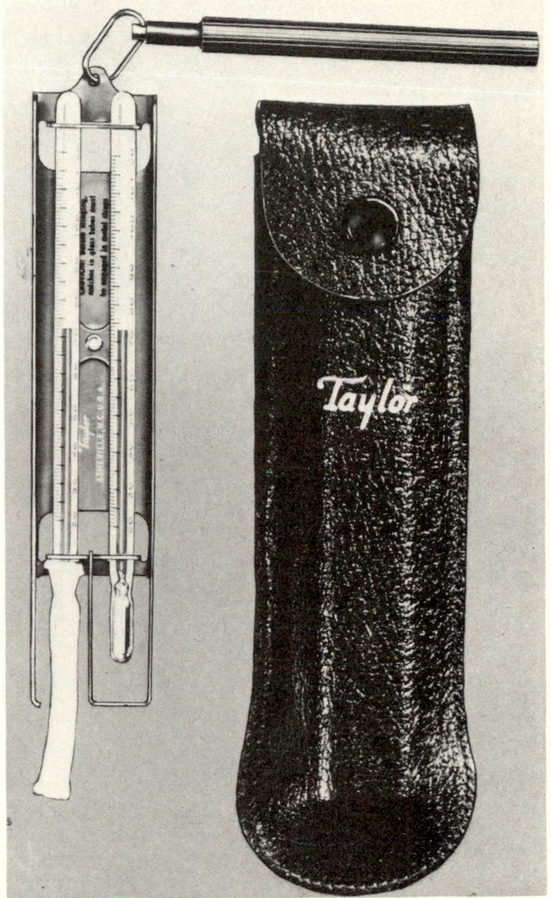

Figure 17-9 A sling psychrometer.

[7]It should be noted that a hygrometer requires frequent recalibration to maintain accuracy.

percent moderate, and 70 percent or greater high humidity. The majority of tropical plants grows best at humidities of at least 50 percent, although most will grow with less.

Raising the Relative Humidity Indoors

A number of techniques are used to raise humidity indoors, with differing degrees of success.

MISTING

Although misting is widely advocated as a means of raising the relative humidity around plants, it is questionable whether it actually does much good. In dry air, water will evaporate very rapidly. A thin film of mist applied to leaves can dissipate into the air in less than half an hour. It is unlikely that such a short period of improved humidity would have much effect.

Misting is, however, beneficial to bromeliads and orchids, which absorb water directly through cells on their leaves.

PEBBLE- AND WATER-FILLED TRAYS

This method works because of the constant evaporation of water into dry air. Trays or saucers are filled with a layer of pebbles or marble chips, which are partially covered with water (Figure 17-10). Plants are then grouped on the tray to benefit from the vapor rising up from the water. The water reservoir is refilled as needed, but never to a depth where the pots would be standing in water.

HUMIDIFIERS

The most reliable way to increase room humidity is by means of a room humidifier. Different models range upward in price from $10.00. Among the most economical are cool vapor humidifiers. In addition to being inexpensive, they use very little electricity.

Figure 17–10 Keeping houseplants on a layer of wet pebbles will help increase the humidity.

Symptoms of Improper Humidity

INSUFFICIENT HUMIDITY

Browning of the tips and margins of leaves on tropical plants is the most common low-humidity symptom. Rapid transpiration causes water to be lost from the leaves before it is able to reach the furthest portions, and browning results. Although unsightly, the browning does little harm to the plant.

Although "tip burn" (Figure 17-11) is the most easily diagnosed symptom of insufficient humidity, others may occasionally appear. Dropping of both new and old leaves may be a symptom, although it is usually associated with other growing problems. Undersized new leaves, a general lack of vigor, and bud drop are others.

EXCESS HUMIDITY

Most indoor plants grow well at humidities of up to 90 percent. However, the disease powdery mildew can develop under humid conditions. It affects roses and begonias among other plants. Another fungus disease, botrytis, is also common at high humidities.

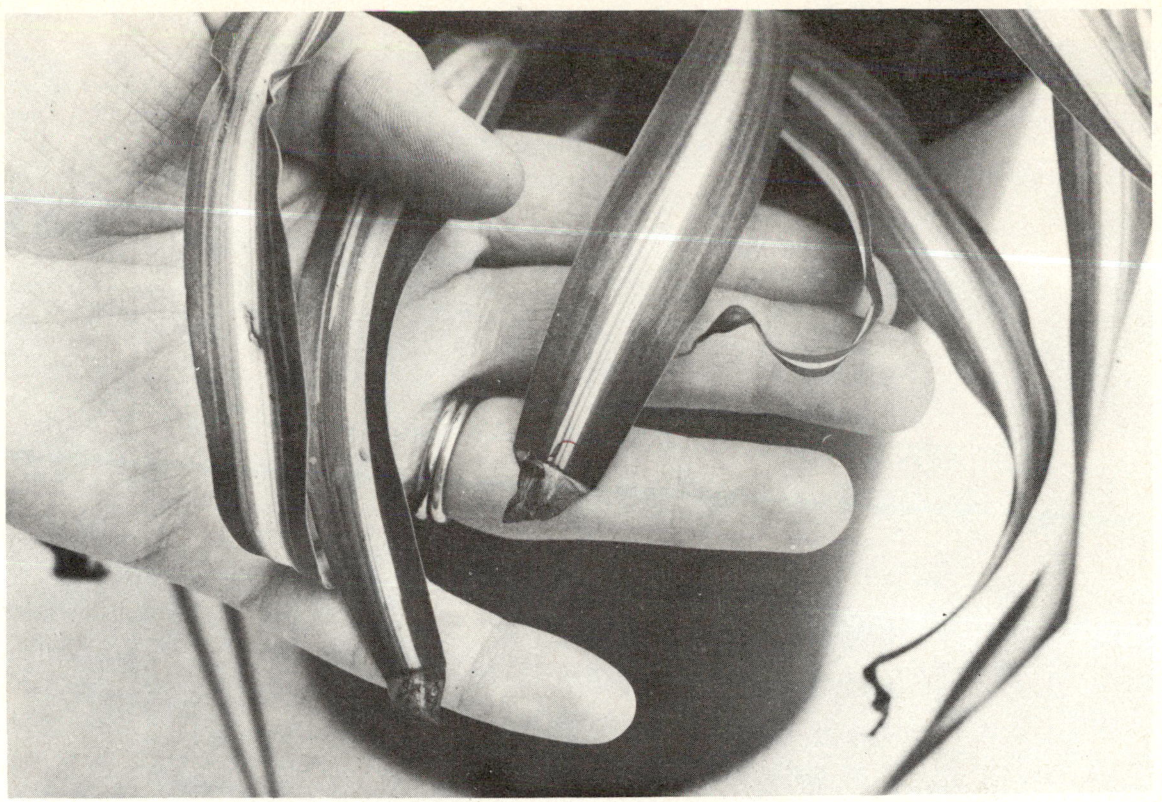

Figure 17–11 Tip burn on a spider plant (*Chlorophytum comosum* 'Vittatum').

Selected References for Additional Reading

American Water Works Association. *Water Quality and Treatment, A Handbook of Public Water Supplies.* McGraw-Hill, New York, 1971.

Behrman, A. S. *Water Is Everybody's Business.* Garden City, NY: Doubleday, 1968.

International Symposium on Humidity and Moisture, 1963. *Humitity and Moisture. vol. 1.* Principles and Methods of Measuring Humidity in Gases. New York: Reinhold, 1965.

Chapter 18

Controlling Houseplant Pests and Diseases

Houseplant problems are almost always attributable to one of two sources: the environment or organisms contained in it. Those caused by improper environmental conditions (light, water, temperature, and the like) are called cultural problems since they are associated with the culture or care of the plant. (These problems are not discussed in this chapter, since the effects of environmental factors were dealt with in the chapters on light, watering and humidity, and potting media.) The organisms harmful to plants can be anything from submicroscopic viruses to inch-long snails. The only characteristic they share is that all are detrimental to plant health, whether by eating foliage, sucking sap, or causing spots on leaves.

HOUSEPLANT INSECTS AND RELATED PESTS

There are insects and also insectlike pests which feed on houseplants. Although it might appear overly technical to separate the two, the distinction can be important if chemicals are used for their control.

Insects, which are six-legged, can be controlled by a single insecticide. Pests like mites, nematodes, and sowbugs belong to other classifications and may require different chemicals for control. Mites, for example, are susceptible to miticides, and nematodes to nematicides. Other than for control purposes these distinctions are unnecessary; all will be discussed under the broad heading of pests (Table 18-1).

Foliage Pests

The easiest pests to detect are often those which damage the aboveground portions of plants—the leaves, stems, buds, and so on. They can be collectively called *foliage pests.* Some foliage pests are more frequently encountered than others; their listing here is ranked from most to least common.

MITES (SPIDER MITES, RED SPIDER MITES, TWO-SPOTTED SPIDER MITES, CYCLAMEN MITES)

As mentioned, mites (Figure 18-1) are not insects. They have eight rather than six legs and are related to spiders; hence the name

Table 18-1 Insects and Related Pests of Houseplants and Their Common Hosts

Pest	Remarks	Common plant hosts
Spider mite	Common on almost all houseplants	African violet (*Saintpaulia ionantha*) Asparagus fern (*Asparagus densiflorus* 'Sprengeri') Azalea (*Rhododendron* hybrid) Cacti Citrus (*Citrus* spp.) English ivy (*Hedera helix* varieties) Prayer plant (*Calathea, Maranta*) Umbrella plant (*Brassaia, Tupidanthus*) Croton (*Codiaeum variegatum*)
Cyclamen mite	Less common than spider mite and attacks fewer species	Begonia (*Begonia* spp.) *Crassula* spp. Cyclamen (*Cyclamen persicum giganteum*) English ivy (*Hedera helix* varieties) Gesneriads (*Saintpaulia, Gloxinia, Episcia*, and other genera) Peperomia (*Peperomia* spp.) Purple velvet plant (*Gynura* spp.) Zebra plant (*Aphelandra squarrosa*)
Scale	Attacks primarily woody plants with some exceptions	Cacti Citrus (*Citrus* spp.) Coffee (*Coffea arabica*) Croton (*Codiaeum variegatum pictum* varieties) Cycad (*Cycas* spp.) Dracaena (*Dracaena* spp.) English ivy (*Hedera helix* varieties) Ferns Figs (*Ficus* spp.) Palms Gardenia (*Gardenia jasminoides*) Kalanchoe (*Kalanchoe* spp.) Orchids Zebra plant (*Aphelandra squarrosa*)

Pest	Remarks	Common plant hosts
Whitefly	Generally uncommon	Begonia (*Begonia* spp.) Citrus (*Citrus* spp.) Coleus (*Coleus blumei* varieties) Ferns Fuchsia (*Fuchsia* × *hybrida*) Gardenia (*Gardenia jasminoides*) Pepper (*Capsicum annum*) Poinsettia (*Euphorbia pulcherrima*)
Mealybug	Common and relatively nonselective	Cacti Coleus (*Coleus blumei* varieties) Croton (*Codiaeum variegatum pictum* varieties) Cycad (*Cycas* spp.) Dracaena (*Dracaena* spp.) Ferns Fuchsia (*Fuchsia* × *hybrida*) Gardenia (*Gardenia jasminoides*) Jade plant (*Crassula argentea*) Palms Poinsettia (*Euphorbia pulcherrima*) Rubber plant (*Ficus elastica*)
Aphid	Occasionally found on the new growth of these and other plants	Ferns Fuchsia (*Fuchsia* × *hybrida*) Purple velvet plant (*Gynura* spp.)
Thrip	Relatively uncommon	Arrowhead (*Syngonium podophyllum* varieties) Citrus (*Citrus* spp.) Fuchsia (*Fuchsia* × *hybrida*) Gardenia (*Gardenia jasminoides*) Gloxinia (*Sinningia speciosa*) Umbrella plant (*Brassaia actinophylla*)
Nematode	Relatively uncommon	African violet (*Saintpaulia ionantha* varieties) Begonia (*Begonia* spp.) Bloodleaf (*Iresine herbstii* varieties, *Calathea* spp.) Cyclamen (*Cyclamen persicum giganteum*) Ferns Fuchsia (*Fuchsia* × *hybrida*)

Table 18-1 (*continued*)

Pest	Remarks	Common plant hosts
		Gardenia (*Gardenia jasminoides*) Rubber plant (*Ficus elastica*)
Root mealybug	Occasionally found in unsterilized media	Arrowhead (*Syngonium podophyllum* varieties) Bromeliads Cacti Dumbcane (*Dieffenbachia* spp.)

Note: Plants listed frequently in this table are not necessarily prone to disease and insect problems but are simply the more thoroughly researched indoor plants.

Figure 18-1 Spider mite damage on a palm leaf.

"spider mites." At first encounter, a mite problem can be among the hardest to diagnose. Mites are so small—about the size of dust specks—they are almost invisible to the naked eye.

Several kinds of mites attack indoor plants. Red or two-spotted spider mites can be red, brown, or creamy white. Their damage is done when they insert their mouthparts into a leaf repeatedly to feed. Each feeding site shows as a pinpoint-sized white dot, and eventually leaves become highly speckled with feeding marks, shrivel, and die. Spider mites attack almost any houseplant having thin leaves, as well as cacti and succulents. They are by far the more common type of mite, and they thrive in low humidity.

The surest diagnosis for spider mites is by examination of the undersides of the damaged leaves in strong light. The mites will be barely visible, but they can be seen clearly with a magnifying glass. In advanced cases of spider mite infestation, webbing will cover leaves. It is to this webbing that the female attaches her eggs.

Cyclamen mites are the second main type of mite affecting houseplants. It is from a main host plant, the florist cyclamen, that their name is derived. Unlike spider mites, which attack old and young leaves indiscriminantly, cyclamen mites cluster on very young leaves and buds for feeding.

Cyclamen mites are nearly transparent and feed by puncturing leaves, but their feeding damage does not show speckling or webbing. Instead it causes curling and cupping of the leaves, giving the young leaves a dwarf appearance.

Cyclamen mites are not as universal as spider mites. They are an unusual problem affecting houseplants, as contrasted to the persistent problem which spider mites pose.

MEALYBUGS

Appearing like specks of cotton, the many types of mealybugs (citrus, long-tailed, and Mexican, for example) are destructive pests. They cause stunting, leaf drop, and eventually the death of a plant.

The adult insects are pink and shaped like sowbugs. They appear white because of the white waxy fibers which they secrete and in which they lay their eggs (Figure 18-2). They are very slow moving and frequently found in the axils of leaves or on the undersides along the veins.

Adult mealybugs are often difficult to kill with sprays because of the protection afforded them by their waxy coats. However, newly emerged young lack this covering and are susceptible to most all-purpose insecticides.

SCALE INSECTS (HARD SCALE, SOFT SCALE)

Scales (Figure 18-3) are relatively immobile and are frequently not recognized as being insects. Like mealybugs, scales secrete a protective covering. The type of covering determines whether they are called *armored* or *soft scales*.

Armored scales are most common on houseplants. They are covered with hard, blisterlike coats and range up to ⅛ inch (3 millimeters) in diameter. Their color can be clear to white, yellow, or brown, depending on the species, age, and sex of the insects.

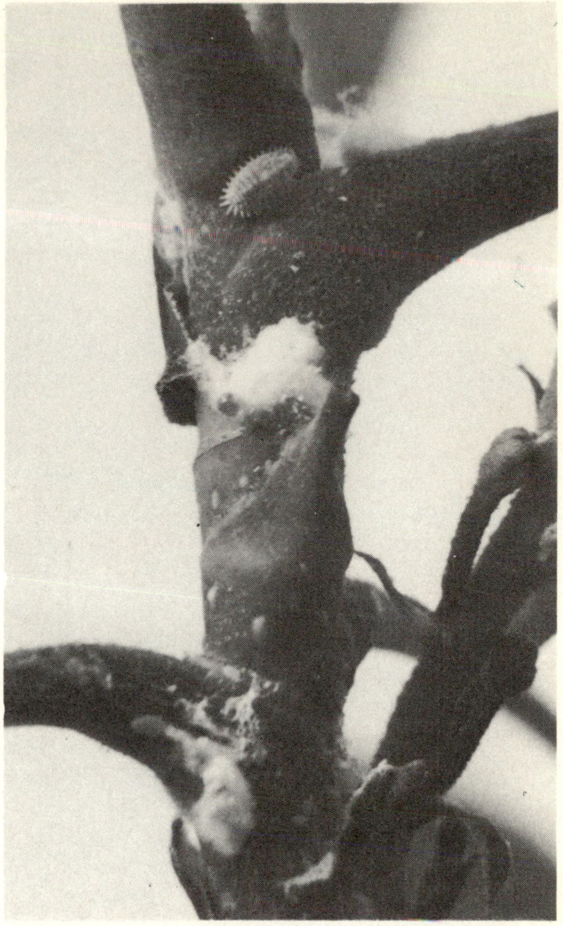

Figure 18-2 Mealybugs. A young adult without protective covering can be seen toward the top. It is flanked by mature insects covered with cottony fibers.

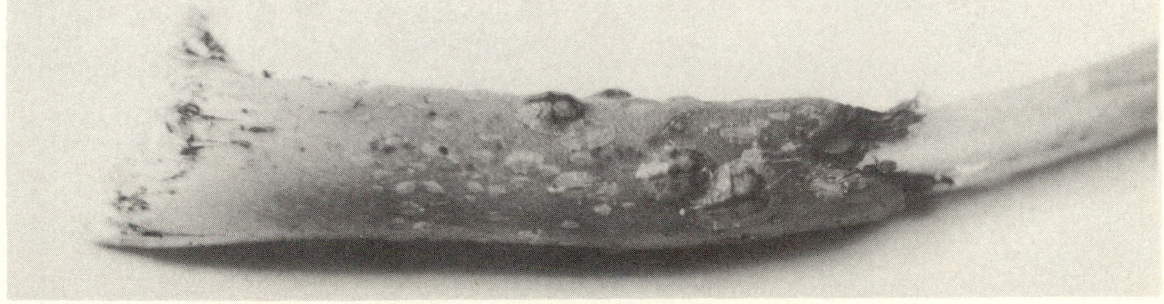

Figure 18-3 Scale insects at the base of a palm frond.

Figure 18-4 Whiteflies.

Soft scales are very much like mealybugs, secreting a soft waxy covering attached to their bodies. They are nearly round in shape and generally larger than armored scales.

Scales of both types do their damage by puncturing plant tissues and extracting sap through their needlelike mouthparts. As a result of this feeding, plants become weakened, and parts or the whole plant will eventually die.

Scales are found on both sides of leaves, on twigs, and on branches, frequently concealing themselves in the leaf axils. When they attach to leaves, a yellow spot will sometimes be found on the reverse side.

In the life cycle of scales, the most visible stage is the adult, who is permanently anchored in one place and grows to a very large size. In the immature crawler stage, scale insects are mobile, and it is at this period when they are most vulnerable to spray insecticides.

WHITEFLIES

Whiteflies (Figure 18-4) are white insects resembling tiny moths. They cluster on the undersides of leaves, flying up when disturbed and resettling almost immediately. Their damage is done by piercing the undersides of leaves to extract plant sap, which causes pale or spotted leaves and weakens the plant.

In a generalized life cycle of a whitefly, the eggs are laid under leaves and hatch in 4 to 12 days to a crawler stage. The crawlers move over the leaf for several hours, then puncture the leaf and feed in one spot. They shed their skin three times, then enter a resting stage and emerge as the familiar white adult. The process takes almost 6 weeks.

When whiteflies are detected, it is almost always in the adult stage. At this point they are bothersome and difficult to spray because of their flying. It is advisable at that point to begin spraying under the leaves to control the crawlers, with the expectation that control will not be complete for several weeks.

APHIDS (PLANT LICE OR GREEN FLIES)

These soft-bodied insects (Figure 18-5) are easily visible and are about 1/32 inch (1 millimeter) across. They can be green, red, black, or any other number of colors and tend to cluster in large numbers on the youngest leaves and buds of plants. Aphids suck plant juices, causing deformed young leaves.

In addition, they secrete a syrupy substance called *honeydew* from their backs. The honeydew makes plant leaves sticky and promotes the growth of sooty black mold, a nonparasitic mold which is nonetheless unsightly. Aphids also transmit virus diseases.

Aphids are easily killed with chemicals during any stage in their life cycle. Unlike other insects, young aphids are born live instead of hatching from eggs. The adult female does not require mating, and she can give birth to up to 100 self-fertile females during her 20- to 30-day lifetime.

Figure 18-5 Aphids.

THRIPS

These insects (Figure 18-6) are just slightly larger than spider mites. They are winged in the adult stage and wingless in earlier immature stages. Foliage damaged from their feeding is silvery-colored, later turning brown. The flowers they attack become streaked, and the buds fail to open.

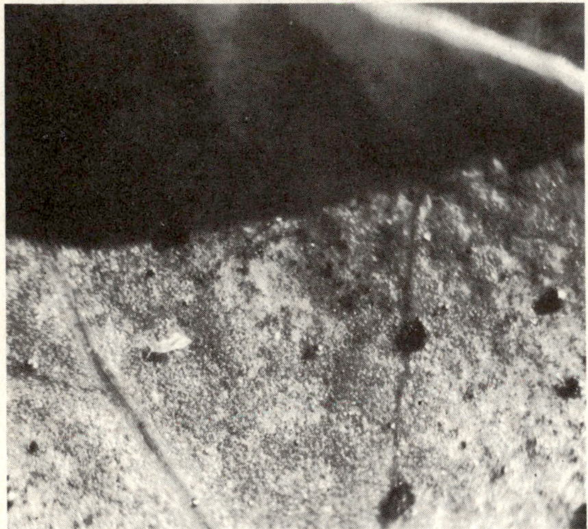

Figure 18-6 Thrip damage.

SLUGS AND SNAILS

These invade from outdoors, often entering hidden in the drainage holes of pots. They chew large irregular holes primarily at night and hide during the day. Baits and sprays designed for outdoor use are both effective.

Root and Media Pests

GNATS (FUNGUS GNATS, MANURE FLIES, MUSHROOM FLIES)

Flying gnats and their immature worm stages (Figure 18-7) are annoying to people and harmful to houseplants. Potting media rich in organic matter provide ideal breeding areas, and a few gnats can multiply to several hundred in a short time.

Figure 18-7 The larva of a fungus gnat.

The adults are brown to black and about 1/8 inch (3 millimeters) long. They are usually seen crawling along the soil surface, where they lay eggs that hatch into tiny white maggots. The burrowing and feeding of the maggots on plant roots cause stunting and root diseases. As the larvae increase in number, the damage intensifies.

Adult gnats can be killed with spray insecticides, but the maggots which do the actual damage must be controlled by drenching.

SPRINGTAILS

Springtail insects derive their name from the sudden jumping motion they go through when they are disturbed, as by watering. The insects spend their entire life cycle in the soil, growing to an adult size of about 1/16 inch (1.5 millimeters).

Unlike parasitic gnat larvae, springtails in all stages eat only dead vegetation, causing no harm to houseplants. They are, however, bothersome and can be eliminated by an insecticidal soil drench.

Nematodes (Eelworms). Nematodes are microscopic or near-microscopic worms which live in the medium and feed on the roots of plants. Certain species live on plant leaves, entering through the stomates to feed. These nematodes are rare on houseplants.

Different species of nematodes inflict different types of damage. The feeding of the root-knot nematode causes irritation of the root and leads to the formation of a knot of tissue at the wound site. Numerous such knots interfere with the root's ability to transport water and nutrients to the foliage, and the top of the plant becomes stunted and pale.

Root-lesion nematode does not cause swellings on the roots but does destroy root cells and leave an entryway for disease. The effects on the top portions of the plant are the same as with root-knot nematode.

Because they are usually microscopic, soil infestations on houseplants are difficult to diagnose. But nematodes can be suspected if the medium was not sterilized before use or if it contains soil. Unpotting the plant and checking for knots will diagnose root-knot nematode. Brown decayed roots could indicate the presence of other types. An exact diagnosis can be made only through a medium and root sample submitted to a soil-testing laboratory. Nematodes are most effectively controlled on growing plants with chemical nematicide soil drenches.

ROOT MEALYBUGS

These root-parasitizing insects look like regular mealybugs but are smaller. Since their feeding weakens plants, they should be controlled with an insecticidal soil drench.

SOWBUGS (PILLBUGS)

These small pests are not insects but are related to the armor-covered crayfish. They live only in damp places and will often be found hiding under pots or living inside the drainage hole. Although they eat mainly dead organic matter, they also feed on roots and small seedlings and should be eliminated by a chemical drench.

EARTHWORMS

Earthworms do not eat live plants and are seldom seen at the surface. During their burrowing they digest potting medium and excrete it behind, leaving a network of tunnels. Unless present in large numbers, they do no damage.

ORGANISMS CAUSING DISEASES OF INDOOR PLANTS

All houseplant diseases are caused by one of four types of microscopic organisms: fungi, bacteria, viruses, or mycoplasmas.

Fungi

Fungi (sometimes called molds) are technically plants, but they do not contain chlorophyll and they are incapable of manufacturing their own carbohydrates. They depend instead on absorbing it from dead or living organisms.

The fungi which cause diseases in plants are parasitic, invading the individual cells of plants with small projections called *hyphae* and extracting nutrition (Figure 18-8). The cells die, and the progression of the disease is begun.

Fungi which attack houseplants usually prey on the roots; foliar fungi-caused diseases are less common. The former are called "soilborne" because plant roots are exposed to the organisms through fungi-contaminated media. Treatment is through modifying environmental conditions (moisture in particular) and chemical fungicides.

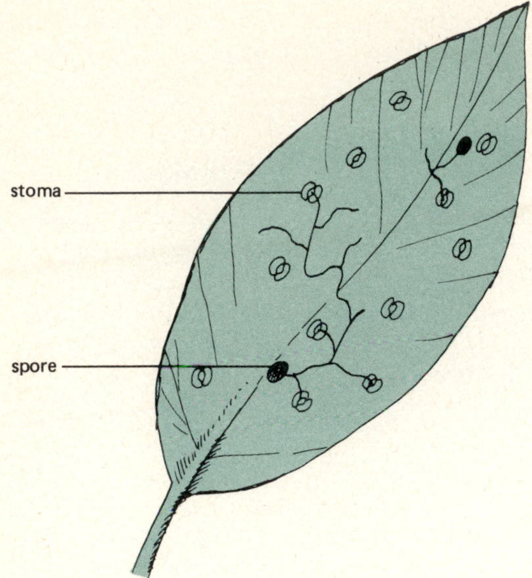

Figure 18-8 Fungus spores germinating and invading a leaf through stomata.

Bacteria

Bacteria are single-celled plants also incapable of manufacturing their food. Some use living plant cells for nourishment and will reproduce rapidly through division under proper environmental conditions. A few bacteria attack indoor plants, causing such disorders as leaf spots and root rots. They can be partially controlled by chemicals.

Viruses and Mycoplasmas

Submicroscopic viruses and mycoplasmas are rarely encountered in houseplants. When they are, it can be assumed that they were present in the plant at the time it was purchased.

Viruses and mycoplasmas are often transmitted between plants through the mouthparts of contaminated insects. Therefore contaminated tools such as pruning tools can also pass the organisms from plant to plant.

SPECIFIC DISEASES OF HOUSEPLANTS

Root Rot

This general term is used for both fungal and bacterial diseases which cause the decay of root tissues. Root rot is usually associated with poor growing conditions such as a heavy medium that excludes air or lacks an outlet for excess drainage water. Initial symptoms are a slowdown in growth and a yellowing of foliage, signs of the inability of the roots to take up water and nutrients. In extreme cases, permanent wilting occurs, and the plants die. An inspection of rot-infected roots will show them to be brown and mushy.

Treatment consists of chemical soil drenches combined with improvements in culture. Repotting in a better-drained medium and a pot bearing a drain hole and less frequent watering are recommended.

Crown or Stem Rot

This disease attacks the stem or crown of a plant at the point where it enters the medium. Numerous bacteria and fungi found in unsterilized potting media often cause the disease, but it is also possible that the organisms may be airborne.

The main symptom of crown rot is brown discoloration of the stem (Figure 18-9). The rotting prevents water passage up through the plant, and it can wilt and break off at the weakened area.

Cacti show a form of stem rot called *dry rot* (Figure 18-10). This disease causes the stem base to discolor to yellow or brown and become sunken. The end effect on the plant is the same as with stem and crown rot.

Once infected, there is little that can be done to cure a plant with this disease. The recommended procedure is to take cuttings of the uninfected portions and root them in sterile medium for new plants. The procedure is the same for cacti; that is, the entire top of the cactus is cut off above the rot area and rerooted.

Mildew

Mildew (Figure 18-11) is a foliar fungus disease which attacks a limited number of susceptible houseplants (see Table 18-2). The first symptom is a gray powdery residue on the leaves. Later the leaves yellow and drop, and the plant eventually dies.

Mildew is controlled by reducing the humidity and applying a fungicide spray. After treatment, keeping the humidity low and all water off the foliage will prevent further problems.

Figure 18-9 Crown rot on a nerve plant (*Fittonia argyroneura*).

Figure 18-10 Dry rot disease at the base of a column cactus.

Figure 18-11 The fungus disease mildew.

Bacterial and Fungal Leaf Spots

A number of organisms occasionally cause leaf spot diseases on foliage plants grown indoors. The symptoms vary with the disease and host, but in general, small pinpoint-sized spots will appear, enlarging and speading slowly from leaf to leaf. The spots will usually show a pattern such as a center black spot surrounded by yellow or an irregular yellow spot ringed in brown.

Figure 18-12 shows a typical fungal leaf spot.

Fungicidal sprays may prevent spread, but often there is no cure. Removing infected leaves as soon as they are detected will slow or stop the spread, and a plant can sometimes be saved in this manner.

Virus and Mycoplasma Infections

The symptoms of a virus infection range widely, depending on the organism and its host plant, but often include yellowing, curled foliage, and stunting.

Although viruses are unlikely to spread between different species of houseplants, no cures are known and diseased plants should be discarded.

PEST AND DISEASE PREVENTION

Prevention is better than any cure that can be administered. It involves avoidance of both

Table 18-2 Diseases of Houseplants and Their Common Hosts

Disease	Common hosts
Leaf spot	Anthurium (*Anthurium* spp.) Arrowhead (*Syngonium podophyllum* varieties) Begonia (*Begonia* spp.) Chinese evergreen (*Aglaonema* spp.) Croton (*Codiaeum variegatum pictum* varieties) Dumbcame (*Dieffenbachia* spp.) Fern German ivy (*Senecio macroglossus*) Gardenia (*Gardenia jasminoides*) Geranium (*Pelargonium* × *hortorum* varieties) Impatiens (*Impatiens* spp.) Jerusalem cherry (*Solanum pseudocapsicum*) Orchids Philodendron (*Philodendron* spp.) Rubber plant (*Ficus elastica*) Snake plant (*Sansevieria trifasciata* varieties)
Mildew	African violet (*Saintpaulia ionantha* varieties) Begonia (*Begonia* spp.) Kalanchoe (*Kalanchoe blossfeldiana* varieties)
Root rot	Cacti Succulents Many foliage plants
Stem and crown rot	African violet (*Saintpaulia ionantha* varieties) Begonia (*Begonia* spp.) Cacti Dumbcame (*Dieffenbachia* spp.) Geranium (*Pelargonium* × *hortorum*) Gloxinia (*Sinningia speciosa*) Orchid Succulents
Virus	Begonia (*Begonia* spp.) Geranium (*Pelargonium* × *hortorum*) Orchid Umbrella plant (*Brassaia actinophylla*) Dumbcane (*Dieffenbachia* spp.) Philodendron (*Philodendron* spp.)

the initial contact with the insect or disease and its spread.

Plants that are healthy and under good growing conditions have been proved to be less susceptible to attacks by insects and diseases than those under "stress." Stress conditions include any factors which slow the normal growth and development of the plant such as poorly drained medium, overly warm temperature, insufficient light, and the like. In addition, plants in a weakened state are less able to survive such disorders.

Figure 18-12 A fungal leaf spot disease on German ivy (*Senecio mikanioides*).

Potting Medium

All potting media used for houseplants should be sterilized prior to use. [8] Unless this is done, fungi, bacteria, and soil pests can be introduced and spread rapidly.

Second, the medium should be composed of ingredients in the proper proportions. This will assure that it drains quickly and thoroughly when watered. A drainage hole or other means for the escape of excess water should be provided in the pot. Media which retain an excessive volume of water or from which water is prevented from escaping have conditions leading to poor root and healthy microorganism growth. Waterlogged media can frequently be traced as the initial cause of root rot diseases.

Finally, a medium should contain adequate nutrients. Poorly nourished plants, like those suffering from any other kind of stress, are commonly attacked by pests and diseases.

Humidity

The high humidities optimal for the growth of most indoor tropicals can be both desirable and undesirable from the standpoint of pest and disease control. Perhaps the worst pests of houseplants, spider mites, are adversely affected by high humidity. Raising the humidity and misting can be used to slow their breeding and control their damage. But high humidity can cause a foliar fungus problem because the spores of the fungus require high humidity to germinate. Overall, unless a plant is commonly susceptible to foliar fungus (see Table 18-2), high humidity is advisable.

Water

Stagnant water, whether in the medium or in the crown, is detrimental to plant health. Crown rot disease, which attacks such houseplants as the African violet, has often been attributed to water left standing in the center of the plant after watering. Barrel-shaped cacti, in which the growing point is in a depressed area on top of the plant, are also susceptible to rot if water is left standing there.

In summation, although misting foliage and showering are recommended on most plants, they should be avoided with a few disease-susceptible species. These plants should be bottom watered or watered with a spouted watering can that will wet the medium and not the plant.

Isolation and Inspection

Isolation of newly acquired plants and older ones showing symptoms is a standard practice for controlling both diseases and insects. Isolation outdoors or in a separate room is preferred since air currents carry disease spores and insects readily.

Weekly inspection, mentioned in Chap-

[8] See Chapter 15 for directions on sterilizing potting media.

ter 14, will help catch disease problems before they become epidemic. Showering (except for disease-susceptible species) is also helpful in washing off insect pests before they become established.

PEST AND DISEASE CONTROL METHODS

Modification of the environment, home remedies, and chemicals all have a use in controlling diseases and pests of houseplants. Modification of the environment for prevention has already been discussed, and the same instructions can be applied to plants which are already infested.

Home Remedies

Home remedies are generally used for pests and not diseases, but used with persistence, they can be effective. In addition, they are completely safe to humans and avoid any danger of accidental pesticide poisoning.

Wiping or washing with soapy water is a home remedy useful for reducing populations of mites or aphids. The pests can be sponged off with a dilute solution, or the entire plant can be held upside down and swished in the soapy water. Soap will not harm the plant, and the agitation will dislodge most of the pests. As a precaution against toxicity and to avoid soap film, rinsing of plants afterward is advisable.

Hand picking of large insects like sowbugs, scales, and mealybugs is another home remedy that works. Toothpicks can be used or cotton swabs dipped in rubbing alcohol and touched to the insects (Figure 18-13). These remedies are most effective if only a few pests are present.

The key to success with both these methods lies in repetition. Probably not all of the pests will be killed, but the population will be reduced to a low level for several weeks.

Figure 18-13 Touching mealybugs or aphids with a cotton swab dipped in alcohol is a reliable control method for small infestations.

Few home remedies are useful for soil insects. An atomizer filled with alcohol will kill many surface pests on contact, but the alcohol should not touch the plant.

Commercial Pesticides

These include the plant-derived "organic" pesticides pyrethrum, nicotine, and rotenone as well as more widely used formulations such as malathion (for insects) and copper oleate (for fungi). All of these chemicals can be dangerous to health (people's, pets', and plants') if not used in accordance with package directions and in observance of safety rules. Plant-derived insecticides should not be considered safer than synthesized ones.

Because of the constantly changing laws regarding pesticide use and the many new chemicals developed each year, it is useless to attempt to prescribe a chemical cure for every houseplant ailment discussed here.

Instead, chemicals currently approved by the Environmental Protection Agency to control diseases will be covered, along with their relative toxicity to humans. The label of the pesticide and recommendations made by government specialists in pest control in the form of Cooperative Extension Service or Department of Agriculture publications should *always* serve as the basis for the application of any pesticide.

Pesticides fall into broad categories depending, first, on what organisms they are designed to control and, second, on the way in which the control is achieved.

The term "pesticide" is often used to refer only to "bug killers." But it actually includes all chemicals used to kill organisms which adversely affect the growth of plants including fungicides, bactericides, insecticides, miticides, nematicides, and herbicides. Insecticides are then further broken down into "contact poisons," which kill when they touch the pests, and "stomach poisons," which must be eaten to kill.

The methods of applying pesticides vary. Sprays are most common for foliage insects, mites, and diseases. But drenches, in which the chemical is mixed in water and applied in place of plain water, are used for soil insects, nematodes, and root diseases.

Systemics are relatively new chemicals applied as sprays, drenches, or soil granules. These chemicals are taken up into the vascular system of the plant and poison insects and disease organisms which prey on that plant. Because a plant to which a systemic has been applied is toxic for a prolonged period, systemics are not often available for home use. Table 18-3 lists some of the pesticides used on houseplants and the pests they control.

SAFETY PRECAUTIONS FOR USING PESTICIDES

Following general safety precautions for pesticide use is the first prerequisite for effectively controlling pests with harm to neither the gardener nor the plant being treated.

Whenever possible, the pesticide with the lowest toxicity to humans should be chosen. Pesticide toxicity is rated by an LD_{50} number, which denotes the number of grams of pesticide per kilogram of body weight that was fatal to 50 percent of a sample of laboratory animals. The smaller the number, the more toxic the pesticide. LD_{50} values of common pesticides are given in Chapter 13.

Personal Safety

For personal safety, all contact with the pesticide should be avoided. Rubber gloves should be used and any skin spills should be washed off immediately with soap and water.

Pesticide sprays, whether aerosol or hand pump type, should be applied outdoors. During cold weather an unoccupied room that can be aired out afterward can be used. The plant to be treated should be placed on newspapers to prevent accidental spraying of carpets or furniture. Particular care should be taken when applying aerosol pesticides. They disperse small droplets into the air, and the airborne pesticide can be inadvertently inhaled.

After spraying is completed, safety precautions should include washing the spray bottle, measuring equipment, and hands with soap. The plant should be placed where it will not be touched by people or animals for several weeks.

Plant Safety and Pesticide Effectiveness

The first rule for plant safety is to follow the label directions. A stronger concentration than recommended may burn foliage severely, whereas a weaker one may fail to kill the pests. Many species of houseplants, including ivies, ferns, and succulents, are easily injured by pesticides; these susceptible plants should be listed on the pesticide label. Test spraying several leaves is always advisable when treating a new species. The plant should be observed for several days for signs of reaction to the pesticide before the remainder of the foliage is treated.

In the application of water-diluted spray pesticides, "spray to runoff" is usually advised. This means that the chemical should be applied heavily enough to drip from the leaves. Coverage should include all aboveground plant parts and the undersides of leaves, even those not presently infested. Aerosol pesticides can present dual hazards to plant health. The propellant used to carry the pesticide may be toxic, and the heat pulled from the leaves when the propellant vaporizes may cause cold injury. To prevent the latter, spray no closer than the directions advise. At a moderate spray distance most of the propellant will vaporize before touching the leaf.

When applying soil drenches, the chemical should be mixed according to directions given for a drench. The pesticide should be

Table 18-3 Selected Houseplant Pesticides[a]

Chemical name	Uses
Rotenone	Plant-derived contact and stomach poison for insects; used for aphids, thrips, whiteflies, mites, and immature scales and mealybugs
Resmethrin	Synthetic pesticide similar chemically to rotenone; used for the same pests
Pyrethrum	Plant-derived insecticide with contact action; effective against the same pests as rotenone
Copper oleate and other copper compounds	Fungicide for powdery mildew; registered for use on fuchsia, ferns, ivy, gardenias, philodendrons, geraniums, begonias, and succulents
Malathion	Insecticide for aphids, red spider and other mites, mealybugs, thrips, whiteflies, and root mealybugs; not recommended for use on crassulas, *Pteris* ferns, *Adiantum* ferns, anthuriums, or kalanchoes
Kelthane	For red spider and other mites; registered for use on gardenias, rubber plants, bracken ferns, Boston ferns, ivy, African violets, and philodendrons
Nicotine sulfate	For aphids, thrips, spider mites, orchid scales, and fungus gnats; registered on philodendrons, fuchsia, ferns, and African violets
Di-Syston	Used for aphids, thrips, mites, and whiteflies; granular systemic applied to the medium and watered in
Orthene	For aphids, thrips, leaf miners, mealybugs, scales, spider mites, and whiteflies; use with caution on gloxinia, hibiscus, and philodendron
Karathane	Miticide and fungicide for powdery mildew, spider mites, and two-spotted mites; registered for use on begonias and cineraria

[a] All pesticides must, by law, be approved and registered with the Environmental Protection Agency for use on any plant species. The above table is not intended as authorized recommendations for pesticide use since the registrations are constantly changing.

applied in sufficient quantity to wet all the potting medium, and any that drains out should be discarded.

Repeat sprays as recommended on package directions are the key to successfully controlling most houseplant pests. Many generations and life stages, from egg to adult, occur at one time, and not all will be chemically susceptible. Repeat sprayings are necessary to do more than temporarily slow reproduction rates.

Resistance of insects and mites to specific pesticides is occasionally encountered in houseplants. Generally all but a few pests will be killed by a chemical, but those remaining continue to breed and eventually build a "resistant population" that cannot be killed by the insecticide. Resistance can be overcome in part by spraying alternately with different pesticides.

Finally, never save leftover mixed insecticides. Mixed pesticides are often inactivated in storage and are a common cause of accidental poisonings.

Selected References for Additional Reading

Pirone, P. P. *Diseases and Pests of Ornamental Plants.* New York: Rodale Press, 1970.

Wellman, F. L. *Tropical American Plant Diseases.* Metuchen. N J: The Scarecrow Press, 1972.

Westcott, C. *The Gardeners Bug Book.* 4th ed., New York: Doubleday, 1973.

Chapter 19

Interior Landscaping

Plants are beautiful creations in themselves. But as with other works of beauty, a little artistic treatment will enhance and focus attention on their most attractive features. A new field of "interior landscaping" has evolved to deal with the selection and arrangement of indoor plants. An interior planting specialist must have not only a knowledge of artistic design but also an understanding of the care requirements associated with particular plants. This person must be an artist *and* a gardener, arranging plants in a manner that is pleasing visually while satisfying their growth requirements.

Most experienced houseplant growers have already tried their hands at interior landscaping, although they may not realize it. For example, when the hostess moves an attractive plant to the coffee table before company comes, she is practicing interior landscaping. Similarly, selecting attractive pots for houseplants is part of the design process. With a bit of artistic sense, a knowledge of plant growth requirements, and sample ideas given in this chapter, anyone can create an interior landscape.

AN HISTORICAL PERSPECTIVE

Using plants for indoor decoration is not a new idea. Indoor plants were much in vogue during the 1800s, and plant explorers undertook lengthy expeditions to secure exotics for wealthy patrons. Unfortunately, the architecture of the period and the lack of central heating made it impossible to grow all but the hardiest species indoors. Cast iron plants, potted palms, and rubber plants were the standard indoor species. Boston ferns also adapted to the low light and cool temperatures and were proudly displayed on ornate fern stands.

Partly because of the inadequate indoor growing conditions, the terrarium and solarium gained favor as plant-growing areas. Terrariums were originally designed as a method of holding plants during sea voyages. From this basically utilitarian structure, they developed into elaborate "parlor cases," highly embellished in the style of the Victorian era. It was not unusual for a terrarium of that period to be made of hand-blown glass, with its supporting structure of filigreed iron or brass or enameled metal.

The solariums or conservatories of that period were the garden rooms of today. They were used to grow and display many exotics, a hobby of the rich during that period.

After World War I, middle-class houses were built with greater window areas and, frequently, with sun porches in the front. By this time more low-light-requiring houseplants had been discovered, and central heating was becoming common. This facilitated indoor plant growing and removed it from being the exclusive province of the rich. But the original plant fervor of the 1800s never returned, and there was a slump in interest until the late 1960s. Architecture reflected this lack of enthusiasm for indoor plants. Ranch-style houses became popular. Although many had large picture windows, these were designed for viewing and were seldom used for plants. The remaining windows were small or, at most, medium-sized, with the emphasis on sleek, horizontal lines.

Currently the pendulum has swung back, and indoor plants are very much in demand for homes, offices, and shopping malls. Houses are being designed with large expanses of glass that provide enough light for growing almost any houseplant. Indoor gardeners are no longer restricted to pots on a windowsill, and large interior plants are flourishing.

From demand comes supply, and the horticulture industry has risen to the occasion. Large specimens of many species are sold today that 10 years ago would have been impossible to find. Species which were previously obscure are now common, and innumerable new varieties have been and are being developed.

All of this presents today's indoor gardeners with a wide variety of plants and accessories from which to choose. Their only problem will be determining how to fit all their favorite plants into their limited living area attractively—the subject of this chapter.

DECORATIVE AND FUNCTIONAL USES OF INDOOR PLANTS

The selection and placement of plants in a room require careful planning to achieve the right effect. Sometimes plants are used as art and enjoyed purely for their aesthetic value. At other times they can be functional in addition to being decorative. Discussion and examples of the decorative and functional uses of indoor plants are given in the sections that follow.

Figure 19-1 A bamboo palm (*Chaedorea erumpens*) used in an office.

Decorative Uses of Plants

SPECIMEN PLANTS

Specimen plants (Figure 19-1) are those displayed purely for their aesthetic appeal. They are often large trees or floor plants, although they may also be hanging baskets, small potted plants, terrariums, or bonsai. Specimen plants are generally used alone and can be thought of as living sculpture. In some instances a specimen plant will become the focal point of the room, attracting more attention than any of the other furnishings. If the plant is large, it can also be used to fill a corner, fitting in a space too narrow for most furniture.

Specimen plants are usually expensive because of their large size and should be considered a major plant purchase. Before you invest in one, try to grow a small plant of the species you intend to buy so that there will be less chance of your losing the plant from improper care. Table 19-1 lists a number of plants available in large sizes which are suitable as indoor specimens.

WALL DESIGNS WITH INDOOR PLANTS

A bright wall near a window can be an ideal spot for a wall grouping of plants (Figure 19-2). The grouping takes the place of a picture or other wall decoration.

Pots can be attached to the wall with the type of clips used to hang plants on fences in outdoor gardens, and a structured design or random pattern can be used. Interest can be added by using unusual potting containers such as antique tins and handmade pottery, although this will make hanging more difficult.

Among the plants which lend themselves more readily to wall grouping are epiphytes. Certain orchids, bromeliads, philodendrons, and ferns can be grown mounted

Table 19-1 Selected Large-Specimen Plants for Indoor Culture

Common name	Botanical name
Norfolk island pine	*Araucaria heterophylla*
Bamboo	*Bambusa* spp.
Umbrella plant	*Brassaia actinophylla*
Cacti	*Cereus, Echinocactus, Ferocactus*
Palms	*Chamaedorea, Howea, Rhapis*
Jade plant	*Crassula argentea*
Cyperus	*Cyperus alternifolius*
Dumbcane	*Dieffenbachia maculata*
Thread-leaf aralia	*Dizygotheca elegantissima*
Corn plant	*Dracaena fragrans* 'Massangeana'
Red-margin dracaena	*Dracaena marginata*
Euphorbia	*Euphorbia millii, Euphorbia lactea, Euphorbia trigona*
Fatsia	*Fatsia japonica*
Weeping fig	*Ficus benjamina*
Rubber tree	*Ficus elastica*
Fiddle-leaf fig	*Ficus lyrata*
Wax-leaf privet	*Ligustrum lucidum*
Avocado	*Persea americana*
Philodendron	*Philodendron, Monstera*
Yew pine	*Podocarpus macrophyllus* or *Podocarpus gracilior*
Ming tree	*Polyscias* spp.

Figure 19-2 A wall arrangement of plants.

on osmunda or bark plaques. These are lightweight and easily hung.

PLANT AREAS

Grouping plants together in a plant area (Figure 19-3) is another striking display method, particularly when combined with spotlights for evening viewing. Houses or apartments with sliding glass doors or floor-to-ceiling windows are particularly adaptable to this idea. First, an area of roughly 2 to 10 square yards (meters) of floor space is selected. The plants are then arranged in random fashion within the area, with taller specimens in the back and shorter ones to the front. Plant stands, sections of logs, weathered crates, and ornamental concrete blocks can all be used to create height differences between plants and move the viewer's eye through the display. Hanging baskets can also be incorporated and add to the variety of form and texture.

In its simplest style, a plant area is situated directly on the floor, and plant saucers are used to keep moisture from the carpet. But an even more attractive display can be created by laying down a sheet of heavy plastic and covering it with a single layer of used bricks or bark chunks. This defines the area and attracts attention. Manufactured plant areas are also available. One style consists of shallow trays filled with chips of stone. The trays can be arranged in any shape and then filled with water. The water increases the humidity around the plant foliage, creating a moist microclimate (Figure 19-4).

Functional Uses of Plants

Few people think about the functional uses of indoor plants. Most are content to use them in a strictly decorative sense and miss the tremendous potential of plants for such uses as room dividers, traffic channelers, and screens. Their humidification, acoustic, and glare-reduction properties are likewise taken for granted. The following examples are just a few of the functional uses of plants in the home.

PLANTS AS ROOM DIVIDERS

Plants are ideal as a casual means of dividing a room into sections. A bushy floor specimen

Figure 19-3 A plant area.

Figure 19-4 Interlocking trays provide a display area for plants, keep the potting medium off the floor, and raise the humidity.

Figure 19-5 An island planter used to separate dining and living areas.

can create a study area in a living room, for example.

In a similar manner, a carefully placed indoor tree can give the feeling of an entryway and create a graceful transition area. In a living/dining area a large, island-type plant area can separate the two sections but still leave a feeling of airiness and spaciousness (Figure 19-5).

PLANTS AS TRAFFIC CHANNELERS

People generally walk the shortest distance from room to room by the route that offers the least resistance. Once these traffic patterns are established, they can be difficult to change, but strategically placed indoor plants can be helpful in this regard. By placing several plants between a chair and sofa, for example, people will naturally be directed to enter the sitting area by another way.

Plants can also be used to channel traffic by means of their attracting rather than obstructing qualities. For example, to direct traffic through a doorway, place plants on one or both sides of the entrance (Figure 19-6). This highlights the doorway and creates "accent plants." The attention-focusing quality of accent plants can be used for other purposes as well. The plants can direct attention to a particular piece of furniture, for example, or call attention to a window view (Figure 19-7).

PLANTS AS SCREENING

Screening is a third functional use of plants. A stragetically placed group of plants can form a visual barrier between kitchen and dining areas or mask the entrance to a bedroom or bath.

If the screening problem is an objectionable window view, plants are the ideal solution. They thrive in the light, yet will not exclude the sunlight as a curtain would. Placing plants on glass shelves (FIgure 19-8) is one way to screen a poor view, and it is suitable for small windows. The glass shelves give maximum light to plants at all levels and permit viewing from all angles.

Figure 19-6 Two large plants used to accent the entrance to the living area.

Figure 19-7 A geranium is used here as a coffee table centerpiece. Palms, a dracaena, and a fig frame the window view.

Figure 19-8 Putting small plants in a window on glass shelves gives them maximum light and decorates the window as well.

One or more hanging plants can also be used for this purpose. They have the advantage of not interfering with dusting and cleaning, and there is no concern about damaging furniture with damp pots, drainage water, or scratches from plant saucers. Not to mention the fact that hanging plants are out of the reach of children and pets.

Not all plants have a growth habit that looks attractive in a hanging display; in fact most plants probably are best displayed in other ways. It is important to consider that a hanging plant is generally viewed from the side and below, so it should be attractive from those perspectives. A cascading or vining growth habit is therefore a prerequisite for a hanging plant. Table 19-2 lists indoor plants suited to hanging culture.

The placement of a hanging plant relative to the window will largely affect its success or failure as a houseplant. The most common mistake is to hang plants too high. Unless artificial light is provided from above, little light will reach the top and an unattractive plant with bare stem bases will result. Ideally, the top of a hanging basket should be no closer than 2 feet (0.6 meter) from the top of the window, and preferably even lower.

Ceiling track hangers (Figure 19-9) represent a considerable investment in money, but they are an especially effective display technique when used in front of a large window. Since they usually hold over a dozen plants, only a gardener who can fill the track with that many healthy specimens should consider this hanging device.

Cord and macrame hangers are popular for suspending a single specimen or a group of hanging plants. By using hangers of varying lengths, hanging heights can be staggered for maximum light availability and screening effect.

PLANT FURNITURE

Furniture specifically designed for growing plants is another way of growing plants indoors.

Clear plastic tables (Figure 19-10) function as terrariums and are available through furniture stores and plastics manufacturers. They provide the high humidity needed by many ferns and orchids and can be a handsome addition to a modern decor.

For the traditional decor, furniture in a more conservative style can be purchased (Figure 19-11). Recessed artificial lighting included in these pieces makes it possible to grow both foliage and flowering plants without natural light.

Hanging baskets also come equipped with artificial lights (Figure 19-12). They can be used for supplemental illumination in low-light areas or strictly for decorative night lighting.

Plant stands popular during the Victorian era are in style again, made from such materials as wood, wicker, and wrought iron. These pedestal-type stands can greatly enhance plants and give an illusion of height.

Table 19-2 Selected Plants Suitable as Hanging Baskets

Common name	Botanical name
Lipstick vine	*Aeschynanthus* spp.
Rattail cactus	*Aporocactus flagelliformis*
Asparagus fern	*Asparagus* spp.
Begonia	*Begonia* spp.
Rosary vine	*Ceropegia woodii*
Spider plant	*Chlorophytum* spp.
Coleus	*Coleus × hybridus*
Cissus vine	*Cissus* spp.
Goldfish vine	*Columnea* spp.
Episcia	*Episcia* spp.
Creeping fig	*Ficus pumila* varieties
Bridal veil	*Gibasis geniculata*
Purple passion vine	*Gynura aurantiaca* 'Purple Passion'
English ivy	*Hedera helix* varieties
Wax plant	*Hoya carnosa* varieties
Pocketbook plant	*Nematanthus* spp.
Fern	*Nephrolepsis*, *Adiantum*, *Davallia*, and other genera
Arrowhead	*Syngonium podophyllum varieties*
Trailing peperomia	*Peperomia* spp.
Heart-leaf philodendron	*Philodendron scandens oxycardium*
Swedish ivy	*Plectranthus australis* and *oertendahlii*
Easter cacti	*Rhipsalidopsis gaertneri* and *rosea*
Strawberry begonia	*Saxifraga sarmentosa*
Devil's ivy	*Epipremnum aureum* varieties
Trailing sedum	*Sedum morganianum × rubrotinctum* and other species
German and wax ivies	*Senecio macroglossus* varieties
Wandering Jew	*Tradescantia* and *Zebrina* spp.
Christmas cactus	*Zygocactus truncatus*

Figure 19-9 A track-type plant hanger for plant display.

Figure 19-10 A Lucite coffee table used as a terrarium.

Instructions for homemade plant stands, utilizing materials such as plywood and sections of clay drainage pipe, are also common in home decorating magazines.

GUIDELINES FOR INTERIOR LANDSCAPING

Interior landscaping is a relatively new field, and there are few absolute design principles to which you must adhere. However, the work of professional interior landscapers makes it possible to give several general guidelines for decorating with plants.

Plant Placement

First, whenever possible, arrange room furniture and plants at the same time instead of placing the furniture and then fitting in the plants. Furniture does not have a light requirement, and if practical, the brightest areas of the room should be left for the plants. Not only will this aid plant growth but it will also minimize bleaching of furnishings from sunlight.

Plant needs should also be considered when buying window coverings. A lightweight curtain to filter midday sunlight might be all that is needed in some rooms. In others, drapes might be rejected entirely in favor of shades which can be rolled up to admit maximum light.

Figure 19-11 An early American étagère equipped with fluorescent lights and humidity trays.

Figure 19-12 A hanging planter equipped with a light.

Take note also of the ease of care when placing plants in the room. Plants in hard-to-reach areas will tend to receive less care than those more easily accessible. For a heavy floor plant, a platform with casters will make moving easy and facilitate cleaning at the same time.

Accessory Selection

Plant accessories (pots, hangers, stands) should be chosen carefully to enhance the plant's beauty. First, the size of the container should be in proportion to the size of the plant to avoid a top-heavy or bottom-heavy look. Double potting frequently creates a bottom-heavy appearance and emphasizes the need for proportion.

Second, select containers and accessories in muted colors and patterns. Simple unglazed clay pots or wicker pot covers are two of the most suitable. They blend with almost any decor and are classic in their appeal.

A bold container will visually overpower a plant, making the pot rather than the plant the focus of attention. However, this is a general guideline and exceptions exist. For example, a room in a Mexican motif might work nicely with bright red or yellow pots. A large specimen might look outstanding in a very elaborate antique container.

Use of Night Lighting

The decorative effects of lighting are well understood by interior designers. Night lighting, for example, is very striking when used with plants. One or two inexpensive spotlights can dramatize houseplants and provide supplemental lighting as well.

The lights can be positioned in many ways for accenting plants. Lights shining up from the floor create interesting ceiling shadows as they pass through plant foliage. Track lights aimed downward from several points on the ceiling can highlight individual plants within a collection.

Impact through Massing

Because many houseplants are small, they tend to go unnoticed unless they are outstanding in some way. Massing plants in groups has thus become a standard design practice used by interior landscapers. Combining plants of varying sizes, forms, and leaf sizes (textures) attracts attention and makes maintenance easier as well. A defined "plant area" is one way of achieving this effect, but just three or four plants can be placed together informally for a similar look.

Selected References for Additional Reading

Elbert, G., and V. Elbert. *The House Plant Decorating Book.* New York: E. P. Dutton, 1977.

Gaines, R. L. *Interior Plantscaping: Building Design for Interior Foliage Plants.* New York: Architectural Record Books, 1977.

Hawkey, W. S. *Living with Plants, A Book of Home Decorating and Plant Care.* New York: William Morrow, 1974.

Helmer, M., and J. Coleman. *Hanging Plants for Modern Living.* Kalamazoo, MI: Merchants, 1975.

Kramer, J. *Indoor Trees.* New York: Hawthorn Books, 1975.

Wallach, C. *Interior Decorating with Plants.* New York: Macmillan, 1976.

Chapter 20

Greenhouses and Related Climate-controlling Structures

A home greenhouse is regarded as the ultimate luxury by many gardeners, conjuring up visions of exotic or high-light-requiring plants which cannot be successfully grown indoors. But then the realities of high construction and heating costs are remembered. With fuel costs rising each year, a greenhouse can be a costly hobby or even a frivolous waste of energy. However, it is possible to have a greenhouse or related structure that is both practical and affordable if the selection is made wisely. The key is to balance the advantages and disadvantages of each of the following structures with regard to range of usefulness, initial construction, labor, materials cost, and heating and cooling expenses. The gardener can then select the structure appropriate for his climate, budget, and hobby interest.

OUTDOOR PLANT-GROWING STRUCTURES

Cold frames, hotbeds, sun-heated pit greenhouses, traditional greenhouses, and shade houses are included in this category. With the exception of shade houses, each structure provides some heat, whether by the sun, decomposing manure, oil, gas, electricity, or a combination.

Cold Frames

A cold frame (Figure 20-1) could be called the "poor man's greenhouse" because it is the

Figure 20-1 A peaked cold frame with the plastic covering rolled up.

least expensive plant-growing structure. Once a common feature of almost every gardener's yard, it is less used now, having been succeeded by more sophisticated structures. But because it is solar heated, it is highly energy efficient, so it may become popular again.

Sunlight heats a cold frame by a phenomenon called the *greenhouse effect,* which occurs in any translucent or transparent structure. Sun passes through the covering, falls on plant leaves or other surfaces, and is changed into infrared (heat) energy. This heat will then be radiated or conducted back into the air and is held in by the covering.

USES OF COLD FRAMES

Cold frames have the most limited usability but are practical for some plants. They can be used to harden off greenhouse-grown transplants, giving cold and wind protection during the transition period and reducing shock after transplanting. They can also be used for raising transplants of cool-season vegetables such as cabbage, cauliflower, and lettuce.

In areas with cool, short, or foggy summers, melons, cucumbers, and sweet potatoes can be grown to maturity in the frame and will produce exceptional yields because of the

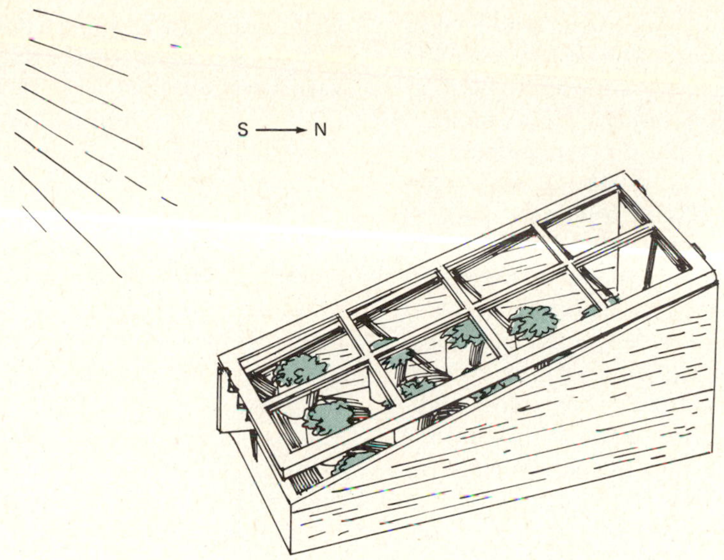

Figure 20-2 A cold frame made from a used window sash. The sash must face south.

extra heat. In fall the growing season can be extended for short-term crops of lettuce, radishes, or beets, and they can be harvested into early winter.

In cold climates seedlings of cool-season flowers like the pansy can be grown to transplant size before being moved into a greenhouse. Hardwood cuttings can also be rooted and overwintered. A cold frame is excellent for cooling bulbs to be forced into bloom in winter and for protecting semi-cold-hardy plants.

COLD FRAME CONSTRUCTION

A cold frame is relatively simple to construct because it does not require electricity or a foundation. Some resemble miniature greenhouses (Figure 20-1), whereas others have only one area of glass or plastic facing south or southeast (Figure 20-2).

The site selected for the cold frame should be shielded from winds, unblocked by shade from trees or buildings, and well drained. The size decided on will depend on the types of plants which will be raised. For a few transplants, cuttings, or potted bulbs, a 3 × 6-foot (900 × 1800-millimeter) frame is adequate and can be readily covered with a 3 × 6-foot (900 × 1800-millimeter) window sash. If film plastic covering is used, size can be flexible because the plastic can be cut to fit. The maximum span of a peaked plastic frame should not, however, be over 7 feet (2.3 meters) and its length 12 feet (3.66 meters) (without additional support) so that water will not collect and cause it to sag. The height of the frame at the tallest point can be 18 inches (45.7 centimeters) to several feet (1 meter) depending on the size of the plants being raised. But its lowest wall should never be less than 12 inches (30 centimeters), as a minimum height for transplants.

The base of a cold frame can be constructed of wood [using 1 × 12-inch (38 × 286-millimeter) boards], block, or concrete. And 2 by 4-inch (38 × 90-millimeter) rafters are sufficiently strong to support either glass or plastic covering.

Because air leakage causes heat loss, the

frame should be tightly constructed and banked with soil for insulation. The covering should fit snugly, with strips of foam padding used as a seal around window sashes.

Beyond the stipulations stated there are no building absolutes, and the gardener can generally build a frame by looking at Figures 20-1 and 20-2. However, books on greenhouse construction or Cooperative Extension Service bulletins will have materials lists and building plans, and provide specific directions.

The term "glazing" refers to the clear or translucent materials used to cover cold frames, greenhouses, and similar structures. It can include a variety of coverings such as glass, fiberglass, or plastic films and also refers to the application of the covering.

Glazing used for cold frames can utilize any of the common commercial materials, but window sashes and plastic are frequently used. Sashes are easier to install (being simply laid over the frame) and permanent, but they are expensive if purchased new. However, salvage yards frequently carry used sashes with broken panes of glass, which are easily replaced. Plastics are inexpensive but are more work to install. Depending on the thickness, plastic or vinyl film may last 1 to 10 years before requiring replacement.

COLD FRAME OPERATION

Temperature is controlled in a cold frame by raising and lowering the glazing to ventilate with cooler outside air. Sash glazings are propped open and plastic coverings propped or rolled up.

Frequency of ventilation will depend on weather. In late winter the days may be relatively cold, and the cold frame should remain closed all day. In spring, sunlight and warmer air temperature will heat the inside of the cold frame rapidly, and daytime ventilation will be necessary. Although many gardeners "guesstimate" when the glazing should be raised, a thermometer in a shaded area of the frame is a better indicator. Anytime the temperature rises above 70°F (21.1°C) (for cool-season flowers and vegetables), the frame should be ventilated. Eighty-five degrees Fahrenheit (29.4°C) is optimum for warm-season plants.

In general the hours of ventilation on sunny spring days are from midmorning (around 10 o'clock) until late afternoon (about 3:30 to 4 o'clock). Closing the frame several hours before sundown allows heat to build up inside to warm the plants through the night. Delaying venting until 10 A.M. makes the bed warm quickly after the cool night.

The question of how low the outside night temperature can drop without injuring frost-tender plants in the frame is not easily answered. In a relatively airtight frame, accumulated heat will normally allow plants to survive down to 25°F (−3.9°C). When temperatures are predicted to fall lower, additional insulation such as burlap, canvas, or carpet should be placed over the glazing. If frost-tolerant plants are being raised, the night temperature can drop as low as 20°F (−6.7°C) before insulation is required. (Banking soil around the edges of the frame will also give extra insulation.)

Hotbeds

A hotbed is structurally and operationally the same as a cold frame. However, a hotbed has a chemical or electrical heat source in addition to heat from the sun. Because of the additional heat, the frame will stay warmer at night, and there is less chance of frost damage. Hotbeds are thus superior to cold frames for growing warm-season plants.

MANURE-HEATED HOTBEDS

The original source of heat for hotbeds was decomposing manure. The decomposition process produced heat which kept the hotbed warm at night. The old method of hotbed heating is not commonly used today because of the inconvenience of hauling manure and its objectionable odor, but it is an economical heating method.

Horse manure is usually used for hotbeds, and it must be fresh. For localities where the temperature is not expected to drop below 12°F (−11°C), enough manure should be hauled to make a layer 12 to 15 inches (30–38 centimeters) thick. For colder areas, add 1 inch additional manure for each degree Fahrenheit lower anticipated temperature (5 centimeters per degree centigrade).

The manure should be laid in a flat pile outside the hotbed for 2 or 3 days until it begins to heat. It should then be turned, sprinkled when dry, and left several more days. By this time it should be heating uniformly and can be forked into a shallow pit dug in the hotbed.

A thin layer should be spread first, with any large lumps shaken out with the pitchfork. The layer should be tamped and water added if needed to make it slightly moist. The layering and tamping process should be repeated until the correct depth is reached, and the top layer covered with 5 to 6 inches (13–15 centimeters) of soil. Additional manure should be banked around the sides of the frame for insulation.

When the manure bed is young, it will generate a large amount of heat and the soil layer may be 90 to 100°F (32–38° C). Planting should be delayed until the temperature falls to 85°F (30°C) or lower. A thermometer buried in the soil at seeding depth will determine this.

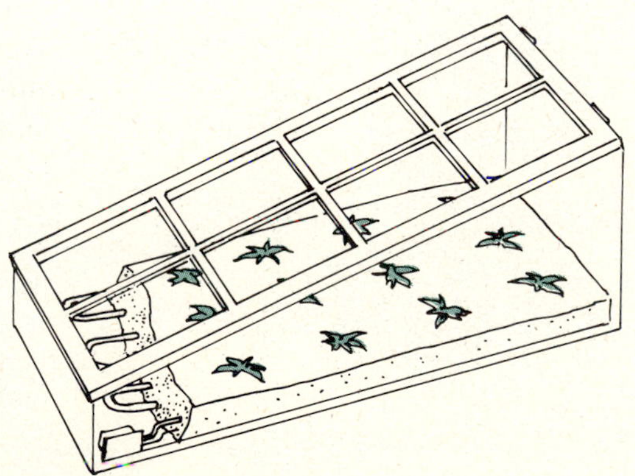

Figure 20-3 An electrically heated hotbed.

CORNSTALK-HEATED HOTBEDS

Although manure is the usual source of chemical heat for hotbeds, cornstalks can also be used. The stalks are chopped, moistened, and packed the same way as the manure. The decomposition will provide heat for 1 to 2 months. Combinations of manure and cornstalks can also be used when neither is available in sufficient volume.

ELECTRICALLY HEATED HOTBEDS

Electrically heated hotbeds use soil-heating cables buried in the bed. A thermostat regulates the amount of heat emitted from the cable (Figure 20-3). Because this type of heating requires electrical power, the hotbed should be located near an electrical outlet. After the current is routed to the bed, the cable is placed on the soil or a 2-inch (5-centimeter) bed of sand. Sand is the preferred material because it will drain water away from the bed and conserve heat. The cables should be spaced according to manufacturers directions without kinks and without touching the frame. About 3 inches (7.6 centimeters) of soil should cover the cable, and thermostat should be set at the recommended temperature (see Table 20-1).

HOTBED OPERATION

Electrically heated hotbeds are easier to operate than manure-fueled beds or cold frames. Because they are thermostatically controlled, there is less chance of over- or underheating. Manure-heated beds, on the other hand, produce a large amount of heat initially but decline rapidly as decomposition is completed. Covering electrically heated hotbeds at night is not essential. However, the heating costs can be reduced by about 30 percent by this method.

Table 20-1 Thermostat Temperatures for Electric Hotbeds

Plant	Thermostat setting: During germination	Thermostat setting: For growing
Cole crops	70°F (21°C)	60–65°F (15–18°C)
Onion	70°F (21°C)	60–65°F (15–18°C)
Lettuce	60–65°F (15–18°C)	60°F (15°C)
Eggplant	70°F (21°C)	65°F (18°C)
Pepper	70–75°F (21–24°C)	65°F (18°C)
Tomato	70–75°F (21–24°C)	62–65°F (16.7–18°C)
Annuals	70–75°F (21–24°C)	65°F (18°C)
Cucumber	75–80°F (24–27°C)	70°F (21°C)
Summer squash	75°F (24°C)	65°F (18°C)

Pit Greenhouses

A pit greenhouse (Figures 20-4 and 20-5) includes features of both a cold frame and a regular greenhouse. You can walk upright in it like in a greenhouse, but it must be covered each night with insulation like a cold frame. As the name implies, it is partially underground, with the lower part constructed in a pit 3 to 4 feet (1–1.2 meters) deep. The

Figure 20-4 A lean-to pit greenhouse. Note the ventilation door on the end and the rolled-up insulation.

glazed portion extends from ground level up to about 4 feet (1.2 meters).

The unique feature of pit greenhouses is their use of the heating and cooling properties of the earth. The depth to which the ground freezes (frost line) averages 24 inches (60 centimeters) over all the United States. In winter unfrozen soil below that level will conduct warmth onto the floor and walls of a pit greenhouse and prevent them from freezing. In summer the temperature of the ground is lower than that of the air, and the earth helps keep the greenhouse cool. A pit greenhouse is the ideal design for cold-winter or hot-summer areas, requiring little or no supplementary heat or cooling and having a moderate construction cost.

Pit greenhouses can be heated exclusively by the sun in most climates, and the temperature will stay above freezing at night and in the range of 40 to 50°F (4.4–10°C) or higher during the day. Plants which can be grown at this range are limited to those classified as "cool-season" species. Most tropical houseplants are excluded, but many flowers such as snapdragons, bulbs, primroses, pansies, and azaleas can be raised along with many common vegetables. Table 20-2 gives a partial listing of these cool-season plants.

With minimal additional heating, the temperature of a pit greenhouse can be raised to a moderate (55–60°F; 12.8–15.6°C) or warm (65–70°F; 18.3–21.1°C) daytime range. At these temperatures the variety of plants which can be grown includes most houseplants and vegetables.

PIT GREENHOUSE STYLES

A pit greenhouse can be considered a full-glass pit or a half-glass pit, depending on whether all or only half the roof is glazed.

A full-glass pit is brightest inside since the roof is entirely glass. However, the glass roof also loses heat rapidly, and supplemental heat might be required to maintain a temperature above freezing. Therefore a full-glass pit is recommended only for climates where the average winter night temperature is not below 15 to 20°F (−9.4 to −6.7°C). Zones 8 and 9 on the Department of Agriculture plant hardiness zone map would be suitable for this style of house.

Figure 20-5 The interior of a pit greenhouse.

A half-glass pit has only the south portion of the roof glazed and could be used in any climate. Direct sunlight will enter from the south, but the north slope of the roof is solid and insulated. Heat loss is reduced substantially with minimal reduction in light. A half-glass pit will remain above freezing without supplemental heat in most climates and could be maintained without heat in zones 6 and 7 and with minimal supplemental heat in zone 5.

More energy efficient still is a lean-to, half-glass pit (Figure 20-4) constructed against a south-facing house wall. Its efficiency is derived from the warmth received from the adjacent heated building and from the wind protection the wall provides.

PIT GREENHOUSE CONSTRUCTION

Site selection is as important to pit greenhouses as to cold frames. In full-glass houses the roof must face north and south. In half-glass houses the glass portion must face south.

No variation in direction can be allowed or the house will not heat properly. Sites with obstructions which would block light are not suitable.

Plumbing is an optional but valuable addition to a pit greenhouse. A single pipe buried below the frost line will enter the greenhouse at a convenient level of about 1 to 2 feet (30–60 centimeters).

Electricity is essential because it is infrequently but indispensibly required for fan ventilation and emergency heating.

The benches built for a pit greenhouse should take advantage of the sunlight entering the house (Figure 20-6), and different bench arrangements can be used depending on the plants being grown. For medium-sized potted flowers, the north bench can be stepped. For overwintering larger potted plants, a bench can be optional, and the plants can be set on the drainage gravel. If a gardener wanted to raise cut flowers, a ground planter 12 inches (30 centimeters) high could be planted with tall-growing carnations or snapdragons.

Benches can be made from wood or wood with pipe supports. Hardware cloth benches are also recommended. Poured concrete, concrete block, or wood is used for ground beds.

The insulation used for nighttime covering of the greenhouse should be thick and

Table 20-2 Greenhouse Temperature Requirements of Selected Ornamental Plants

Botanical name	Common name	Temperature requirement	Code[a]
Aphelandra squarrosa	Zebra plant	Warm	Fl, Fo
Justicia brandegeana	Shrimp plant	Intermediate	Fl
Begonia semperflorens hybrids	Wax begonia	Cool-warm	Fl
Mammillaria, Opuntia, Rebutia, and other genera	Cacti	Cool	Fl, Fo
Calceolaria herbeohybrida	Pocketbook plant	Cool	Fl
Celosia cristata	Cockscomb	Cool	Fl
Senecio × hybridus	Cineraria	Cool	Fl
Cyclamen persicum	Cyclamen	Cool	Fl
Fuchsia × hybrida	Fuchsia	Cool	Fl
Pelargonium × hortorum	Zonal geranium	Cool	Fl, Fo
Sinningia speciosa	Gloxinia	Warm	Fl
Impatiens spp.	Impatiens	Cool-warm	Fl
Chrysanthemum frutescens	Marguerite daisy	Cool	Fl
Euphorbia pulcherrima	Poinsettia	Warm	Fl
Primula spp.	Primrose	Cool	Fl
Saintpaulia ionantha	African violet	Warm	Fl
Salvia splendens	Salvia	Cool	Fl
Schizanthus pinnatus	Butterfly flower	Cool	Fl
Solanum pseudocapsicum	Christmas cherry	Warm	Fl
Matthiola incana	Stock	Cool	Fl
Schlumbergera truncata	Christmas cactus	Cool	Fl, Fo
Fatsia japonica	Japanese fatsia	Cool-warm	Fo
Begonia × rex-cultorum	Rex begonia	Warm	Fo
Chlorophytum comosum	Spider plant	Cool-warm	Fo
Cissus species	Kangaroo vine, grape ivy, and other	Warm	Fo
Coleus × hybridus	Coleus	Warm	Fo
Codiaeum variegatum	Croton	Warm	Fo
Dieffenbachia maculata varieties	Dumbcane	Warm	Fo

Botanical name	Common name	Temperature requirement	Code[a]
Ficus spp.	Fig, rubber plant	Warm	Fo
Hedera helix varieties	English ivy	Cool-warm	Fo
Maranta leuconeura varieties	Prayer plant	Warm	Fo
Peperomia spp.	Peperomia	Warm	Fo
Pilea spp.	Pilea	Warm	Fo
Sansevieria trifasciata	Snake plant	Warm	Fo
Saxifraga stolonifera	Strawberry plant	Cool-warm	Fo
Selaginella spp.	Irish moss	Warm	Fo
Tradescantia, Zebrina	Wandering Jew	Warm	Fo
Iris spp.	Bulbous or Dutch iris	Cool	Fl
Freesia × hybrida	Freesia	Cool	Fl
Agapanthus africanus	Lily of the Nile	Cool	Fl,Fo
Crocus spp.	Crocus	Cool	Fl
Hyacinthus orientalis	Hyacinth	Cool	Fl
Tulipa spp.	Tulip	Cool	Fl
Narcissus spp.	Daffodil	Cool	Fl
Lathyrus odoratus	Sweet pea	Cool	Fl
Viola × wittrockiana	Pansy	Cool	Fl
Viola odorata	Violet	Cool	Fl
Dianthus caryophyllus	Carnation	Cool	Fl
Abutilon megapotanicum	Trailing abutilon	Cool	Fl
Calendula officinalis	Pot marigold	Cool	Fl
Camellia japonica and *Camellia sasanqua*	Camellia	Cool	Fl, Fo
Citrus spp.	Citrus	Cool	Fl, Fo
Daphne spp.	Daphne	Cool	Fl
Osteospermum ecklonis	Osteospermum	Cool	Fl
Gazania ringens	Gazania	Cool	Fl
Gerbera jamesonii	Transvaal daisy	Cool	Fl
Lantana camara	Lantana	Cool	Fl
Antirrhinum majus	Snapdragon	Cool	Fl
Plumbago ariculata	Cape honeysuckle	Cool	Fl
Petunia × hybrida	Petunia	Cool-warm	Fl
Rosmarinus officinalis	Rosemary	Cool	Fl, Fo
Anemone coronaria	Anemone	Cool	Fl
Clivia miniata	Clivia	Cool	Fl, Fo
Ranunculus asiaticus	Ranunculus	Cool	Fl
Cypripedium spp.	Lady slipper orchid	Cool	Fl
Cymbidium spp.	Orchid	Cool	Fl
Adiantum spp.	Maidenhair fern	Cool-warm	Fo
Asplenium spp.	Mother fern, birdnest fern	Cool-warm	Fo
Cyrtomium falcatum	Holly fern	Cool-warm	Fo
Nephrolepis spp.	Sword fern	Warm	Fo
Phyllitis scolopendrium	Hart's-tongue fern	Cool	Fo
Polistichum spp.	Shield fern	Cool	Fo
Muscari spp.	Grape hyacinth	Cool	Fl
Lobularia maritima	Sweet alyssum	Cool	Fl
Tropaeolum majus	Nasturtium	Cool	Fl
Oxalis spp.	Bulbous oxalis	Cool	Fl

[a]Fl—plant grown primarily for its flowers; Fo—plant grown primarily for its foliage.

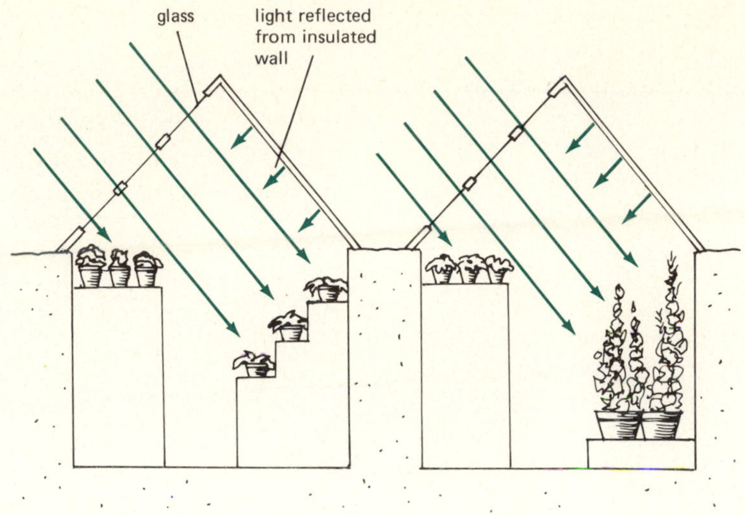

Figure 20-6 Two styles of benches for a pit greenhouse. The ground bed on the right enables the gardener to grow tall vegetables and flowers.

efficient. Filled canvas pads laid on the glass and covered with a tarpaulin are one solution. Fiberglass insulation with a paper or foil backing which can be rolled up with pulleys during the day (Figure 20-4) can also be used.

PIT GREENHOUSE OPERATION

Daily maintenance of a pit greenhouse involves covering and uncovering the glazing and ventilating by opening the doors on either end of the house. As with a cold frame, the covering should be taken off by early morning when the sun is shining and replaced in late afternoon. On cloudy days with below-freezing temperatures the coverings should not be removed, and the house may remain covered for several days during long stretches of cloudy weather.

Watering may be required daily, although plants lose water slowly in the cool humid atmosphere. Careful watering with a narrow spout is recommended to minimize spillage. Cool temperatures and high humidity make the fungus disease mildew (Chapter 13) a constant threat, but one that can be lessened by cautious water use.

One advantage of a pit greenhouse is that it can be left unattended for several days, if necessary, by leaving the insulation over the glass. The temperature will normally stay above freezing.

Summer care of the house includes covering the glazing with a "shading material" to decrease the light entry and heat buildup. Lath, bamboo blinds, and Saran (a plastic, netlike covering) can be used for this purpose. Because of its southern exposure, a pit greenhouse can overheat easily, so all vents should be left open except on very cool nights. Air circulation with a fan may also be necessary.

Conventional Greenhouses

A conventional greenhouse offers its owner the advantage of almost complete environmental control. Although partially heated by the sun, gas, electricity, or oil is the main heat source, and the heat is thermostatically controlled to keep the house cool, moderate, or warm. Depending on your budget, venting, watering, cooling, shading, and fertilizing can also be automated.

Of the structures discussed, conventional greenhouses are the most expensive to operate because of heat loss through the large glazed area. However, conventional greenhouses are the most common for commercial and home use because of the ease of construction and superior light penetration.

Many options in shape, glazing and framing materials, and heating, watering, and cooling systems are available when choosing a greenhouse. The gardener should think of his greenhouse as a component system and select each component according to climate, budget, convenience, plant, and aesthetic requirements.

Most greenhouses are ordered from catalogs, and with competition encouraging lower prices, high quality cannot always be assumed. Some inexpensive greenhouses are essentially toys, having only sheets of fiberglass screwed to a flimsy wood frame. Such greenhouses are not suitable for use in harsh climates. Cold air would enter through the poorly designed junctions of the glazing and the first snowfall would collapse the roof. Look for guaranteed quality and durability when selecting a greenhouse. Information given by the manufacturer should include:

- Frame construction
  - Framing material (wood or metal)
  - If wood, type and dimensions of framing members
- Glazing
  - Type of glazing material (glass, fiberglass, PVC film, and so on)
  - Guarantees of glazing against breakage and discoloration, and expected life
  - Options of buying replacement glazing
  - Airtightness of glazing
- Ventilation
  - Type and location

GREENHOUSE SHAPES

Span Roof Freestanding. These are the most common and are used frequently by commercial growers. Their design is angular, with a pitched roof and straight or sloping sides. Three variations on the span-roof greenhouse are the standard, Dutch light, and curvilinear (Mansard).

Standard greenhouses (Figure 20-7) have straight sides and a two-sided roof that slopes down to the sides of the house. It may be glazed to the ground or have sidewalls up to 30 inches (76 centimeters) high under the benches. Sidewalls decrease the rate of heat loss and give greater stability but shade the area under the benches.

A Dutch-light greenhouse (Figure 20-8) is similar to a standard span roof except that the sides are glazed to the ground and are constructed at a slight angle rather than being vertical. The angle of the sides causes winter sun to strike the glass at close to 90°, the optimum angle for complete light penetration. Due to the absence of sidewalls and because of the angle of the glass, Dutch-light houses are slightly brighter than standard houses.

The mansard or curvilinear greenhouse (Figure 20-9) takes this principle a step further. Each section of glass attaches to the next to form a tunnellike structure. Because the sun's angle changes with the seasons, each set of panes will eventually be at the optimum angle for light penetration. Mansard greenhouses are not readily available and their light transmission is not significantly superior to that of full-glass standard or Dutch-light houses.

Lean-to. Lean-to greenhouses are split, span-roof greenhouses of one of the types discussed. Their appeal lies in convenience and aesthetics. Since they can often be built to cover a sliding glass door (Figure 20-10), they can be warmed with the household heat. A lean-to can also extend the living area to form a garden room or hobby area.

Figure 20-7 A standard greenhouse with brick sidewalls.

Although most lean-to greenhouses are attached to the south side of the house for maximum light, they can be situated on another side for convenience. East- or west-facing greenhouses receive moderate light and grow almost any plant, and even north-facing greenhouses can be used for shade-loving ferns and many houseplants.

Quonset. These greenhouses are usually framed of bent pipe and glazed with one or more layers of plastic film (Figure 20-11). Although less attractive than some styles, they are inexpensive and can be recommended for home use.

Dome. Although they appear faddish, geodesic domes (Figure 20-12) constructed of interlocking triangles are actually very well designed. The angles of the glazing permit excellent light penetration, and hanging plants can be hung from the structure to make maximum use of the light.

GREENHOUSE FRAMING MATERIALS

The frame of a greenhouse, the part which supports the glazing, can be of almost any material. A strong, weather-resistant material is essential for even a temporary structure. The relative strengths of greenhouse framing materials are discussed below.

Figure 20-8 A Dutch light greenhouse. Note that the sides are entirely glass and are not quite vertical.

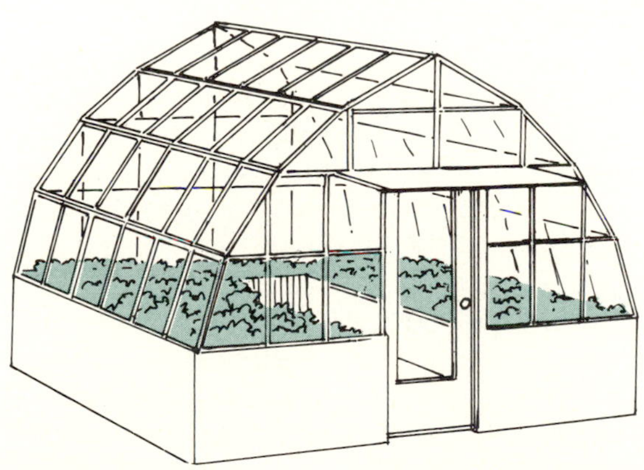

Figure 20-9 A cross section of a mansard greenhouse showing the differing angles of the panes of glass.

Wood was once the main greenhouse framing material. It is a good insulator, moderately durable, but requires some maintenance such as painting. Redwood, western red cedar, and pressure-treated softwoods are all satisfactory. Creosoted wood should never be used in a greenhouse; the creosote will vaporize into the greenhouse and injure the plants.

Aluminum alloy is lightweight and moderate in price. It is durable and low maintenance but loses heat more rapidly than wood.

Steel bars or pipe can be used to frame a greenhouse. The material is strong and relatively inexpensive but unless galvanized, it must be painted regularly to prevent rust. Steel also transmits heat rapidly, making it a poor insulator.

Figure 20-10 A lean-to greenhouse.

FOUNDATIONS AND SIDEWALLS

In order to withstand wind without glass breakage or tipping, a greenhouse should have a foundation. A foundation will also keep the house level. Many prefabricated greenhouses state that no foundation is needed. However, unless the greenhouse is to be used only as a temporary structure, a foundation is recommended. As a minimum, cinder block can be laid on level soil. But a concrete slab is preferable, poured with bolts to anchor the walls and the water and electrical pipes in place.

If the design includes sidewalls, a concrete base to support them should go at least 1 foot (30 centimeters) below the frost line. The walls can then be built of the chosen material.

GLAZING MATERIALS AND METHODS

For many years, wooden sashes with panes of glass puttied into place were the main glazing material. Glazing used today includes a wide range of plastics and fiberglass as well as glass.

Glass. Still preferred for looks, durability, and ease of cleaning, glass is and will remain a popular glazing material. Its light transmission ability is excellent, and although breakage can occur, it will not be a problem in a well-constructed house. The expense is moderate to high but is compensated for by its long life.

Modern methods of glazing with glass utilize several systems. In "dry glazing," the glass is inserted into a channel in wood or

Figure 20-11 Quonset style plastic greenhouses. These are built exceptionally tall to house citrus trees.

between metal bars. Glass applied by dry glazing is easy to install, but air leakage can be a problem.

A second glazing system places the glass on a strip of putty and holds it in place with a clip. The junction is then covered with a cap to weatherproof the seal (Figure 19-17).

Film Plastic (Polyethylene). These thin (2- to 6-millimeter thickness) plastics can be stapled to a wood frame or run in a continuous piece over a quonset house. They are inexpensive but last only 1 to 2 years in most climates. When signs of tearing or discoloration appear, they should be replaced.

Greenhouses covered with plastic film are very airtight. Consequently, the humidity can climb very high, causing condensation runoff from the plastic and creating conditions favoring disease development. Proper ventilation is therefore essential in a plastic house.

Vinyl Plastic (Polyvinyl Chloride or PVC). Vinyl plastics are thicker than film plastics (3–12 millimeters) but still moderately flexible. They are generally stapled onto a wooden frame.

Vinyl plastics are moderate in cost and

Figure 20-12 A geodesic dome greenhouse for home use.

usually have a quaranteed life of up to 10 years. Both transparent and translucent types are sold.

Fiberglass. There are many brands and grades of corrugated fiberglass used for glazing greenhouses. The durability of some is excellent, and they are guaranteed for 10 to 20 years. Most require resealing of the surface with a plastic resin every 1 to 2 years to prevent abrasion of the protective coating. The manufacturer should provide information regarding the durability and maintenance of the glazing material.

Double Glazing. The loss of heat through all glazing materials is relatively rapid, from both conduction through the material and air leakage at junctions. In double glazing, two layers of glazing are applied to the greenhouse, eliminating air exchange and creating a "dead air" space.

The easiest double glazing for home greenhouses involves supplementing the present glazing with a layer of plastic film. The film can be paplied either inside or outside the house and will result in up to a 50 percent heat saving.

HEATING SYSTEMS

Self-contained forced-air heating units are the best for home greenhouses. They can be oil or gas fired or electric. Electric units do not require venting and are easy to install. They are installed at one end of the greenhouse and aimed so the fan circulates warm air throughout the house.

Gas and oil heaters set up inside the greenhouse require venting. In several models the heater fits into the wall of the greenhouse. The combustion chamber is outside, and no venting is required.

Heating by household heat is used frequently with lean-to greenhouses. However, a fan will still be required to circulate air between the two if the greenhouse is very large.

COOLING SYSTEMS

Cooling of greenhouses is accomplished mainly by venting and shading. Many inexpensive greenhouses are constructed to be ventilated only by opening the doors; however, vents on the roof or at the eve are more efficient because the warmer air will rise and flow out through them. Greenhouse models with these vents are preferable.

Shading of a greenhouse in summer will restrict direct sunlight and decrease heating of the interior. Bamboo blinds, cheesecloth, or Saran placed over the glass will function very effectively. A coating of whitewash can be used on glass greenhouses and can be removed in fall with water and a brush.

In areas with very hot summers an evaporative cooler can lower the temperature of the greenhouse to less than the outside air temperature. Evaporative coolers are widely used in commercial greenhouses, and small units are available from home greenhouse suppliers.

In this cooling system, a fiber pad is installed at one end of the greenhouse with water trickling through it at a slow rate. A fan at the opposite end of the house pulls outside air in through the pad. The water evaporates into that air, absorbing heat in the process and reducing the air temperature.

Figure 20-13 A double-layer bench arrangement in a greenhouse. Note that low-light-requiring ferns are being grown on the lower bench.

INTERIOR GREENHOUSE EQUIPMENT

Most greenhouses are equipped with benches which support plants at a convenient working level. For greenhouses without sidewalls, double layers of benches (Figure 20-13) are practical, with lower-light plants grown on the bottom level.

Benches framed with wood and covered with galvanized hardware cloth have much to recommend them. They are rustproof, permit free air and heat circulation, and allow light to pass through to plants growing in the area beneath.

When large potted plants must be accommodated, a section of bench can be left out and the pots placed on the floor of the greenhouse. If vegetables such as tomatoes or trellised cucumbers are being raised, a ground bed (Figure 20-14) will be needed.

Other equipment used for operating a greenhouse is minimal. A thermometer that specifies the minimum and maximum temperature reached is necessary and should be placed out of direct sunlight and at the same level as the plants.

A humidity chamber for rooting cuttings is almost indispensable and can be easily made by draping vinyl or plastic film tent fashion over wire supports (Figure 20-15).

A mist nozzle (Figure 20-16) is useful although not essential. Likewise, a fertilizer proportioner (Figure 20-17) will enable the greenhouse gardener to fertilize plants while watering by mixing concentrated fertilizer into the water supply.

Automatic systems for shading and vent-

Figure 20-14 A ground bed for raising large foliage plants.

ing are relatively expensive; however, an automatic watering system can be easily and inexpensively installed. Commercial greenhouses use this "spaghetti system," consisting of a flexible ½-inch (13-millimeter)-diameter plastic pipe inserted at various points with "microtubes" about the thickness of spaghetti. The water flows at low pressure through the main line and out each of the microtubes to a pot (Figure 20-18). The system is identical to the trickle irrigation discussed in Chapter 11.

Growing Plants in a Greenhouse

ORNAMENTALS

Growing ornamental plants in a greenhouse is much like growing them in the house. They require the same essentials of water, fertilizer, and potting medium and the same maintenance. However, in the greenhouse they receive higher light and humidity conditions, and consequently they will grow faster and may be healthier. But a greenhouse should not be expected to guarantee success with plants. The person who has successfully grown plants indoors will probably be even more successful in a greenhouse, but one who has had a number of problems with plants indoors will have the same problems in the greenhouse. The greenhouse will not substitute for knowledge or experience.

Greenhouse environments can be divided into two types by temperature, although some plants can be grown in both temperature ranges. A cool greenhouse maintains a night temperature above freezing and generally in

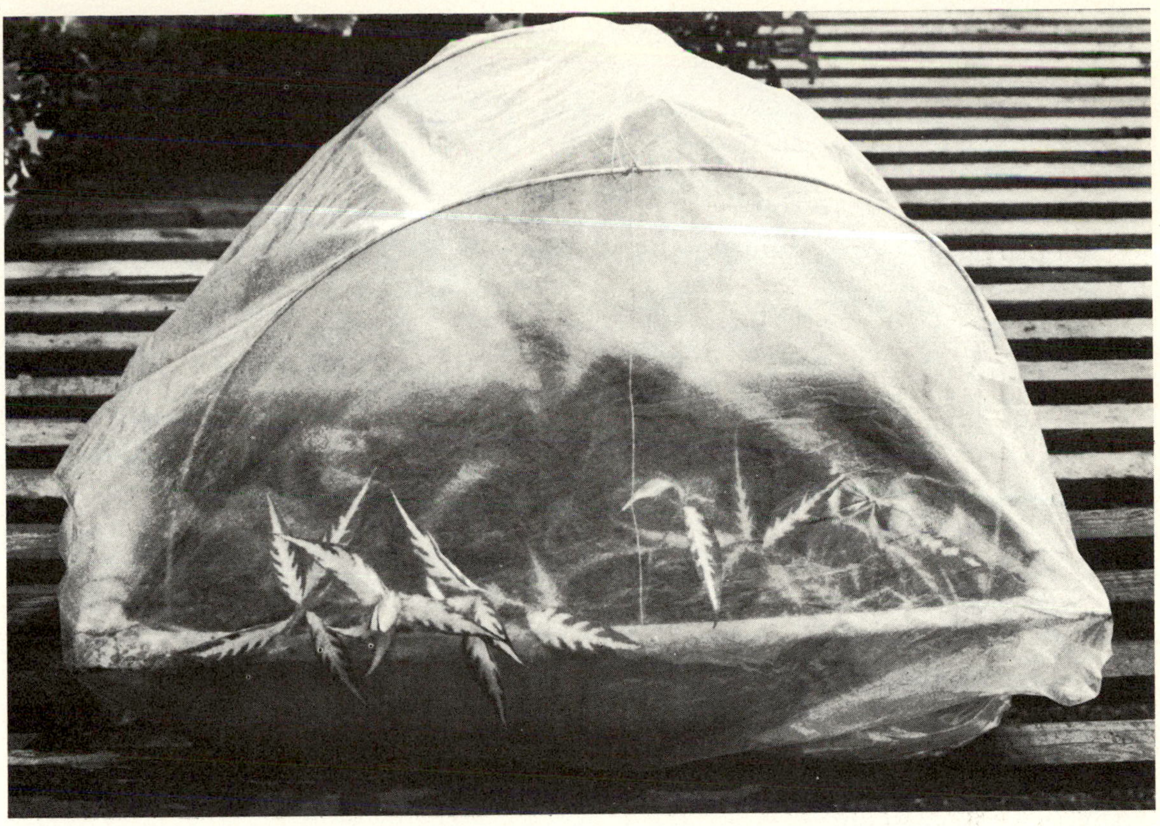

Figure 20-15 A rooting chamber made from plastic draped over wire supports.

the range of 40–55°F (4.4–12.8°C). Daytime highs can reach up to 60°F (15.6°C). In a tropical or warm greenhouse the minimum night temperature is 55°F (12.8°C) with daytime around 65–70°F (18.3–21.1°C). The drop to a lower nighttime temperature is a standard practice to promote healthy growth by slowing respiration during the night hours.

Table 20-2 gives a listing of ornamental plants which are commonly grown in greenhouses and their suggested temperature ranges. Because of the improved growing conditions which a greenhouse provides, many beautiful flowering plants from semitropical climates can be grown, most of which will not be familiar to temperate-climate gardeners. As temperature above freezing and generally in with any unknown plant, it is worthwhile to research the specific growing requirements of a species to assure the best performance.

VEGETABLES AND FRUITS

For some people, a greenhouse is seen as the means for growing fresh produce out of season. A number of vegetables and fruits can be successfully grown in a greenhouse, although their growth rate and taste may not equal what they would be outdoors due to the reduced light intensity and shorter days. Vegetables which can most successfully be greenhouse grown are listed here, along with a brief discussion of their growing requirements.

Beans. Snap beans should be grown in a warm greenhouse and can be seeded in February or March in ground beds for a May crop. Pole bean varieties grown on a trellis work

Figure 20-16 A mist nozzle.

well, but bush types can be used if space is limited.

Beets. Beets can be grown for both their tops and their roots. They are a cool-greenhouse crop and can be grown throughout winter.

Carrots. Short varieties such as 'Oxheart' or 'Short 'n Sweet' can be grown in ground beds from a February sowing and harvested in early spring. Carrots for fall and winter use are more easily obtained by mulching in the remaining carrots in the vegetable garden and pulling them as needed. See Chapter 6 for details.

Cucumbers. Cucumbers should be grown on trellises and can be harvested through most of the winter if managed correctly. Seed sown in November, January, and March should yield a nearly continuous supply of fruits.

Varieties of cucumbers which do not require pollination to set fruit (called *parthenocarpic*) are best for greenhouse culture. Growing temperature should be at least 60°F (15.6°C) and preferably 70°F (21.1°C) for best production.

Grapes. Grapes take up a large amount of room and can be grown only in a large greenhouse. They cannot be produced in midwinter, but only a month or two before their regular season.

Use a 12-inch (30-centimeter)-diameter tub for each vine and train it flat against the wall. The vine will break dormancy and flower early in a cool house. The flowers will need to be hand pollinated with a soft brush.

Herbs. A number of leafy herbs can be grown successfully year-round in either a cool or a warm greenhouse, including mint, par-

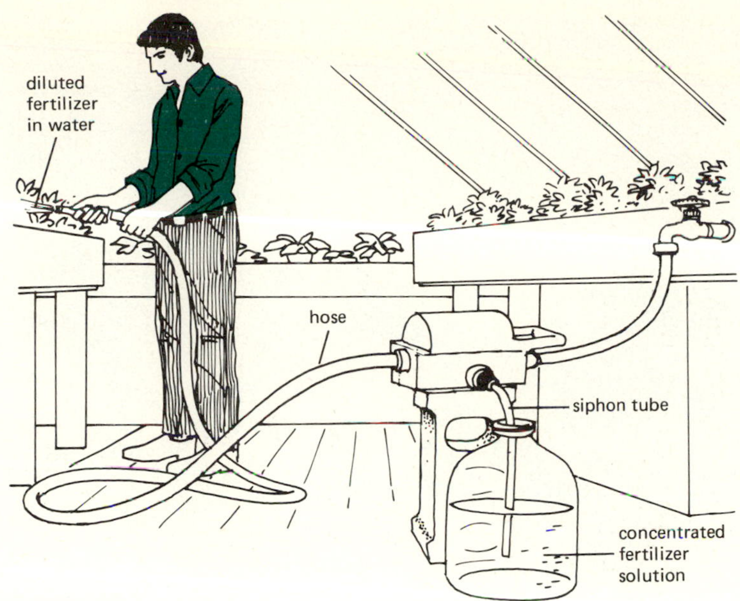

Figure 20-17 A simple fertilizer proportioner system that works by water pressure.

sley, rosemary, chive, thyme, basil, and sage.

Lettuce. Either 'Bibb' or leaf lettuce will thrive under cool conditions. A continuous winter harvest can be expected from November from seed sown about once per month.

Onions. Green onions can be grown easily from seed in a cool greenhouse. Two to three sowings starting in September will provide onions through spring.

Radishes. This quick-maturing vegetable will be ready 1 month after the seed is sown and can be grown throughout winter in a cool house.

Tomatoes. Tomatoes can be propagated from seed or cuttings and grown in large pots or a ground bed (Figure 20-19). A warm greenhouse (60°F (15.6°C) or higher) is best. As the plants grow taller, they should be staked upright. Hand pollination is necessary, and lightily tapping the flower clusters during the bloom period will assure good fruit set. Choosing tomato varieties bred for greenhouse growing will help ensure success.

Hydroponic Greenhouse Gardening

Hydroponic gardening involves supplying water and minerals to plants by flooding their roots with a liquid nutrient solution. No soil or potting medium is used. Instead the plant roots are supported in beds of gravel, sand, or similar unreactive material, which is filled and drained several times per day with the water-based solution.

Hydroponic gardening is considerably more complicated than conventional culture. However, the practice achieved considerable popularity recently as a hobby and even as a commercial enterprise. There are many books dealing with the science of hydroponics which offer growing instructions and recipes for nutrient solutions.

Shade Houses

Shade houses (Figure 20-20) provide a growing area of filtered shade for plants which are

Figure 20-18 A "spaghetti" watering system for potted plants in a greenhouse. The lead weight at the end of the microtube keeps it from falling out of the pot.

Figure 20-19 Tomatoes growing in a greenhouse ground bed.

not tolerant of full summer sunlight. Fuchsias, impatients, and many begonias and ferns grow well only in a shade house during the summer months but can tolerate full sun in the greenhouse during the winter months when the sun is dim. A shade house can also be used for hardening transplants and for summering houseplants outdoors. If large enough, it can be used as an outdoor living area.

Shade houses are usually constructed with a wooden frame and covered with lath or Saran. The covering may or may not extend to include the sides of the structure. Lath houses shaded by pieces of lath or 1 × 2-inch (25 × 50-millimeter) wood strips form an attractive, durable structure. The spacing of the lath will depend on the intensity of the sunlight in the area and the amount of shade required by the plants. North-to-south orientation of roof lath is suggested because it casts alternate sun and shade on plant leaves through the day. However, orientation is not crucial, and many gardeners prefer to orient lath east-west or in decorative crisscross patterns.

Plastic Saran can also be used to cover a shade house. It is less durable, lasting 1 to 2 years, but provides an even shade throughout the structure. The depth of shade will be determined by the weave of Saran selected.

Figure 20-20 A lath-covered shade house for the home gardener.

INDOOR PLANT GROWTH UNITS

Greenhouse Windows

A greenhouse window is one way to expand the indoor growing area for houseplants or to raise garden transplants. Although the heat cannot be controlled as in a regular greenhouse, the light is better and the humidity often higher than in an ordinary window.

When buying or constructing a greenhouse window, a major criterion is ventilation. A side or top vent should be included in the design for excess heat dissipation in summer. Beyond this requirement, the considerations of price and aesthetic appeal become important. A large window is not usually correspondingly higher in price than a small one, but it will provide valuable extra space for additional plants at the same time that it becomes an attractive addition to a room.

Environment-controlling Growth Units

These indoor units (Figure 20-21) monitor and regulate heat, humidity, and light continuously for optimal plant growth. They are useful in homes with insufficient natural light and for species which are difficult to grow out of a greenhouse. They are also attractive for displaying and maintaining collections of exotic or miniature plants. However, for grow-

Figure 20-21 An environment-controlled plant growth unit for indoor use.

ing ordinary houseplants their expense is usually prohibitive.

Terrariums

The primary plant-growing benefit of terrariums is their ability to hold in humidity, the lack of which causes many houseplant problems. But the glass that holds in the moisture also excludes some light, making terrariums suitable mainly for medium- and low-light plants.

Almost any container, aquarium, fish bowl, water carboy, or bottle can be used as a terrarium. Those with narrow openings are more difficult to plant, but the main consideration is that the container be made of clear or only slightly tinted glass. Colored glass containers will exclude light, and plants growing in them will eventually die.

In making a terrarium, first provide a layer of drainage material at least 1 inch (3 centimeters) deep in the bottom of the container. This layer will be a reservoir for excess water if the terrarium is accidentally overwatered. Most prepackaged or homemade potting media will be satisfactory and should be moistened, if necessary, to slight dampness.

Assembly of an open-mouth terrarium can be done quickly using your hands, but narrow-necked terrariums require special tools and more time. The drainage layer and potting medium should be poured in through a funnel to avoid dirtying the inside of the glass. Each layer is then smoothed out with a tool fashioned by taping a spoon to a piece of wooden molding, a dowel, or a straightened coat hanger.

Plants are inserted by soaking the media from the roots, folding the leaves upward, and sliding them through the opening. They are then positioned and planted with the spoon tool and a plain dowel. Unrooted cuttings can also be used and will root easily in the humid atmosphere.

The plants chosen for the terrarium should be medium- or low-light-requiring and slow growing. Table 20-3 lists several dozen recommended species. Cacti and succulents are not suitable for terrarium planting. The stagnant air and restricted light provide poor

Table 20-3 Plants for Small and Moderate-sized Terrariums

Botanical name	Common name	Remarks
Adiantum spp.	Maidenhair fern	Many species suited to terrarium culture
Araucaria heterophylla	Norfolk island pine	Seedlings make attractive miniature trees
Begonia spp.	Begonia	Miniature species are hard to find but work well, including *Begonia foliosa*
Chamaedorea elegans	Parlor palm	Several seedlings are usually sold in a small pot and can be separated and used for small trees
Dracaena surculosa	Gold-dust dracaena	Formerly known as *Dracaena godseffiana;* shrubby plant with deep green leaves irregularly spotted yellow
Episcia dianthiflora	Lace-flower vine	Fuzzy-leaved relative of the African violet bearing white tubular flowers
Ficus pumila	Creeping fig	Small-leaved creeping vine; the varieties 'Minima' and 'Quercifolia' are smaller foliaged but rarer
Fittonia verschaffelti 'Minata'	Miniature nerve plant	Slow growing and low-light-requiring; leaves are about 1″ (2.5 cm) long
Maranta leuconeura varieties	Prayer plant	Low-light-requiring; young plants are ideal but may outgrow the terrarium in 6 months
Pellaea rotundifolia	Button fern	Deep, blue-green pea-sized leaves on a flat-growing plant
Peperomia caperata	Emerald ripple peperomia	One of the smaller and more attractive peperomias; can be propagated by a single leaf pushed into the soil
Peperomia rubella		Small-leaved peperomia with red stems and leaf undersides; grows up to 6″ tall
Podocarpus macrophyllus	Southern yew	Slow growing and useful as a small tree; deep green ribbon-type foliage
Pteris spp.	Table fern	Most species are excellent; some are plain green, others ruffled or lined in silver
Saxifraga stolonifera	Strawberry geranium	Small plants will root readily in the terrarium; produces interesting runners
Schlumbergera spp. varieties	Christmas cactus	Normally sold as *Zygocactus;* Christmas, Thanksgiving, and Easter cacti; all prefer moist soil and will thrive in a terrarium
Selaginella spp.	Irish moss	Many species are excellent, either creeping or forming small mounds
Sinningia pusilla	Slipper plant	Tiny flowering plant only 3″ across; flowers produced sporadically in pastel shades
Soleirolia soleirolii	Baby tears	Commonly sold as *Helxine soleirolli.* Has tiny 1⁄16″ (2-mm) leaves and a matlike growth habit; becomes tall and stringy in insufficient light

growing conditions for these and other xerophytic plants.

When assembly of the terrarium is complete, the new plants should be watered in to assure contact of the medium with the roots. Trickle several teaspoons (10–15 milliliters) of water onto each plant by pouring it down the dowel. Rinse any media off the glass with a trickle of water or by wiping with a wet paper towel attached with a rubber band to a dowel.

Terrarium maintenance involves pruning and occasional watering. Fertilizer is not used because it encourages rapid growth, and the plants will outgrow the container. Even without fertilizer, pruning may be needed to control vigorous species and to encourage branching. If the container has a wide mouth, pruning can be done with a scissors and the prunings taken out and discarded. In a narrow-neck terrarium a razor blade taped to a dowel can be used, and a pickup tool will remove the trimmings and any dead leaves or flowers.

How frequently a terrarium will require watering depends on the size of the opening. In a sealed terrarium no water is lost to the outside air, and watering should never be required. A narrow-neck terrarium will have minimal water loss through the opening and will require infrequent watering, on an average of every 2 to 6 months. Wide-mouth terrariums, however, have less restricted air exchange and may require watering every 1 to 4 weeks.

Overwatering causes the death of terrarium plants in many instances, so caution is advisable. In wide-mouth terrariums the medium can be checked with the fingers and should always feel slightly damp. In narrow-neck containers water should be applied only when the top of the medium begins to appear dry and never as long as there is condensation on the glass or water in the pebble reservoir. Watering by teaspoonfuls (5 milliliters) is suggested to lessen the danger of overwatering.

Garden Rooms

Garden rooms combine the features of a greenhouse and living room, creating an area useful for both growing plants and dining or relaxing. Solariums, Florida rooms, atriums, and conservatories are all variations of garden rooms.

Unlike a greenhouse, a garden room is normally not completely glazed. A completely glass room becomes too hot during the day and cools off too rapidly at night to be comfortable for people. Instead it is partially glazed, perhaps with large windows and several skylights. Ventilation and electricity are also present along with furniture and, usually, a waterproof floor. Plumbing and drainage facilities are optional.

Location of the garden room should be governed by convenience rather than orientation. A large garden room designed for family use will be most conveniently situated near kitchen, living, or dining areas. A more intimate room could be placed off a master bedroom or bath. If several locations seem agreeable, orientation can be considered. An east-facing garden room is ideal in many ways. It provides morning sunshine yet remains relatively cool and shady the remainder of the day. A south-facing room might be ideal for cold or low-light climates since it receives direct sunlight all day. However, a north room would be better suited to warm areas of the country. A west-facing room is brightest from midday until sunset and would be a cheery retreat late in the afternoon.

The cost of a garden room is approximately equal to that of a regular room on a cost per square foot (square meter) basis. Since most garden rooms are additions, the services of an architect will probably be required. Consult, if possible, an architect with experience in designing garden rooms, and bring pictures from magazines or books to help explain the project and what you have in mind.

Selected References for Additional Reading

Abraham, G., and K. Abraham. *Organic Gardening under Glass.* Emmaus, PA: Rodale Press, 1975.

Jones, J. L. *Home Hydroponics . . . And How to Do It.* Paradise Valley, AZ: Beardsley, 1975.

Kramer, J. *Garden Rooms and Greenhouses.* New York: Harper and Row, 1972.

Neal, C. D. *Build Your Own Greenhouse.* Radnor, PA: Chilton, 1975.

Sunset Books Editors. *Greenhouse Gardening.* Menlo Park, CA: Lane, 1976.

Taylor, K. S., and E. W. Gregg. *Winter Flowers in Greenhouse and Sun-Heated Pit.* New York: Charles Scribner's Sons, 1976.

Walls, I. G. *The Complete Book of Greenhouse Gardening.* New York: Quadrangle/New York Times, 1973.

Appendix of Plant Societies

African violets and other Saintpaulias	*The African Violet Society of America, Inc.*, Box 1326, Knoxville, TN 37901. "African Violet Magazine" (published 5 times annually). *Saintpaulia International,* Box 1064, Knoxville, TN 37919, c/o Alma Wright. "Gesneriad Saintpaulia News" (published bimonthly).
amaryllis	*The American Plant Life Society,* The American Amaryllis Society Group, Box 150, La Jolla, CA 93037, c/o Dr. Thomas W. Whitaker. "Plantlife-Amaryllis Yearbook."
arils	*The Aril Society, International,* 11500 Versailles Ave. NE, Albuquerque, NM 87111, c/o Mrs. Richard A. Wilson. Newsletters (published several times annually) and "The Aril Society International Yearbook."
begonias	*American Begonia Society, Inc.*, 6333 West 84th Pl., Los Angeles, CA 90045, c/o Jacqueline Garinger. "The Begonian" (monthly bulletin).
bonsai	*The American Bonsai Society,* 953 South Shore Dr., Lake Waukomis, Parkville, MO 64151, c/o Herbert R. Brawner. "Bonsai Journal" (published 4 times annually) and newsletter "ABStracts." *Bonsai Clubs International,* 455 Blake St., Menlo Park, CA 94025, c/o Horace Hinds, Jr. "Bonsai Magazine" (published 10 times annually). *The Bonsai Society of Greater New York, Inc.*, Box E, Bronx Park, Bronx, NY 10466, c/o S. Dreilinger. "The Bonsai Bulletin" (published quarterly). *Bonsai Society of Texas,* Box 11054, Dallas, TX 75235, c/o George Gray.

boxwood	*The American Boxwood Society,* Box 85, Boyce, VA 22620, c/o Mrs. Andrew C. Kirby. "The Boxwood Bulletin" (published quarterly).
bromeliads	*Bromeliad Society, Inc.,* P.O. Box 3279, Santa Monica, CA 90403, c/o Kathy Door. "The Bromeliad Journal" (published 6 times annually).
cacti and succulents	*Cactus and Succulent Society of America, Inc.,* 1593 Las Canos Rd., Santa Barbara, CA. "Cactus and Succulent Journal" (published bimonthly). *International Cactus and Succulent Society,* Box 1452, San Angelo, TX 76901. "ICSS Newsletter" (published 3 or more times annually).
camellias	*The American Camellia Society,* Box 212, Fort Valley, GA 31030, c/o Milton H. Brown. "The Camellia Journal" (published quarterly).
carnivorous plants	*Carnivorous Plant Society,* 329 Helen Way, Livermore, CA 94550, c/o J. A. Mazrimas. "Carnivorous Plant Newsletter" (published quarterly).
chrysanthemums	*National Chrysanthemum Society, Inc.,* USA, 394 Central Ave., Mountainside, NJ 07092, c/o Mrs. W. A. Christoffers. "NCS Journal" and "The Chrysanthemum" (both published quarterly).
daffodils	*The American Daffodil Society, Inc.,* 89 Chichester Rd., New Canaan, CT 06840, c/o G. S. Lee, Jr. "The Daffodil Journal" (published quarterly).
dahlias	*The American Dahlia Society, Inc.,* 345 Merritt Ave., Bergenfield, NJ 07621, c/o Irene B. Owen. "Quarterly Bulletin" and "Annual Classification List."
delphiniums	*The Delphinium Society,* 7540 Ridgeway Rd., Minneapolis, MN 55426, c/o Philip H. Smith. Yearbook and seed.
elms	*Elm Research Institute,* Harrisville, NH 03450, c/o John P. Hansel.
epiphyllums	*Epiphyllum Society of America,* 218 E. Greystone Ave., Monrovia, CA 91016, c/o Dr. Herbert S. Irwin. "Epiphyllum Bulletin" (issued irregularly).
ferns	*The American Fern Society,* Dept. of Biology, Vanderbilt University, Nashville, TN 37235, c/o Dean P. Whittier. "American Fern Journal" (published quarterly) and "Fiddlehead Forum" (a newsletter published 5–6 times annually). *Los Angeles International Fern Society,* 2423 Burritt Ave., Redondo Beach, CA 90278, c/o Wilbur W. Olson. Monthly fern lessons, newsletter, and annual magazine.

fuchsias	*American Fuchsia Society,* Hall of Flowers, Golden Gate Park, San Francisco, CA 94122. "American Fuchsia Society Bulletin" (published monthly). *National Fuchsia Society,* 10934 East Flory St., Whittier, CA 90606, c/o Martha Rader. "The National Fuchsia Fan" (published monthly).
fruit	*American Pomological Society,* 103 Tyson Bldg., University Park, PA 16802, c/o Dr. L. D. Turkey. "Fruit Varieties Journal" (published quarterly). *Dwarf Fruit Trees Association,* Dept. of Horticulture, Michigan State University, East Lansing, MI 48824, c/o Dr. Robert F. Carlson. "Compact Fruit Trees" (published bimonthly) and "Compact Composit" (proceedings of annual conference). *North American Fruit Explorers,* 10 South 55 Madison, Hinsdale, IL 60521, c/o Robert Kurle. "Pomona" (published quarterly). *Southern Fruit Council,* Rt. 3, Box 40, Summit, MS 39666, c/o James J. Anding. "The Southern Fruit Garden" (published 3 times annually).
geraniums	*International Geranium Society,* 11960 Pascal Ave., Colton, CA 92324. "Geraniums Around the World" (published quarterly).
gesneriads	*American Gesneria Society,* Box 549, Knoxville, TN 37901. "Gesneriad Saintpaulia News" (published bimonthly). *American Gloxinia and Gesneriad Society, Inc.,* Box 174, New Milford, CT 06776, c/o Charlotte M. Rowe. "The Gloxinian" (published 6 times annually).
gladiolas	*North American Gladiolus Council,* 30 Highland Pl., Peru, IN 46970, c/o Bob Dorsam. "NAGC Bulletin" (published quarterly).
gourds	*American Gourd Society,* P.O. Box 274, Mount Gilead, OH 43338, c/o John Stevens. "The Gourd" (published 3 times annually).
hemerocallis (daylily)	*American Hemerocallis Society,* Signal Mountain, TN 37377, c/o Mrs. Arthur W. Parry. "The American Hemerocallis Journal" (published quarterly).
herbs	*The Herb Society of America,* 300 Massachusetts Ave., Boston, MA 02115. "The Herbarist" (published annually).
hibiscus	*The American Hibiscus Society,* Box 98, Eagle Lake, FL 33839, c/o James E. Monroe. "The Seed Pod" (published quarterly).
hollies	*The Holly Society of America, Inc.,* 407 Fountain Green Rd., Bel Air,

MD 21014, c/o B. C. Green, Jr. "Holly Letter" (published 3 times annually) and "Proceedings" (published biannually).

hostas — *The American Hosta Society,* 114 The Fairway, Albert Lea, MN 56007, c/o Nancy M. Minks. "The American Hosta Society Newsletter" and "Bulletin of the American Hosta Society."

hydroponics — *Hydroponic Society of America,* Box 516, Brentwood, CA 94513, c/o Paul W. Droll. Monthly newsletter and quarterly journal.

irises — *The American Iris Society,* Missouri Botanical Garden, 2315 Tower Grove Ave., St. Louis, MO 63110, c/o Clifford W. Benson. "AIS Bulletin" (published quarterly).

Sections of the American Iris Society

The Society for Japanese Irises, 17225 McKenzie Hwy., Rt. 2, Springfield, OR 97477, c/o Lorena Reid. Quarterly bullentins of The American Iris Society as well as membership to and bulletins of The Society for Japanese Irises.

Median Iris Society, 10 South Franklin Circle, Littleton, CO 80121, c/o Harry B. Kuesel. "The Medianite" (published quarterly).

Reblooming Iris Society, 903 Tyler Ave., Radford, VA 24141, c/o Dr. Lloyd Zurbrigg. "Reblooming Iris Recorder" (published triannually).

The Society for Siberian Irises, South Harpswell, ME 04079, c/o Dr. Currier McEwen.

Spuria Iris Society, Rt. 2, Box 83, Purcell, OK 73080, c/o Mrs. Joseph P. Crawford. "Spuria Newsletter" (published quarterly).

Society for Louisiana Irises, Box 175, University of Southwestern Louisiana, Lafayette, LA 70501, c/o Barbara F. Nelson. "Newsletter" (published quarterly).

ivies — *American Ivy Society,* 128 West 58th St., New York, NY 10019, c/o Mrs. J. Pierot. "American Ivy Society Bulletin" (published 3 times annually).

light gardening — *Indoor Light Garden Society of America,* c/o The Horticultural Society of New York, 128 West 58th St., New York, NY 10019. "The Light Garden" (published bimonthly).

lilacs — *International Lilac Society, Inc.,* Box 315, Rumford, ME 04276, c/o Walter W. Oakes. "Lilacs" (published quarterly) and "Pipeline" (published monthly).

lilies — *The North American Lily Society, Inc.,* Rt. 1, Box 395, Colby, WI 54421, c/o Betty Clifford. Yearbook and quarterly bulletin.

magnolias	*The American Magnolia Society,* RD 5, Box 532, Asheville, NC 28803, c/o Virginia Melnick. Newsletter published quarterly.
native plants	*California Native Plant Society,* Suite 317, 2490 Channing Way, Berkeley, CA 94704, c/o Mrs. V. F. Rumble. *New England Wild Flower Society, Inc.*, Hemenway Rd., Framingham, MA 01701, c/o Mrs. Bigelow Green. "New England Wild Flower Notes" (published quarterly).
nuts	*Northern Nut Growers Association, Inc.*, 4518 Holston Hills Rd., Knoxville, TN 37914, c/o Spencer B. Chase. "Annual Report" and "Newsletter" (published quarterly).
oleanders	*National Oleander Society,* 5127 Ave. 01/2, Galveston, TX 77550, c/o Mrs. Cortus T. Koehler. "National Oleander Society" (published annually).
orchids	*American Orchid Society, Inc.*, Botanical Museum of Harvard University, Cambridge, MA 02138, c/o Gordon W. Dillon. "American Orchid Society Bulletin" (published monthly). *Cymbidium Society of America, Inc.*, 6787 Worsham Dr., Whittier, CA 90602, c/o D. G. Saurenman. "Cymbidium Society News" (published monthly).
palms	*The Palm Society,* 1320 South Venetian Way, Miami, FL 33139, c/o Mrs. T. C. Buhler. "Princepes" (published quarterly).
penstemons	*The American Penstemon Society,* 1547 Monroe St., Red Bluff, CA 96080, c/o Howard A. McCready. "Bulletin" (published annually).
peonies	*American Peony Society,* 250 Interlachen Rd., Hopkins, MN 55343, c/o Greta M. Kessenich. "American Peony Society Bulletin" (published quarterly).
primroses	*The American Primrose Society,* 7100 SW 209th, Beaverton, OR 97005, c/o Mrs. John Genheimer. "Quarterly of the American Primrose Society."
rhododendrons	*American Rhododendron Society,* 2232 NE 78th Ave., Portland, OR 97213, c/o Mrs. B. J. Lamb. "The Quarterly Bulletin of the American Rhododendron Society."
rock gardens	*American Rock Garden Society,* Office of the Secretary, 90 Pierpont Rd., Waterbury, CT 06705, c/o Milton S. Mulloy. "American Rock Garden Society Bulletin" (published quarterly).

roses	*American Rose Society,* P.O. Box 30,000, Shreveport, LA 71130, c/o Harold S. Goldstein. "The American Rose" (published monthly).
sempervivums	*The Sempervivum Society,* 11 Wingle Tye Rd., Burgess Hill, Sussex, England. "Sempervivum Society Journal" (published quarterly) and "Yearbook."
terrariums	*The Terrarium Association,* 57 Wolfpit Ave., Norwalk, CT 06851, c/o Robert Baur. "Terrarium Topics" (published 6 times annually).

Glossary

abscisic acid a growth-inhibiting chemical that induces abscission and dormancy and inhibits seed germination, among other effects.

abscission the dropping of leaves, fruits, flowers, or other plant parts.

absorption the uptake of any material by a plant or seed, such as water by the roots or pesticide through the leaves.

accent plant a landscape plant used to call attention to a particular feature of an area, for example, an attractive shrub planted near the front door of a house.

acid-forming a material that reacts to acidify the soil or acidifies the soil as it decomposes; for example, peat moss, agricultural sulfur, and oak leaves.

acid-loving describes plants which grow best in acidic potting media or soil.

active growth a phase of the plant life cycle characterized by rapid stem lengthening and leaf production and sometimes flowering or fruiting.

adhesion the attraction of solids for liquids, for example, the attraction of soil for water.

adventitious bud a bud produced on a part of a plant where a bud is not usually found, such as on a leaf vein or root.

adventitious root a root produced by other than primary root tissues, such as from the stem.

aeration a method of increasing the water and air penetration into compacted soil by the removal of slices or cores of soil throughout the compacted area.

aggregation the clumping together of soil particles into large groups.

air layering the rooting of a branch or top of a plant while it is still attached to the parent.

allelopathy the injury of one plant by another through excretion of toxic substances by the roots; the black walnut has allelopathic properties.

amendment a material added to soil that renders it more suitable for plant growth; for example, sand, lime, compost, and peat moss.

anatomy the internal structure of an organism.

angiosperm the botanical designation of a plant that produces its seeds in ovaries.

annual a plant that grows to maturity, flowers, produces seed, and dies during one season; often refers to frost-tender flowers.

anther the top, pollen-producing portion of the stamen.

anthocyanin a red pigment found in many plants, including the fruits of cherries and strawberries.

antidesiccant a spray used on transplanted plants to slow transpiration and prevent burning of the foliage.

aphid a small, soft-bodied insect that feeds on young leaves and buds and sucks plant juices, causing curling and distortion of new growth.

arboretum (arboreta) a place where trees are grown—either alone or with other plants—for scientific and educational purposes.

arboriculture the branch of horticulture that deals with the care and maintenance of ornamental trees.

artificial medium a growing medium that does not contain soil.

ash content the amount of mineral matter present in a soil amendment.

atomizer an instrument that breaks a liquid into small droplets through the use of forced air, for example, a pesticide applicator.

auxin a plant hormone involved in dormancy, abscission, rooting, tuber formation, and other activities.

axil the angle formed where a petiole joins a stem.

axillary bud a bud located in a leaf axil.

bactericide a chemical used to control bacterial disease.

bacterium (bacteria) a single-celled, non-chlorophyll-containing plant; some bacterial species assist in the decomposition of organic matter and the fixation of nitrogen, whereas others cause such plant diseases as fireblight.

balanced fertilizer a fertilizer containing equal proportions of nitrogen, phosphorus, and potassium.

balled and burlapped describes plants (usually trees) grown in the ground, dug, and sold with the root ball wrapped in burlap.

bare root refers to deciduous plants grown in the ground, dug, and sold while dormant with no soil around the roots.

bearing age the age at which a tree produces its fruit.

bed a soil area prepared for planting.

bench a raised platform in a greenhouse for growing plants.

berm a mound or ridge of soil created to add interest or screening in a landscape.

biennial a plant that produces vegetative growth during one season, flowers during the second season, and then dies; examples are carrots and parsley.

binomial system the system of naming plants developed by Carolus Linnaeus whereby plant names are given in two parts, the genus and species.

biological control the use of a predatory organism to kill a plant pest, for example, the use of *Bacillus thuringiensis* disease to kill caterpillars.

black frost a frost that kills plant tissue, so named because of the resulting blackening of injured plants.

blanch to exclude light from a vegetable to make it more tender, less strong tasting, or more attractive.

blight a bacterial or fungal disease that kills portions of a plant and may eventually destroy the entire plant.

block garden a form of vegetable gardening that involves planting vegetables in groups or blocks rather than rows.

borer any insect that burrows into the stems, roots, leaves, or fruits of a plant.

botanical insecticide an insecticide derived from a plant; examples are pyrethrum, rotenone, and ryania.

bottom watering a technique for watering potted plants in which water is poured into the drainage saucer and absorbed through the drainage hole by capillary action.

bract a small, underdeveloped leaf that functions as a flower petal; a flower often arises in the axil of a bract.

brambles the collective name given to the fruits in the genus *Rubus;* examples are raspberries, blackberries, dewberries, and loganberries.

branch spreader a piece of wood inserted between the trunk and branch of a fruit tree to widen the crotch and create a stronger branch.

breaking dormancy causing a plant or seed to change from a dormant to an active growth phase.

broadleaf weed a dicot weed.

bud a rudimentary shoot or flower; may be either vegetative (containing leaves, stem, and so forth), flowering (producing one or more flowers), or mixed (containing both vegetative and reproductive parts).

budding implanting a bud of one plant onto another so that the resulting shoot bears the characteristics of the bud parent on the roots of the host.

bulb an underground storage organ composed of a shortened stem covered with modified leaves called *scales*.

burn browning of a plant due to lack of water, exposure to high temperatures, or exposure to a toxic chemical.

cactus a member of the plant family Cactaceae.

candle the succulent new growth of a needle evergreen such as spruce or pine.

cane the long shoots bearing fruits which are produced by grapes and brambles.

cane cutting *see* **stem section cutting.**

canker a disease particularly affecting woody plants which destroys the cambium and vascular tissue of a generally localized area of a plant.

capillary action the sideways and upward movement of water through a growing medium owing to the adhesion of water molecules to the medium.

capillary water the water that moves in soil or another growing medium owing to the forces of adhesion and cohesion.

carbohydrates the sugar and starch required for the growth, structure, and metabolism of a plant.

carotene an orange plant pigment that gives color to oranges and persimmons, among other plants.

cation-exchange capacity (CEC) the ability of a growing medium or soil to attract and hold nutrients.

Cell-pack® a plastic container of transplants in which each transplant has its own root packet.

cellulose a chemical compound that forms a main part of plant cell walls and provides strength to stems and other parts.

central leader the main, upright shoot of a tree.

central leader form a training method for fruit trees that leaves the central leader intact (as opposed to modified central leader and vase forms); also refers to the natural form of many ornamental trees.

chelated nutrient an essential element for plant growth chemically formulated to be nonreactive in soil or potting media.

clay one of the three primary soil particles: clay particles are platelike, extremely small, and retain nutrients well.

clod a hard lump of clay soil that is difficult to break.

coarse aggregate a material such as sand or perlite, for example, made up of relatively large particles and added to a growing medium to improve its drainage.

cohesion the attraction of identical molecules for each other, for example, the cohesion of water molecules.

cold hardiness or tolerance the minimum temperature at which a plant can survive.

cole crop vegetables of the genus *Brassica,* including broccoli, kale, cauliflower, and mustard.

colloids an interspersed system of finely divided soil particles.

compaction excessive packing or settling of a potting medium or soil.

companion planting the planting of insect-repellent plants next to susceptible ones to repel pests.

compatible describes two species that will cross-pollinate or that can be grafted together.

compensation point the point at which the amount of light a plant receives photosynthesizes sufficient carbohydrate for respiration but does not produce enough carbohydrate for growth.

complete fertilizer a fertilizer containing nitrogen, phosphorus, and potassium; *see also* **balanced fertilizer.**

compost an organic soil amendment made from decomposed garden refuse.

composting the piling of garden refuse to allow it to decompose.

compound leaf a leaf composed of more than one leaflet; *see also* **simple leaf.**

container gardening the growing of fruits, vegetables, and ornamentals outdoors in containers; also called *pot gardening.*

container-grown refers to landscape plants grown to selling size in containers such as cans or plastic pots.

cool-season vegetable a species of vegetable that grows best at daytime temperatures ranging from 50 to 65°F (10–18°C) and is tolerant of frost; includes peas, lettuce, broccoli, and turnips.

cork cambium the layer of cells that produces bark on woody plants.

corm a thickened underground storage organ (stem) found on such plants as gladiolus and crocus.

Cornell Mixes the Cornell University potting media.

corner planting a group of trees, shrubs, or groundcovers used to visually soften the corner of a house.

cotyledon part of an embryo of a seed made up of one or more simple leaves; serves as a storage organ in some cases, for example, in peas and beans.

cover crop a grass or similar fast-growing plant sown over a vegetable garden in fall and turned under in spring to supply organic matter to the soil.

critical photoperiod the specific dark/light ratio necessary for various plant responses to occur, particularly flowering and germination.

crop rotation planting a different species in an area of the garden each year to prevent buildup of diseases or insects associated with particular crops.

crusting a soil surface condition typified by a smooth, glazed appearance and common on silty soils or after rain.

cucurbit a group of warm-season vegetables having similar cultural requirements; included are melons, cucumbers, squash, and pumpkins.

cultivar a combination of the words "cultivated variety," designating a plant variety.

cultural control alterations in plant care (watering, fertilizing, and so forth) for the purpose of controlling disease, insect, or weed problems.

cut flower a flower grown specifically for cutting and displaying in a vase.

cuticle the thin, waxy layer on the exterior of stems and leaves that retards water loss.

cutting a vegetative plant part that regenerates roots and forms new plants.

cytokinin a naturally occurring plant hormone involved in stem elongation, bud formation, breaking dormancy, and other activities.

damping off a disease in which a soilborne fungus attacks seedlings at the soil line, constricting the stem and killing the plant.

day-neutral plant a species that flowers irrespective of the dark/light ratio.

deciduous a plant that drops its leaves in winter; the opposite of evergreen.

decline the general decrease in vigor of a previously healthy plant.

decomposition the breakdown or rotting of plants and animals.

defoliation the dropping of most or all leaves from a plant.

desuckering removal of the suckers.

determinate describes tomato varieties which remain short and bushy; *see also* **indeterminate.**

dethatching removal of thatch from a turf area by a machine or by hard raking.

dicot a subdivision of angiosperms characterized by the presence of two cotyledons, net-veined leaves, and the like; most cultivated plants are included in this group; *see also* **monocot.**

dioecious a species in which the male and female organs of reproduction are produced on separate plants; examples are holly, dates, and bittersweet.

direct seeding sowing seed in the area where the plants will grow to maturity.

disease a plant malady caused by a microorganism or an environmental factor; the latter diseases are called *physiological* diseases.

division a method of vegetative propagation involving separation of a plant into two or more pieces, each containing a portion of the roots and crown.

dormancy a phase of a plant life cycle characterized by slowed or stopped growth; leaf drop and death of the aboveground portions may also occur.

dormant oil a lightweight oil sprayed on dormant plants for control of insects such as scale; the oil coats the insects to exclude air, suffocating them and their eggs.

double glazing the application of two layers of glazing to a greenhouse to reduce air leakage and to conserve heat.

double leader two competing central leaders in a tree.

drainage the rate at which water flows through a soil or potting medium; also, the percentage of the applied water which the material retains; poor drainage is recognized by slow water flow and high retention, whereas good drainage is recognized by fast water flow and low retention.

drain tile interlocking clay tiles buried in an area with poor drainage to carry excess water away from plants.

drill hole fertilization method a technique for fertilizing deep-rooted plants such as trees by pounding holes into the ground throughout the root zone of the tree, filling them with granular fertilizer, and covering with peat moss or topsoil.

drip irrigation *see* **trickle irrigation.**

drought tolerance the ability of a plant to live without much water in the soil or to absorb the small quantity of capillary water remaining in a dry soil.

dry rot (of cacti) the stem rot of cacti characterized by discolored and sunken stem areas.

dwarf the miniature varieties of a species; *see also* **semidwarf** and **standard.**

dwarfing rootstock a rootstock that makes the scion grafted onto it grow less vigorously.

eelworm a nematode.

embryo the immature plant in a seed.

encapsulated fertilizer a slow-release fertilizer composed of beads coated with a plastic-like resin that is permeable to water, which enters through the coating, dissolves the fertilizer, and carries it back out through the coating.

espalier the stylized training of a tree or shrub to grow flat against a wall or trellis, sometimes with a precise branch pattern.

essential elements the 16 micro- and macronutrients required for plant growth.

established describes a seedling or other plant whose roots have penetrated well into the soil.

ethylene a gas produced by senescing plants or plant parts and ripening fruits; used commercially in pineapple production and to induce flowering of bromeliads (see Chapter 3).

everbearing describes strawberry and raspberry varieties which bear two crops annually: one in early summer and one in fall.

fall vegetable gardening the practice of seeding cool-season vegetables in summer to mature and be harvested in fall.

family a unit of plant classification grouping related genera.

fan and pad cooling a technique for cooling a greenhouse whereby a fiber pad is installed at the end of the greenhouse with water trickling through at a slow rate; a fan at the other end pulls outside air through it into the house; water then evaporates into the air as it passes through the pad, absorbing heat in the process and reducing the air temperature.

fast-release fertilizer a granular, powdered, or liquid fertilizer containing nitrogen only in nitrate form—the form immediately usable by plants.

fertilizer analysis the proportions of essential plant elements contained in a fertilizer which are listed on the container.

fertilizer burn an injury to plant roots owing to an excess concentration of fertilizer; wilting and browning of leaf margins are symptoms resulting from water deficiency.

fertilizer salt a chemical compound that includes one or more of the essential elements for plant growth.

fibrous (netlike) roots a fine, branched root system such as that found in many grasses.

filament the stalk portion of the stamen which supports the anther.

filler material a material such as sawdust added to a fertilizer to increase its volume inexpensively.

flat a shallow plastic, wood, or metal tray used to grow seedlings, carry plants, and the like.

fleshy roots roots thickened with stored carbohydrates or water.

floriculture the science and practice of flower growing and arrangement.

flower border a long, narrow flower bed.

foliar feeding a method of supplying nutrients to plants by spraying them with fertilizer, which is absorbed through the leaves.

foot-candle a measure of light intensity; superseded by the metric lux.

fork a narrow angle between two branches.

form the overall shape of a plant, such as vase-shaped, pyramidal, or mounded.

formal flower bed a symmetrically designed bed for annual and perennial flowers.

foundation planting shrubs or groundcover planted against the base of a house.

freestanding bed a flower, vegetable, or shrub planting designed to be accessible from all sides.

frond a fern leaf.

frost-hardy describes plants which are able to withstand frost.

frost pocket a low-lying area more susceptible to frost than the surrounding vicinity.

frost-tender describes plants that are killed at temperatures below freezing.

fruit a swollen ovary.

fruit load the amount of fruits that a plant is bearing; a "heavy load" indicates a large quantity of fruits.

fruit set the development of fruits after flowering; few fruits develop in "poor fruit set," as opposed to "good fruit set."

fungicide a chemical used to control a fungus-caused disease.

fungus (fungi) a single-celled or multicelled plant lacking chlorophyll that may be saprophytic (living on dead organic matter) or parasitic (existing on living plants and other organisms); saprophytic fungi are useful in the garden because they decompose compost and release the nutrients in organic matter, but parasitic fungi are the cause of many plant diseases.

fungus gnat a minute, winged insect whose wormlike larvae are found occasionally in unsterilized potting media and damage plant roots through feeding.

gall an abnormal swelling on a plant part; leaf, stem, and root galls are typical examples.

genetic dwarf a fruit tree that is a dwarf owing to its genetic makeup rather than to grafting on a dwarfing rootstock.

genus (genera) a group of related species; one or more genera are grouped in a family.

germination the sprouting of a seed, spore, or pollen grain.

germination rate the percentage of seeds that will germinate under favorable conditions.

gibberellin a naturally occurring plant hormone involved in increasing flower, fruit, and leaf size, and elongation; also influences vernalization and other activities.

girdle to restrict the function of the xylem or phloem of a dicot plant; often occurs when strings or wires are left on young plants; as the plant grows, the wire becomes embedded in the wood, and the portion above dies; girdling by cutting through an area of phloem is also used to encourage flowering.

glazing the clear or translucent material used to cover a greenhouse or related plant growing structure.

girdling root a root, usually of a tree, that becomes embedded in the trunk of the tree at ground level and will eventually kill part or all of the tree.

grading changing the slope level of the soil in an area.

grafting implanting a branch or bud from one plant onto another.

granular fertilizer a fertilizer in a pelletized form.

greenhouse effect heating the air in a translucent or transparent structure owing to energy from the sun; as light energy passes through the glazing, infrared rays are trapped by the glazing.

green manure crop a grass or similar fast-growing plant sown over a future garden area and turned under before it reaches maturity to provide organic matter in the soil.

groundcover low-growing plants which form matlike growth over an area.

ground bed a planter constructed on the floor of a greenhouse for growing tall vegetables or flowers.

growing medium (media) a material used for rooting and growing plants which may contain sand, peat moss, soil, or any other number of ingredients.

grub the soil-inhabiting larva of many beetles.

gymnosperm a botanical group of evergreen plants with generally needlelike foliage and usually bearing cones; includes pines, juniper, and spruce.

harden off to gradually acclimatize a greenhouse- or house-grown plant to outdoor growing conditions.

hardpan a layer of compacted soil that slows or stops movement of water.

hardwood cutting a cutting made from the mature growth of a woody plant, usually taken in late fall or early spring.

head the leafy portion of a tree.

heading back a pruning technique to control size and shape of shrubs by cutting back shoots to the parent branch to encourage new growth.

heal in to temporarily cover the roots of a plant with soil to prevent their drying until the plant can be properly planted.

heavy soil a soil composed predominately of clay particles.

herbaceous not producing woody growth.

herbaceous cutting a cutting of a nonwoody plant.

herbicide a chemical used to control weeds; a "weed killer."

hill a group of seeds planted close together; hill planting is common for melons and squash.

hoar frost a frost that leaves ice crystals on soil and plants owing to condensation of humidity at a decreased temperature.

holdfast a suction-cup-like organ on some climbing vines, anchoring them to walls and other surfaces.

home horticulture the growing of ornamental and edible plants in and around the home.

hose-end sprayer a hose attachment for applying pesticide or fertilizer; the sprayer contains a concentrated chemical that is diluted with water through a siphon mechanism as it is applied.

hot cap a paper or plastic dome set over warm-season vegetables in early spring to protect them against frost and to increase the daytime growing temperature.

humic acid an acid formed by the decomposition of organic matter.

humidity the amount of water vapor in the air.

hydroponics a method of growing fruits, vegetables, and flowers in which the plants are planted in beds of sand, gravel, or a similar material and the roots are supplied with nutrients in a water-based solution.

incompatible describes two plants that will not graft together or that form a weak and unsatisfactory graft union; also refers to two species which will not cross-pollinate.

indeterminate describes tomato varieties which grow into long vines; *see also* **determinate.**

integrated pest management (IMP) a system of insect and disease control stressing thorough understanding of the life cycle of the pest as well as cultural and biological controls; chemical control is used only when cultural methods fail.

internode the portion of plant stem between two successive nodes.

interplanting sowing seeds of a quick-maturing vegetable with seeds of a slower one or between transplants; the fast-maturing vegetable is then harvested before it begins to crowd the main crop.

interstock a section of stem inserted between the stock and scion in a graft union; the interstock may contribute a characteristic to the tree to make an otherwise incompatible stock and scion graftable.

invasive roots a vigorous root system that can clog sewer lines and wells.

irrigation watering.

juvenility the first phase of plant development in which the plant is strictly vegetative.

lankiness unattractive lengthening of stem internodes, usually owing to insufficient light.

larva (larvae) the immature stage of an insect; larvae are frequently wormlike, whereas the adult may be winged.

latent heat of the earth the relatively uniform, above-freezing temperature of the soil; a pit greenhouse makes use of this latent heat by having its walls belowground.

lateral a side branch arising from a vertical shoot.

lath house a garden structure covered with thin strips of wood and used for growing shade-loving plants.

lava rock a coarse aggregate made from ground volcanic rock.

lawn substitute a groundcover plant tolerant of foot traffic and planted in lieu of grass in a lawn area.

layering a method of propagating shrubs or vines with a trailing growth habit or flexible branches: a shoot is bent to the ground, held in place with a wire loop or stone, and covered with soil; after it generates roots, it is then severed from the parent plant.

leaching a succession of waterings used to dissolve and flush out excess fertilizer or other chemicals from a soil or potting medium.

leaf apex the tip of a leaf.

leaf base the section where the leaf blade joins the petiole.

leaf blade the flattened portion of a leaf which is responsible for the majority of photosynthesis.

leaf bud cutting a cutting that includes a short section of stem with a leaf attached; leaf bud cuttings are taken after the stem tip is used.

leaf cutting a cutting made from a leaf and its attached petiole.

leaflet one of the expanded, leaflike parts of a compound leaf.

leaf margin the edge of a leaf blade.

leaf miner a wormlike insect larva that tunnels between the upper and lower surfaces of a leaf, leaving a scribble pattern.

leaf mold a soil amendment or potting medium component composed of partially decayed leaves.

leaf vegetable a vegetable raised for its edible leaves, such as lettuce, chard, or Chinese cabbage.

lean-to greenhouse a greenhouse constructed with one wall in common with a house or other building.

LD_{50} an indication of pesticide toxicity established by experimentation on laboratory animals; abbreviation for "lethal dose that will kill 50% of a population" and measured in mg/kg of body weight: the higher the LD_{50}, the lower the assumed toxicity of the pesticide.

light duration the total number of hours of light a plant receives per day.

light intensity the brightness of light.

light soil a soil composed predominately of sand.

limbing up the removal of the lower branches of a tree.

lime any of several calcium-containing materials used to raise soil pH.

loam a soil having large proportions of both sand and clay particles.

long-day plant a species that flowers only with a short daily dark period.

low-maintenance landscape a landscape that is planned for minimal upkeep.

lux a metric measure of light intensity, equal to the illumination on a surface 1 m from a standard candle.

macronutrient one of the essential elements needed in relatively large amounts for plant growth; these include nitrogen, phosphorus, calcium, magnesium, and sulfur.

macropore space a large pore space found between large particles or aggregations of particles; *see also* **micropore space.**

maleic hydrazide a growth-inhibiting chemical occasionally used to slow the growth of lawns and hedges.

maturity the phase of plant development characterized by flowering and fruit production.

mealybug a destructive plant pest that is about ⅛–¼ in. (3–6 mm) long and covered at maturity with white waxy fibers which form a protective covering.

medium (media) a material used for rooting or supporting the roots of plants.

meristem an area of rapidly growing and dividing cells.

microclimate a small area with a climate differing from that of the surrounding area because of such factors as exposure, elevation, and sheltering structures.

micronutrient one of the essential elements needed in minute quantities for plant growth; these include iron, copper, zinc, boron, molybdenum, chlorine, and cobalt.

microorganism an organism that can only be seen through a microscope, for example, fungi, bacteria, and mycoplasmas.

micropore space a small pore space found between minute particles such as clay; *see also* **macropore space.**

mildew a foliar fungus disease characterized by a white powdery layer on plant leaves.

mite an eight-legged, near-microscopic plant pest that damages foliage by sucking cell contents.

miticide a chemical used to control mites.

modified central leader form a training method for fruit trees in which the central leader is removed back to the nearest scaffold branch to restrict the height of the tree and facilitate fruit harvesting.

monocarp a plant that grows vegetatively for more than one year but only flowers once and then dies.

monocot a subdivision of angiosperms that includes plants with a single cotyledon in the embryo and mostly straplike leaves with parallel venation; most are nonwoody and have underground storage organs; *see also* **dicot.**

muscadine grape *Vitis rotundifolia,* an American grape species native to the South.

mutation a spontaneous change in the genetic makeup of a plant, for example, variegated shoots on an otherwise green plant.

mycoplasma a microscopic organism intermediate between a virus and a bacterium that causes several plant diseases.

nematicide a chemical used to control nematodes.

nematode a microscopic or near-microscopic wormlike plant pest that damages roots and occasionally burrows in foliage.

needle feeding a method of fertilizing and watering deep-rooted plants such as trees: consists of a container of fertilizer connected to a pipe with a pointed tip; a garden hose is in turn connected to the pipe; the pointed tip is pushed into the ground around the tree, and when the water is turned on, dissolved fertilizer is forced into the ground around the roots.

nitrate a nitrogen-containing compound and the form in which most plants absorb nitrogen.

node the point of leaf attachment to a stem.

nonselective herbicide an herbicide that injures or kills plants regardless of species.

nutrient an essential element for plant growth.

nutrient deficiency the lack in a plant of a sufficient quantity of one or more of the essential elements.

offset a young plant produced at the base of a parent plant; in the case of cacti and a few other plants such as pineapple, the offset may develop on top of the parent plant.

olericulture the science and practice of vegetable growing.

open-center form *see* **vase form.**

organic defined by the laws of chemistry as a chemical that contains carbon atoms; also, a fertilizer or pesticide derived from a naturally occurring material.

organic matter the decomposing bodies of dead plants and animals.

ornamental a plant grown for its beauty.

ornamental horticulture the branch of horticulture that deals with the cultivation of plants for their aesthetic value; this designation includes landscape horticulture, floriculture, indoor plant growing, and sometimes turfgrass growing.

ovary the bulbous base of the pistil which contains egg cells; the ovary enlarges to become the fruit.

overwatering applying an excess volume of water at too-frequent intervals for healthy plant growth.

parthenocarpy the growth of a fruit without fertilization of the included egg; also, the continued development of a fruit after the abortion of the embryo.

patio tree a small ornamental tree planted near a patio or deck for shade and beauty.

peat moss any of several bog plants, particularly sphagnum moss, which are harvested, dried, and used as components in growing media or for soil amending.

peat pot a pot made of compressed peat moss and used for growing seedlings, which are then planted in the garden in their container.

pelleted seed a vegetable seed coated with a dissolving material that makes it larger and easier to sow.

perennial a plant that lives for an indefinite number of years; often refers to herbaceous flowering perennials.

perfect flower a flower with both male and female parts.

perlite a course aggregate made from expanded volcanic rock.

pesticide a chemical used to control an undesirable organism; *see also* **bactericide, fungicide, herbicide, miticide, nematicide.**

petiole the stalk that attaches the leaf blade to the stem.

pH a measure of the acidity or alkalinity of a soil, potting medium, or liquid.

phloem the portion of the vascular system in which carbohydrates are moved throughout the plant.

photoperiod the dark/light ratio in a day.

photoperiodism the effect of the daily dark/light ratio on the growth and flowering of plants.

photosynthesis the chemical process by which green plants manufacture carbohydrates in the presence of light.

phototropism the hormone-induced leaning of a plant toward light.

physiological disease a plant malady due to insufficient amounts of essential nutrients or excessive (toxic) amounts of certain elements such as sodium in the soil.

pinching removal of the terminal bud of a plant to promote branching.

pistil the female parts of a flower: style, stigma, and ovary.

plant breeding the science of controlled pollination of plants to develop new varieties.

planter a defined area, usually raised and constructed of wood or concrete, for growing plants.

plant food a lay term for fertilizer.

plastic mulch a layer of clear or black plastic used as a mulch in vegetable or ornamental plantings.

plugging a method of lawn establishment involving transplanting small cores of sod into a prepared area.

pollarding a formal training method applied to deciduous trees whereby branches of the tree are pruned back almost to the main trunk every year.

pollination the deposition of pollen on the flower stigma.

pomology the science and practice of fruit growing.

pony pack a plastic package of four to eight flower or vegetable seedlings.

pore spaces the gaps between the particles of soil or other growing media; the gaps may be filled with either air or water.

postemergence herbicide an herbicide that kills established weeds.

pot-bound describes a containerized plant allowed to grow in a pot too long; the overabundance of roots in the pot jeopardizes the health of the plant.

pot up to dig up a seedling and transfer it to a pot.

potting compost British term for a growing medium.

preemergence herbicide an herbicide that kills weed seedlings as they germinate but has no effect on established plants.

propagation plant reproduction by sexual or vegetative methods.

pruning removal of parts of a plant for size control, health, or appearance.

public area the area of a landscape in front of a house and viewable from the street; includes the front door, driveway, and walkway.

pup popular name given to the offset of a bromeliad.

radiation frost a frost caused by the loss of heat from plants by radiation during the night.

recommended variety a variety of fruit or

vegetable well adapted to growing in a particular region and recommended by the local Cooperative Extension Service.

refoliate to grow new leaves after being leafless; for example, a deciduous tree refoliates each spring.

relative humidity the amount of water vapor present in the air compared to the total amount the air could hold at its present temperature.

renovation pruning severe, heavy pruning used to rejuvenate a neglected plant; used on fruit trees, grapes, strawberries, ornamentals, and sometimes houseplants.

resistance nonsusceptibility of a plant to insects or disease.

respiration the breakdown of carbohydrates to yield energy for growth and cell activities.

rhizome a usually thickened, horizontal, underground stem.

rock garden a combination of rocks and plants that creates a natural-looking area for the cultivation of plant species native to rocky or alpine regions.

roguing eliminating certain plants from a group, usually those considered the least desirable; may also refer to the removal of diseased or infested plants to prevent the spread of the condition.

root-bound *see* **pot-bound.**

root cutting a cutting made from sections of roots alone.

root hair a tubelike outgrowth of a root epidermal cell through which water and dissolved nutrients pass.

root mealybug root-parasitizing insects resembling foliar mealybugs.

root rot the death and decay of plant roots owing to unfavorable growing conditions and resulting infections by microorganisms.

root tuber a thickened root that stores carbohydrates and is used in propagation.

root vegetable a vegetable grown primarily for its edible root, such as radishes, turnips, and carrots.

root zone the volume of soil or growing medium containing the roots of a plant.

rooting hormone a plant hormone used in promoting rooting of cuttings.

rootstock the plant that bears the roots of the new plant in a budding or grafting operation.

rosette a plant form in which the plant has a shortened stem with leaves that overlap and radiate outward from a central point; examples are African violets and strawberries.

runner an aboveground stem produced as a natural means of vegetative reproduction of plants such as ferns and strawberries; one or more plants grow on the stem, generate roots, and become independent of the parent.

rust a fungus disease characterized at one point in the life cycle by rust-colored pustules on the leaves of the affected plant; rust diseases require two host plant species for a complete life cycle.

Saran® a plastic, screenlike material used for covering a shade house to reduce light penetration.

scaffold branch a main branch growing from the trunk of a tree.

scale a destructive plant pest that is attached to a plant and exudes a shell-type covering.

scarification the mechanical scraping or damaging of a seed coat to facilitate germination.

scion the twig (or bud) to be grafted onto the roots of another plant, the stock, in a budding or grafting operation from which the bud or branch for grafting is taken.

scorch the death of plant tissue from excess heat or insufficient water.

scoria a coarse aggregate made from ground volcanic rock.

screening plant a plant used to block a view or to separate areas in a house or landscape.

scurf the white or silver scales that cover some leaves, for example, bromeliads.

seed coat a covering that helps prevent injury and drying of the seed.

seed leaves the first leaves (cotyledon) produced by a seedling; they generally do not resemble the mature leaves of the species; *see also* **true leaves.**

seed potato a section of a potato tuber containing at least one bud which is planted to start a potato plant.

seed tape a strip of water-soluble plastic tape embedded with vegetable seeds.

selective herbicide an herbicide that harms some species but not others, for example, 2,4-D, which is selective for dicot weeds in turfgrass.

self-incompatibility the inability of pollen to fertilize eggs of the same plant.

semidwarf a plant (usually a fruit tree) intermediate in size between dwarf and standard (full size).

semihardwood cutting a cutting made from the partially matured new growth of a woody plant.

senescence the aging and death of a plant or any of its parts.

sepal the often green, leaflike outermost part of a flower that encloses and protects the unopened flower bud.

service area the area of a landscape used for storage, laundry lines, pet runs, and other utilitarian purposes.

shade house a garden structure covered with Saran or lath and used for growing shade plants such as ferns.

shading material (whitewash, lath, or bamboo blinds, for example) applied to glazing to decrease light entry and heat buildup in a greenhouse or related structure; also, the act of applying the material.

shock wilting and leaf drop owing to root loss during transplanting; in houseplants, shock may result from a decrease in light density.

short-day plant a species that flowers only with a long daily dark period and a short daily light period.

sidedressing a method of fertilizer application that involves sprinkling a narrow band of fertilizer on both sides of a plant row over the root area.

sidewall a concrete block or wood wall extending partway up the side of a greenhouse in lieu of glazing and used to reduce heat loss.

silt one of the three basic soil particles; *see also* **clay.**

simple leaf a leaf with a single blade; *see also* **compound leaf.**

single element fertilizer a fertilizer that contains one of the elements essential for plant growth; urea is a single-element fertilizer.

slip a synonym for cutting.

slow-release fertilizer a fertilizer chemically formulated or encapsulated to release its nutrients over a period of several weeks or months.

small fruits the collective name given to the plants which bear edible fruits of a relatively small size; included are grapes, strawberries, and raspberries.

softwood cutting a cutting made from the new growth of a woody plant.

soil ball the soil and the contained roots of a balled and burlapped plant.

soilborne refers to a disease or insect pest that lives in the soil; many wilt diseases and nematodes are soilborne.

soil drench a liquid poured over the root zone of a plant to kill soil insects or to cure root rot.

soil layer the topsoil or subsoil portion of soil.

"source" to "sink" movement the translocation of material (hormones, metabolites, and so forth) from their point of origin or manufacture to the sink, generally meristems, roots, and storage tissues.

spaghetti system an automatic watering system involving a flexible plastic pipe inserted at intervals with spaghetti-sized microtubes which lead to plants and apply water at a slow rate.

species a group of individuals with many characteristics in common and usually interbreeding freely.

specific epithet the second word in a botanical name.

specimen plant a showy ornamental plant grown solely for its beauty.

spore a simple reproductive cell that develops into a plant; usually used with reference to fungi and ferns.

spp. abbreviation for species (plural) designating collectively all species within the genus; for example, *Quercus* spp. refers to all oaks.

springtail a minute, harmless insect characterized by an unusual jumping motion.

stamen the male parts of a flower: anther and filament.

standard a tree or shrub with a tall stem that stands alone; also, the normal size of a species; *see also* **dwarf, semidwarf.**

starter solution a liquid fertilizer high in phosophorus content and applied to seedlings to hasten growth.

stem section cutting a cutting made from a short piece of thickened leafless stem containing at least one node.

stem tip cutting a cutting made from the top of a growing stem.

stigma the knoblike top of the pistil that receives pollen.

stolon *see* **runner.**

stomate (stomata, stomates) an opening between leaf cells which functions in gas exchange.

stone fruit a fruit that contains a single, large pit; included are plums, peaches, apricots, and nectarines.

stratification the chilling of seeds to aid in breaking their dormancy.

stress a harmful growing condition that inhibits healthy plant growth, for example, drought (water stress), root boundness, and excess heat or cold (temperature stress).

stringiness unattractive lengthening of stem internodes, usually due to insufficient light.

style the portion of the pistil connecting the ovary and stigma.

submersion watering a watering technique for potted plants in which the container is submerged in water for several minutes and then allowed to drain; useful for water-repellent potting media and sphagum-lined hanging baskets.

subsoil the layer of soil directly beneath the topsoil which is characterized by a lower proportion of organic matter and less fertility than the topsoil.

subsurface irrigation an irrigation technique in which water is applied to an area through seepage from underground pipes.

succulent a plant having thickened leaves or a stem adapted for retaining water; also the tender new growth of plants.

sucker a vigorous, vertical-growing shoot produced at the base of a plant; also, a young plant produced around the base of a mature one.

sucking insect an insect that damages plants by puncturing the leaves and sucking the sap.

summer bulb a bulbous-type plant that is not cold-hardy; includes tuberous begonia, canna, and gladiolus.

sun scald an injury to a plant from excess sun or a sudden increase in light intensity.

surface root a plant (usually a tree) root that grows along the surface of the soil; also, the very shallow roots of certain plants.

systemic pesticide a pesticide absorbed into a plant, killing the organisms that prey on it.

taproot a single, fleshy root that grows vertically into the soil; for example, carrots and parsnips.

taxonomy the scientific classification of plants and animals according to their presumed natural relationships.

teepee training tying stakes together in pyramid fashion as a trellis for vining vegetables.

terminal bud a bud located at the tip of a stem.

terrarium a transparent glass or plastic container retaining high humidity used for displaying and growing plants indoors.

texture the visual impact of a plant owing to the size of its leaves: "coarse-textured" plants have large leaves, and "fine-textured" plants have small leaves; also refers to the proportions of sand, silt, and clay in a soil.

thatch a layer of stems and roots (both dead and living) that accumulates between the sur face of a turf and the soil.

thinning a pruning technique for shrubs which involves removing the oldest stems to promote new growth; also, the selective removal of the excess fruits on a fruit tree to improve the size and quality of the remaining fruits; also, removal (usually by pulling out) of excess seedlings spaced too closely for optimum growth.

thrips minute, winged insects that damage plants by rasping the surface of the leaves and lapping the exuded sap.

tip burn the death and browning of leaf tips, common on indoor plants in low-humidity environments.

tomato cage a wire mesh cylinder used to support a tomato plant without tying.

topdressing a method of fertilizer application that involves the sprinkling of fertilizer over the soil surface.

topiary the stylized training of a shrub by pruning or shearing to grow in a controlled shape.

topping removal of the central leader of a tree to make the head fuller (not a recommended pruning practice).

topsoil the uppermost layer of soil, usually characterized by a higher quantity of organic matter and nutrients than the subsoil.

trace element a micronutrient.

training a system of pruning and tying plants to conform them to a particular size or shape; used on fruits to aid harvesting and on ornamentals for decorative effect.

translocation the movement of carbohydrates, water, minerals, and other materials within the vascular system of a plant.

transpiration the loss of water from a plant, usually through leaves, in vapor form.

transplant a seedling plant grown in a cold frame, in a greenhouse, or on a windowsill for later planting outdoors; also, to dig up and move a plant to another location.

trap crop the planting of an expendable insect-attracting plant near a crop plant to draw the pests away from the desired crop (not proved to be an effective pest control method).

trickle irrigation an irrigation technique for outdoor plants involving flexible water pipes inserted at intervals with small microtubes that run to the base of nearby plants.

true leaves the second set of leaves produced by a seedling; these leaves resemble the normal leaves of the species, whereas seed leaves do not.

tuber a thickened underground stem in which carbohydrates are stored; for example, the potato; *see also* **root tuber.**

utility area *see* **service area.**

U-C Mixes the University of California potting media.

underwatering applying water in insufficient quantity or at insufficient frequency for healthy plant growth.

variegation the genetic patterning of leaves with white or yellow markings.

variety a subdivision of a species; usually differs from the species in one or more minor ways.

vascular bundle the bundle consisting of xylem and phloem; the conducting system in plants.

vascular cambium the layer of cells that gives rise to xylem and phloem cells.

vascular system the network of cellular pathways that moves water, minerals, and carbohydrates within a plant; includes the xylem and phloem.

vase form a training method for fruit trees in which from two to five scaffold branches grow from a short trunk and the central leader and other scaffold branches are removed to keep the tree short and make harvesting easier; also refers to any tree that grows naturally in this fashion.

vase-shaped a tree or shrub without a central leader and having a V- or Y-shaped form.

vegetative describes the parts of a plant not involved in the sexual reproduction process.

vermiculite a coarse aggregate made from expanded mica and having a high cation-exchange capacity.

vernalization the subjecting of a plant to cool temperatures for the promotion or enhancement of growth or flowering.

virus a microscopic noncellular organism dependent on living hosts for its nutrition and reproduction and the cause of many plant diseases.

volunteer a seedling produced by natural reseeding of a garden plant.

warm-season vegetable a vegetable that grows best at daytime temperatures ranging from 5 to 90°F (18–32°C); includes tomatoes, eggplant, peppers, and corn.

water holding capacity the ability of a growing medium to retain water.

water sprout a vigorous, vertical-growing shoot produced on the trunk or branches of a tree.

wet-dry watering cycle a technique for watering houseplants whereby the plants are only watered when the medium becomes dry and just before the plants wilt.

wetting agent a chemical added to pesticide spray, peat moss, or soil to decrease surface tension and increase water adhesion or penetration.

whiteflies white insects resembling tiny moths which cluster under plant leaves and cause damage by extracting sap.

wick watering a watering technique that moves water from a reservoir to the potting medium through a fabric wick to keep the potting medium constantly moist.

winter annual a frost-hardy flower planted in the fall in mild-winter climates for winter bloom; for example, stock and calendula.

wrapping the practice of covering the trunk of a young tree with paper tape or burlap to prevent sun scald.

xylem the portion of a plant vascular system in which water and minerals, taken in by the roots, move throughout the plant.

yield the amount of produce obtained from a fruit, nut, or vegetable variety; "good yield" indicates heavy production.

zone of elongation the section of a root, just behind the apex, in which cells produced in the meristem enlarge and lengthen the root.

Illustration Credits

CHAPTER 1

Chapter opening photo: courtesy of American Airlines; **1-1:** courtesy of Vocational Education Productions; **1-4:** courtesy of R.E. Partyka; **1-5:** USDA map; **1-6:** with permission of J.B. Edmond et al., *Fundamentals of Horticulture*, Philadelphia, Blakiston, 1951, pp. 165-168.

CHAPTER 2

Chapter opening photo: courtesy of The New York Public Library Picture Collection; **2-1:** modified from Wilson et al., *Botany*, 5th ed., New York, Holt, Rinehart and Winston, 1971, p. 648; **2-15:** modified from Wilson et al., p. 208; **2-16, 2-17c:** from *Botany*, 5th ed., by Carl E. Wilson and Walter E. Loomis. Copyright 1952 © 1957, 1962, 1967, 1971 by Holt, Rinehart and Winston, Inc. Reprinted by permission of Holt, Rinehart and Winston.

CHAPTER 3

Chapter opening photo: Pamela Harper; **3-2:** modified from Wilson et al., p. 326; **3-4:** courtesy of Claud L. Brown; **3-6:** courtesy of A.A. DeHertogh and H.P. Rasmussen; **3-8:** California Cooperative Extension Service.

CHAPTER 4

Chapter opening photo: Elinor S. Beckwith; **4-1, 4-3:** USDA; **4-4:** courtesy of George J. Ball, Inc.; **4-6:** courtesy of R.E. Partyka; **4-7:** USDA; **4-9:** from *Botany*, 5th ed., by Carl E. Wilson and Walter E. Loomis. Copyright 1952© 1957, 1962, 1967, 1971 by Holt, Rinehart and Winston, Inc. Reprinted by permission of Holt, Rinehart and Winston; **4-14:** George Taloumis; **4-19, 4-20, 4-21:** USDA; **4-22 a, b:** courtesy of Vocational Education Productions; **4-23:** courtesy of Robert F. Carlson.

CHAPTER 5

Chapter opening photo: Soil Conservation Service; **5-2:** courtesy of Vocational Education Productions; **5-7:** Soil Conservation Service; **5-12:** USDA; **5-14:** adapted from USDA Yearbook of Agriculture; **5-16:** courtesy of Vocational Education Productions; **5-18:** George Taloumis.

CHAPTER 6

Chapter opening photo: USDA; **6-1:** U.S. Department of Commerce; **6-2:** USDA maps; **6-5:** George Taloumis; **6-10, 6-11:** USDA; **6-12:** courtesy of J.C. Allen and Son; **6-14, 6-15:** USDA; **6-17:** Michigan State University Cooperative Extension Service; **6-18, 6-19, 6-20, 6-21, 6-22:** USDA; **6-23, 6-24:** Pamela Harper; **6-25:** USDA.

CHAPTER 7

Chapter opening photo: USDA; **7-2, 7-6:** George Taloumis; **7-10, 7-11, 7-12:** USDA.

CHAPTER 8

Chapter opening photo, 8-2, 8-10: USDA; **8-15:** George Taloumis.

CHAPTER 9

Chapter opening photo: courtesy of Marge Coon; **9-1, 9-2, 9-3:** courtesy of Park Seed Company; **9-7, 9-8, 9-9, 9-10, 9-12:** George Taloumis; **9-13, 9-14, 9-15:** courtesy of Marge Coon; **9-16:** Pamela Harper.

CHAPTER 10

Chapter opening photo, 10-6: USDA; **10-10:** George Taloumis; **10-16:** Pamela Harper.

CHAPTER 11

Chapter opening photo: George Taloumis; **11-6:** USDA; **11-9:** courtesy of Vocational Education Productions; **11-21:** courtesy of Monrovia Nursery Company; **11-22, 11-23:** courtesy of Chapin Watermatics; **11-24, 11-25:** USDA; **11-27:** George Taloumis.

CHAPTER 12

Chapter opening photo: courtesy of J.C. Allen and Son; **12-2b:** courtesy of Vocational Education Productions; **12-4:** courtesy of The Lawn Institute; **12-5:** courtesy of Vocational Education Productions; **12-6:** USDA; **12-7:** courtesy of The Lawn Institute; **12-8:** California Cooperative Extension Service.

CHAPTER 13

Chapter opening photo, 13-2, 13-3, 13-4, 13-5, 13-6: USDA; **13-7:** California Cooperative Extension Service; **13-8:** USDA; **13-9:** courtesy of R.E. Partyka; **13-10:** USDA; **13-11:** courtesy of Vocational Education Productions; **13-12:** courtesy of R.E. Partyka; **13-13:** courtesy of O.W. Barnett, **13-14:** courtesy of Vocational Education Productions, **13-15:** courtesy of Marge Coon; **13-17:** George Taloumis; **13-20:** USDA.

CHAPTER 14

Chapter opening photo: courtesy of Vocational Education Productions; **14-4:** photo © 1977 CBS Publications; **14-8, 14-9:** courtesy of Marge Coon; **14-10:** courtesy of Marge Coon; **14-13:** George Taloumis; **14-15:** courtesy of Vocational Education Productions.

CHAPTER 15

Chapter opening photo: Laura Rice; **15-3:** USDA; **15-5:** courtesy of Vocational Education Productions.

CHAPTER 16

Chapter opening photo: courtesy of General Electric Company; **16-5:** U.S. Department of Commerce.

CHAPTER 17

Chapter opening photo: Sam Dasher/DPI; **17-9:** courtesy of Nasco International.

CHAPTER 18

Chapter opening photo, 18-4, 18-5: USDA; **18-6, 18-7, 18-11, 18-12:** courtesy of Marge Coon.

CHAPTER 19

Chapter opening photo: Elinor S. Beckwith; **19-1:** courtesy of Marge Coon; **19-7:** courtesy of Ethan Allen, Inc.; **19-8:** George Taloumis; **19-10:** courtesy of Dome Industries; **19-11:** courtesy of Earthway Products.

CHAPTER 20

Chapter opening photo, 20-1: USDA; **20-4, 20-5:** courtesy of George Taylor; **20-7, 20-8, 20-10, 20-11:** USDA; **20-12:** J and D Solar Products; **20-16:** courtesy of Vocational Education Productions; **20-18:** courtesy of Chapin Watermatics; **20-19:** courtesy of Vocational Education Productions; **20-20:** courtesy of George Taylor; **20-21:** courtesy of General Aluminum Products.

Index